STRANGE BEAUTY

STRANGE BEAUTY

*Murray Gell-Mann and the Revolution
in Twentieth-Century Physics*

by GEORGE JOHNSON

ALFRED A. KNOPF   NEW YORK   1999

THIS IS A BORZOI BOOK
PUBLISHED BY ALFRED A. KNOPF, INC.

Copyright © 1999 by George Johnson
All rights reserved under International and Pan-American
Copyright Conventions. Published in the United States by
Alfred A. Knopf, Inc., New York, and simultaneously in
Canada by Random House of Canada Limited, Toronto.
Distributed by Random House, Inc., New York.

www.randomhouse.com

Knopf, Borzoi Books, and the colophon are registered
trademarks of Random House, Inc.

Grateful acknowledgment is made to the following for
permission to reprint previously published material:
Harold P. Furth: "Perils of Modern Living" by Harold P. Furth.
Originally published in *The New Yorker* in 1956. Reprinted
by permission of the author.
Alfred A. Knopf, Inc.: "Cosmic Gall," from *Collected Poems
1953–1993* by John Updike, copyright © 1993 by John
Updike. Reprinted by permission of Alfred A. Knopf, Inc.

Library of Congress Cataloging-in-Publication Data
Johnson, George.
Strange beauty: Murray Gell-Mann and the revolution in
twentieth-century physics / by George Johnson. — 1st ed.
       p.   cm.
   Includes bibliographical references and index.
   ISBN 0-679-43764-9 (alk. paper)
   1. Gell-Mann, Murray.   2. Nuclear physicists—
United States—Biography.   I. Title.
QC774.G45J65   1999
530'.092—dc21
[B]   99-19952   CIP

Manufactured in the United States of America
First Edition

For Nancy

There is no excellent beauty that hath not some strangeness in the proportion.

—Francis Bacon
(Quoted by Murray Gell-Mann in an article explaining his theory of cosmic-ray particles whose behavior seemed to defy the laws of physics)

In our work we are always between Scylla and Charybdis; we may fail to abstract enough, and miss important physics, or we may abstract too much and end up with fictitious objects in our models turning into real monsters that devour us.

—Murray Gell-Mann, in a 1972 lecture on quarks in Schladming, Austria

CONTENTS

*16 pages of illustrations will
be found following page 212*

STRANGE BEAUTY

ON THE TRAIL
TO LA VEGA

It was Memorial Day weekend of 1996, in the middle of what turned out to be one of New Mexico's worst droughts of the century. The seemingly endless dry spell reminded many of the climatic disaster said to have driven the Anasazi, the original inhabitants of this land, from their stone settlements around Mesa Verde, causing the collapse of a civilization. To escape the heat, I left my house in Santa Fe and drove as high as you can go into the nearby Sangre de Cristo Mountains. After leaving my Jeep in the ski basin parking lot, already some 10,000 feet above sea level, I began walking higher. My destination, La Vega, "the meadow," lay at the base of Santa Fe Baldy, an 11,600-foot peak of Precambrian granite that juts above the timberline.

Almost as soon as I reached the trail head, I realized that, once again, I had misjudged the perversity of New Mexico weather. Looking out across the Rio Grande Valley, I could see the next mountain range, the Jemez, where just weeks earlier a fire had devastated fifteen thousand acres of one of my favorite places, the wilderness backcountry of Bandelier National Monument. Now storm clouds were boiling up over the Jemez and sweeping toward the Sangre de Cristos. The temperature began dropping, and before long snow flurries, of all things, were swirling around me.

I was wishing I had worn a jacket and long pants instead of khaki shorts and a T-shirt, when, as I rounded a corner on the trail, I heard a familiar voice. "Well, hello," a man in a floppy cotton hat and a windbreaker called out enthusiastically. He was walking toward me from the opposite direction. "How are *you?*" he said. It took me a few seconds to realize that I had randomly encountered the subject of this biography, my Santa Fe neighbor Murray Gell-Mann, hiking with his stepson, Nick Levis.

For weeks now I had been trying to pin down Gell-Mann for another interview. He had been running hot and cold ever since I had told him, two years earlier, that I intended to write his life story. Lately he had been more helpful. But now I was worrying that his second thoughts were being followed by third and fourth thoughts, and I had no idea what stage our relationship was in. I was relieved that he seemed genuinely pleased to see me. And I was struck again by how much, contrary to so many of the legends, Gell-Mann liked people and conversation, the easy camaraderie of encountering someone familiar on a mountain trail. The physics lore is filled with stories of Gell-Mann cutting down a colleague with a withering remark, of the mocking names he assigned to people whose ideas he didn't respect. Particle physics is the most competitive of intellectual sports, and faced with a theory or a theorist he didn't like, Gell-Mann could be merciless. But up in the mountains, in New Mexico, he seemed almost able to relax.

He introduced me to Nick, who like me was shivering without a jacket. When I said I was headed for La Vega, Gell-Mann was delighted at the coincidence. "La Vega," he said, his mouth stretched wide to mimic as perfect a northern New Mexican accent as you might hear in the villages of Chimayo or Truchas, down the other side of the mountain. He and Nick had also been heading to La Vega when the drop in temperature caused them to turn around, a little way up the trail, at Nambe Creek—"*Nam*-be," Murray said, with just the right amount of padding around the *b*. Now they were heading home.

If Gell-Mann was disappointed about not reaching this particular goal, he didn't show it. His eyes sparkled, and he seemed happy just to be out in the woods again. A few weeks earlier, the cardiologists had stuck a catheter in his chest, checking on his progress since a recent heart attack. They were relieved to find that the artery they had scraped out—a Roto-rooting, Gell-Mann called it—was still open. There was another, less threatening obstruction further downstream, but the doctors decided to leave it alone.

I was tempted to turn around and join Murray and Nick on the hike back. But somehow it seemed improper. This was not Murray Gell-Mann, the Nobel laureate, the discoverer of the quark and the Eightfold Way, but simply a man on a holiday with his stepson. My strategy all along had been to avoid making him feel cramped. I was in this for the long haul. After a few minutes, we parted ways. I made it about a mile past Nambe Creek. Then, just before the descent into the meadow, the clouds went black and I also decided to save La Vega for another day. Heading back down the mountain, I thought about how much I had come to like this brilliant, complicated, always fascinating, and often exasperating man.

When we visit the ruins of ancient civilizations, we reserve a peculiar fascination for those giant, elaborate structures that seem to serve no practical purpose whatsoever: the pyramids built by the Egyptians on the Nile and the Maya in Mexico, or the large circular kivas of Chaco Canyon in northwestern New Mexico. They stand meaningless now, rock-solid projections long outlasting whatever ideas they were meant to represent. Catholicism still survives, so we can understand some of the rationale behind Chartres, St. Peter's, and the other great cathedrals and basilicas of Europe. But we have barely a hint of the ideas that motivated the construction of the Sphinx.

It is sometimes said that the cathedrals of the late twentieth century are the giant particle accelerators, monuments to the belief—far from obvious on its face—that buried beneath the rough surface of the world we inhabit is a crystalline order so beautiful and subtle the mind can barely grasp it. Engaging in a fantasy, we can imagine, centuries and centuries from now, archaeologists (from this planet or perhaps from beyond the solar system) perplexed and captivated by the remains of the seventeen-mile-circumference particle accelerator being constructed at CERN, the European Center for Nuclear Research, near Geneva, or the four-mile ring at Fermilab in Illinois. These "atom smashers" are among the largest, most powerful machines ever built by the human race—not for the purpose of generating power, like the dams and nuclear reactors, or for predicting the weather or simulating nuclear explosions, like the supercomputers. Their sole purpose is intellectual: to find the faintest glimmers of evidence that, despite so many appearances to the contrary, we live in a mathematically symmetrical universe. How is it that a civilization long ago became so obsessed with this idea? That will be the riddle of these twentieth-century sphinxes.

If our parchments and our data banks survive along with the wreckage of our great machines, the archaeologists will learn a remarkable story: How the elders of the church of science came to believe that, despite what we perceive, matter is not continuous; it is made of invisible particles linked together in a beautiful architecture.

As the atomists would show over the years, the seemingly infinite variety of the world is generated by some one hundred elements, neatly arranged in the Russian chemist Dmitri Mendeleev's periodic table of the elements.

Viewed from the heavens, any hint of geometry on the earth— land divided into rectangles and circles, rock cut into blocks and piled straight and high—is usually a sign of intelligent creatures imposing order on an irregular world. But surely, the scientists believed, this harmony we find so soothing runs deeper. Beneath the world's confusion of forms is a scaffolding built according to a geometry as pleasing to the mind as a Gothic cathedral.

Since no one could directly see this geometry, the best one could hope for was to study its shadows. And so the physicists began to build the machinery they believed would provide an indirect glimpse. At first these devices were as simple as a jar enclosing gold foil leaves that seemed to waft in the wind of an invisible essence called electricity. By the early twentieth century, scientists were making gas-filled tubes that glowed in the dark with what they took to be mysterious beams of positive and negative charge. By studying and measuring these weird emanations, the physicists reached a powerful consensus: The world was even more elegant and symmetrical than Mendeleev and the atomists dared imagine. The variety of atoms found on the earth and in the sky were made up of combinations of just three particles: the proton, the electron, and the neutron.

But this newfound simplicity was short-lived. Not content with their instruments, the scientists built bigger and bigger machines. With the first particle accelerators, small enough to fit on a tabletop, they began smashing their elementary particles into each other and discovered that they weren't so elementary after all. They could be shattered into fragments. When they built bigger accelerators to smash the pieces even harder, they were left with fragments of fragments. Placing carefully designed detectors on mountaintops or sending them aloft in balloons, they found traces of still other particles, the cosmic rays bombarding the planet from space. Soon, there were so many of these "elementary" constituents that they threatened the very desire for order that had driven the search. The physicists were in despair.

And then, leading them out of the confusion, came the young scientists whose string of discoveries would do so much to make sense of it all, to find pattern hiding beneath the confusion. Viewed through these magicians' wonderful new lenses, the clouds lifted and order shone through. But it came at a curious price. To restore beauty to the core of creation, humanity was asked to believe in truths stranger than any that had come before.

The most remarkable of these wizards was Murray Gell-Mann. Graduating from Yale University at age eighteen, by the time he was twenty-one he had earned a Ph.D. from the Massachusetts Institute of Technology. Less than three years later, he began his revolution with an astonishing theory explaining the unlikely behavior of certain cosmic rays—the so-called "strange particles" that bombarded the earth from space. The legend was born. From then until a decade later, when he proposed the existence of quarks, Gell-Mann dominated particle physics. He is sometimes called the Mendeleev of the twentieth century, for what he provided was no less than a periodic table of the subatomic particles. In a fanciful allusion to Buddhist philosophy, Gell-Mann called his organizing scheme the Eightfold Way. While the periodic table shows that the plenitude of atoms can be generated by combining just three particles— the proton, electron, and neutron—the Eightfold Way shows that the hundreds of subatomic particles are made up of a handful of the elements Gell-Mann named quarks. Complexity was reduced to simplicity again.

But there is an important difference between the architecture of Mendeleev and the architecture of the Eightfold Way. And it is here that one can glimpse the enormity of the intellectual upheaval brought on by Gell-Mann and his colleagues. The periodic table, now a commonplace in any high school chemistry course, classifies the elements according to properties we can perceive with our senses. Every element is characterized by a unique mass and charge. Mass is something we feel when we pick up a rock; we generate charge when we shuffle across a carpet and touch a doorknob. Classified according to these commonsense qualities, the elements miraculously arrange themselves into columns—the rare earth metals, the noble gases, and so forth—whose members share similar characteristics.

In its ability to sift pattern from chaos, the Eightfold Way is at least as powerful, but tantalizingly more subtle. The qualities Gell-Mann used to arrange the subatomic particles were far more abstract than charge and mass. In his scheme, particles were classified according to elusive qualities called isospin and strangeness, which have no counterpart in the world of everyday experience. To describe the invisible patterns said to underlie the material world, Gell-Mann's strangeness was soon followed by more new qualities with names like charm, truth, and beauty. They "exist" not within the familiar world of three dimensions (four, if you include time), but within artificially constructed mathematical spaces, imaginary realms of pure abstraction.

Was this world stuff or mind stuff? To say that Gell-Mann "discovered" the quark is not quite right. All of his great breakthroughs came from playing with symbols on paper and chalkboards. His most important tools, he liked to say, were pencil, paper, and wastebasket. His discoveries were not of things but of patterns—mathematical symmetries that seemed to reflect, in some ultimately mysterious way, the manner in which subatomic particles behaved. But then "*invented* the quark" is not quite right either—implying some kind of postmodern relativism in which science is pure construction, just another philosophy. When Mendeleev drew his table, he left blank spaces for unknown elements that were discovered only years later. This manmade artifice was predicting truths about the real world. And so it was with the Eightfold Way. New kinds of particles demanded by Gell-Mann's abstract invention showed up in the experimenters' atom smashers.

The conflicting views of the nature of scientific ideas—are they discovered or invented?—are starkly laid out in the titles of two books: *The Hunting of the Quark* by Michael Riordan and *Constructing Quarks* by Andrew Pickering. Are quarks real particles (whatever that means) or mathematical contrivances? It's a debate that Gell-Mann refused to engage in. Philosophy, he thought, was a waste of time. But the puzzling questions about the reality of quarks—particles that cannot in principle be independently observed—quietly churned in his mind. One can see the struggle in the words he wrote and the lectures he gave. Ultimately he and just about everyone stopped worrying about it. Whether invented or discovered or something in between, it was Gell-Mann's quarks and his Eightfold Way that laid the foundation for the explanation physicists have given for how the world is made. For years particle physicists argued over who was the smartest person in their field: Richard Feynman or Murray Gell-Mann.

This idea of breaking the world into pieces and then explaining the pieces in terms of smaller pieces is called reductionism. It would be perfectly justified to consider Gell-Mann, the father of the quark, to be the century's arch-reductionist. But very early on, long before mushy notions of holism became trendy, Gell-Mann appreciated an important truth: While you can reduce downward, that doesn't automatically mean you can explain upward. People can be divided into cells, cells into molecules, molecules into atoms, atoms into electrons and nuclei, nuclei into subatomic particles, and those into still tinier things called quarks. But, true as that may be, there is nothing written in the laws of subatomic physics that can be used to explain higher-level phenomena like human behavior. There is no way that one can start with quarks and predict that cellular life would emerge and evolve over the eons to produce physicists. Reducing downward is vastly easier than explaining upward—a truth that bears repeating.

In the last decade, what aspires to be a new branch of science has sprung up to try and come to grips with complex phenomena—organisms, economies, ecosystems, societies, the thunderstorms that sweep through the Rockies. Gell-Mann, some fifteen years after winning a Nobel Prize for his reductionist tour de force, reversed direction and helped found the Santa Fe Institute, a world center for studying complexity. Part of his motivation was political. An ardent conservationist, he hoped to find scientific ammunition to support his environmental causes. He wanted to understand the complexity of the rain forests and convince the world that they must be preserved. But he also hoped to deepen the world's understanding of the relationship between the unseen particles science understood so well and the unruliness of the world that confronts us every day. Sitting in his small office, with its pictures of the particles he had discovered hanging on the walls like family portraits, he would look out at the Sangre de Cristo Mountains, at all this rich biology and geology begging to be understood. And, though some of his Santa Fe colleagues would beg to differ, he believed he had come close to figuring it out.

For all his accomplishments, Gell-Mann hadn't always, or often, been as self-assured and easygoing as he appeared that afternoon hiking in the mountains. I shouldn't be fooled, some of his old colleagues told me, by the newer, mellower Murray. As I explored his past, I found that his reputation as an intellectual show-off was well earned. He had long been interested in almost everything—classical history, archaeology, linguistics, wildlife ecology, ornithology, numismatics, French and Chinese cuisine—and he was always ready to lure people into conversations where he could display the depth of his knowledge and, it sometimes seemed, the shallowness of their own. The breadth of his learning had become legendary. He had taken visitors to dinner at Chinese restaurants, ordering in what seemed like passable Mandarin. He had strolled the streets of Kathmandu and Chinatown, translating the signs out loud. These performances could be exhilarating, and it was hard for people not to succumb to the pure delight he took as he reveled in the linguistic diversity of the world. But in his worst moments he could come off as a bully, someone who assumed that anyone who disagreed with him simply hadn't understood the argument. Everyone who knew him had seen the classic pose: Say something wrong or ask an ignorant question and he would raise his eyebrows in mock astonishment, then groan, holding hand to wrinkled brow as if his head were about to explode from the sheer weight of your misapprehension. He would sigh wearily, contemplating the effort he was about to expend setting you straight.

Over the years colleagues had been left dumbfounded by how self-centered the man could be. After accepting an invitation from the eminent journal *Nature* to speak at the most important event in its existence, the centenary celebration in London in 1969, Gell-Mann had abruptly canceled with three days' notice. He had an earache, he said. (As it turned out, the lecture was scheduled for the day after he would receive the early-morning call notifying him that he had won the Nobel Prize in physics—a coincidence, he would later insist; he hadn't been tipped off that the call was coming. And he really did have an earache, he pleaded to *Nature*'s editor, John Maddox, though, in the end, not a serious one.)

After charming his hosts by speaking in Swedish at the elegant Nobel Prize dinner, Gell-Mann had puzzled and then profoundly offended them by failing to submit his official lecture for publication in the annual celebratory volume. Seized with a pernicious case of writer's block, something that has plagued him all his life, he fended off one urgent telegram after another with abject apologies, finally conceding—months after the deadline was extended again and again for his benefit—that he wouldn't be submitting a lecture after all. Among the rows of volumes commemorating each year's prizes, one will find an empty page for Murray Gell-Mann.

It was difficult to know what to make of his behavior. He always seemed to feel genuinely sorry when he let people down. And it was hard for them not to forgive him, especially those who had seen his other side. This was also a man capable of taking time from his research to write a long reply to a high school student seeking advice about a career in physics, or to a mother worrying over how to raise her young prodigy to be as emotionally solid as he was intellectually precocious. (It was a balance Gell-Mann regretted he had never himself achieved.) And for every physicist he had cut down, there was another whose career he had promoted. When it came to writing recommendations for students and colleagues, no one was more generous. When two younger physicists, George Sudarshan and George Zweig, received only scant recognition for independently discovering some of the same phenomena that led to Gell-Mann's fame, he tried to make amends, providing glowing testimonials, nominating them for awards. Sudarshan, he wrote, apparently came up with the long-sought theory of the weak nuclear force, which drives many forms of radioactive decay, before anyone else, including Gell-Mann and Feynman. Proposing Zweig for the prestigious Majorana Prize in physics, for "seminal work" on the quark model, Gell-Mann added, "his contributions to this go far beyond mine." But he didn't always follow through on his good intentions. When an editorial in the *New York Times* erroneously referred to the Israeli physicist Yuval Ne'eman, not Zweig, as co-discoverer of the quark, Gell-Mann made a note to write to the editor and set things straight. But he never got around to it.

He always felt overwhelmed by all the things he wanted, or was expected, to do. And he was terrible at organizing his time. Every day, he would sheepishly tell people, he fell eight years behind. As the years went by, he became a worse and worse correspondent. "I'm getting to be as bad as Dick Feynman at answering letters," he apologized to a colleague in 1960. Feynman had long been the standard by which one measured such things. Eventually, Gell-Mann stopped answering mail at all. The advent of e-mail only made matters worse. One day I walked into his office and found him sadly staring at the screen of his Powerbook, crushed by the sheer glut of electronic epistles to process. Each hitting of the delete key was a decision he would rather not have to make.

In Santa Fe, where he had moved after retiring from Caltech, both sides of the old Murray remained. The charming conversationalist and apologetic procrastinator alternated with the unrepentant dispenser of acid remarks. And as in the past, he was sometimes subject to volcanic eruptions. A new secretary at the Santa Fe Institute once made the mistake of mentioning that she had just seen a television show about Feynman, who, since his death, had become such a celebrity. Murray erupted, attacking his old colleague's reputation and leaving her stupefied and wondering what she had done wrong. If Murray had mellowed from his days as the enfant terrible of physics, then the distinction was lost on her. He seemed to miss Feynman sometimes—he had once considered him among his closest friends—but he resented the way he had become enshrined like a dead rock star, with tapes of every lecture he had ever recorded dragged out of a closet somewhere and sold on cassette tapes and CDs to adoring fans.

Gell-Mann could be especially short with science writers, as I learned when I first met him, in 1992, at a conference on complexity in Santa Fe. The meeting was held at Sol y Sombra, the magnificent estate (the name means "sun and shade") on Old Santa Fe Trail where the artist Georgia O'Keeffe had gone to die. I walked into the meeting and got my first glimpse of Gell-Mann: his full head of tightly packed white hair, his styleless glasses with black plastic frames. He was wearing a bolo tie with a turquoise clasp and a jacket with an emblem of the Nature Conservancy, one of the environmental organizations he champions. The field of complexity is intimately related to the phenomenon called chaos, and Murray was loudly complaining about a popular book on the subject written by my former *New York Times* colleague James Gleick. I had admired *Chaos* immensely and was a little shocked when Murray denounced "this Gleick person," as he called him, for supposedly undermining the public's understanding of science. He conceded that Jim's book was beautifully written, but that somehow just made it worse. And Gleick's biography of Feynman made Murray livid. (Later he met Jim's brother, a scientist then visiting Santa Fe. They hit it off well, and from then on, Murray called him "the good Gleick.")

When the meeting broke for lunch, I carried my plate to one of the long wooden tables and sat down. I felt a mild adrenaline jolt when I saw Gell-Mann walk in my direction and, quite by accident, sit down across from me. He put out his hand and said in his deep, nasal voice, "Hi. I'm Murray Gell-Mann." I apprehensively introduced myself as an editor for the *New York Times*. "Oh, the *Times*," he said, smiling with amusement. "That's the place that employs that— what is his name?—that *Wilford* person."

It seems that John Noble Wilford, the dean of American science journalism, had once written a story that Gell-Mann didn't like. In the mid-1980s, some scientists at Purdue University were arguing that Galileo had got it wrong: A feather and a brick dropped inside a vacuum would not land simultaneously after all. A fifth force of nature—beyond gravity, electromagnetism, and the strong and weak nuclear forces—would cause some objects to accelerate faster than others. Wilford had called Gell-Mann to ask his opinion of what might conceivably have been a monumental discovery. After subjecting Wilford to a five-to-ten-minute oration on everything that was wrong with science-writing today, Gell-Mann tried to dissuade him from writing the piece. No one had heard of these scientists, Murray told him. Their analysis was shaky and would doubtless turn out wrong. As I listened to Gell-Mann tell the story, I could empathize with the frustration Wilford must have felt. Right or wrong, the fact that some card-carrying physicists were publishing this theory—now long forgotten—in *Physical Review Letters* was certainly newsworthy. Getting a quote from Gell-Mann would help put the story in perspective. I could imagine the clock ticking above Wilford's head, the deadline approaching, and Gell-Mann stubbornly refusing to cooperate. Looking for a good quote, Wilford apparently did what any of us might have done: He asked a leading question, something like, "Well, if the theory *does* turn out to be right, would it be important?" "Well, yeah, of course," Murray had replied. He was appalled to read the next morning on the front page of the *Times* that "Dr. Murray Gell-Mann, a theoretical physicist at the California Institute of Technology, said that if the conclusions of the study were correct, it was fair to speculate on the existence of a fifth force. . . ." Never mind that Wilford had taken all the care in the world to point out in his story how very tentative the research was. This had happened six years before, and Gell-Mann was not about to forgive him.

The story, given just the right spin by Gell-Mann, set off a round of laughter at the lunch table, which had filled with other physicists. I could see that it was going to be open season on science writers. And Gell-Mann was on a roll. "Things used to be worse," he said. He told about another science writer, a two-time winner of the Pulitzer Prize, who infuriated him by refusing to believe in the existence of the famously elusive particle called the neutrino. He was, Gell-Mann declared, "a man of impenetrable stupidity unmatched even by science writers today." This was getting to be a bit much. I had heard that Gell-Mann, the perfectionist and procrastinator, was having a huge amount of difficulty trying to write his own book, *The Quark and the Jaguar,* explaining complexity to a general audience.

The manuscript was late, and the publisher was ready to demand that he return his rather considerable advance. I couldn't resist. "How is your book coming along?" I asked. "Umm. Not very well," he admitted. He turned away and began talking to some physicists about string theory.

On the last day of the conference, when each scientist was giving a summary statement, I was asked (to my surprise and dismay) to tell what I thought of the affair. I had no idea what to say, but my encounter with Gell-Mann had been gnawing at me and I found myself describing our conversation at lunch that first day. The audience laughed at what they recognized as vintage Gell-Mann. And Murray stared straight ahead with a pained expression on his face. Sure that I had made an enemy, I avoided him for weeks. Then, at a Halloween party at one of his former student's houses (we were carving jack-o'-lanterns), he sat down next to me, as cordial as could be. "You know, I'm finding that this book-writing business isn't as easy as it looks," he said a little sheepishly. I took it as a touching concession, and before I knew it he was giving me chapters of his manuscript to read.

Though we had become friendly, he wasn't exactly thrilled when I told him two years later that I had decided to write his biography. He was a little flattered, I think, and he flattered me by saying I was one journalist he would trust to tell his story. But he wasn't sure he would be able to cooperate. His agent, John Brockman, was hoping to follow *The Quark and the Jaguar,* which had been sold to a dozen publishers around the world for more than a million dollars, with the *auto*biography of Murray Gell-Mann. He tried to convince Murray that this was a zero-sum game: Any good stories Gell-Mann told me would be money down the drain, material he couldn't use in his memoirs. I shouldn't have been surprised when, back in Manhattan, Brockman invited me for lunch in his office. Gell-Mann would tell all his friends and colleagues not to talk to me, he warned. I said I didn't believe for a minute that Gell-Mann had that kind of influence (he had made his share of enemies along the way). The agent struck again from a different direction. A biography about Gell-Mann written by a journalist is probably worth $100,000, he told me. But an autobiography—*coauthored* perhaps with a journalist— would be worth a million. But then it wouldn't be my book, I said. I had no intention of being amanuensis even for someone as intriguing as Gell-Mann. Though there are exceptions, the most honest biographies are usually unauthorized ones.

A little shaken (I had already signed a contract), I convinced myself I could write Gell-Mann's life story even if he refused to talk to me ever again. After all, the subjects of most biographies are dead. I started interviewing people Gell-Mann knew and struggling through the papers he had written. When he heard that I was planning a trip to Caltech, he called ahead to ask his colleagues not to tell me any "funny stories" about him. His agent had convinced him there was a market for a book of Gell-Mann anecdotes, like Feynman's *"Surely You're Joking, Mr. Feynman!"* a surprise best-seller. "I don't know any *funny* stories about Murray," one of his former collaborators grumbled, and proceeded to give me an earful of anecdotes Murray wouldn't have been tempted to put in his own book.

Soon after that, I moved from Manhattan to Santa Fe, where I would see Gell-Mann at scientific conferences or driving around town in a gold Range Rover with a license plate that read QUARKS. I would occasionally encounter him at restaurants and parties. Once my wife ran into him shopping alone at the Albertson's supermarket. (It was his turn to buy the groceries, he told her.) Slowly, I became a familiar presence and not so much of a threat.

Often we would see him with his wife, Marcia, a lovely poet he had met several years earlier in Aspen, Colorado, where he owned an old Victorian house. Gell-Mann was part of a group of physicists who had started going there in the 1960s for summer gatherings at the Aspen Center for Physics. Some of them, like Murray, had stretched a little and used their extra income from government and corporate consulting to buy homes for prices that seemed extravagant then—$100,000 for something that might cost $20,000 in the real world. Now they were real estate millionaires. Who would have guessed that particle physics would turn out to be so profitable, at least for some members of the generation that flourished after World War II?

Murray had been alone for a decade since his first wife, Margaret, the light of his life, died of cancer in 1981. Marcia had saved him, and Murray wanted her to know it. Only the biggest diamond earrings, the finest restaurants, the most expensive bottles of wine, were good enough. She drove around town in her own Land Rover. Sometimes the age difference—two decades—grated. He would complain about the loud rock music she liked to play or the slumber parties she would organize with students from the writing class she taught at the University of New Mexico. It became a standing joke, Murray affectionately complaining about his "expensive wife."

You could hear the anxiety in his voice as well as the pride. With houses in Aspen, Santa Fe, and, for a while, Pasadena, it was taking a huge cash flow to sustain his existence. He always seemed to be off on another expensive adventure, traveling to China, South America, Mexico, Antarctica, Cuba, the Galápagos, driven to where he could see the most exotic birds, the most beautiful wildlife. And to help pay for the trips and the other expenses, he was hitting the lecture circuit, always on the run.

Whenever our paths crossed in Santa Fe, he was unfailingly friendly but still not quite convinced he should talk to me. I didn't press but let him know that I was quietly working away. But he hinted that his plans of writing a memoir were fading. He had so little time and writing was such agony for him.

Finally, a year after I had begun immersing myself in his life, he sat down for what would become a series of regular interviews. And toward the end of the project, he surprised me by allowing unrestricted access to his personal archives. One by one, he let go of his valuable anecdotes. One evening, early on, we went to dinner at a Chinese restaurant, one where Gell-Mann could get the low-sodium food Marcia and his doctor had insisted on since the heart attack. Then we got in the Range Rover and drove up a winding dirt road—so rutted we had to slow to five miles per hour—to a place he owned in the foothills above Tesuque, a rural (or now semirural) community just north of Santa Fe. He was about to leave on a trip abroad and he needed to find his field glasses for bird-watching. With its west-facing wall of plate glass windows, the view from the house was magnificent. Looking out, one could see Sandia Mountain way down by Albuquerque, Tsichoma Peak, over in the Jemez range, Canjilon Peak, up north beyond Abiquiu. Behind the house, Lake Peak was glowing in the last minutes of sunlight. Together these comprised the four sacred mountains of the Tewa Indians, who still lived in pueblos down along the Rio Grande. Murray had learned a couple of the mountain's original names from his pueblo friends, and now he started reciting them. As he gazed out at the sunset, it was clear how much he loved this house and how sad he was that he had to sell it. Marcia didn't like the spartan feel of the place—with its brick floors and unfinished wooden roof beams, it was more like a large, very nice cabin—or the long drive into town. They had recently sold the home in Pasadena where he had lived for years and had just bought a house with an indoor swimming pool in Santa Fe's expensive museum district. The house itself was like a museum, with Murray's collections of indigenous American pottery, African art, rare books, and ancient weapons—an Eskimo harpoon, a North African mace, a blowgun complete with poison darts, a Sumatran dagger, a Chinese beheading sword. "For keeping Marcia in line," he said sardonically. He could get away with comments like that because it was so clear how devoted to her he was.

Traveling from one exotic place to another, surrounding himself with historical treasures—it seemed like a wonderful life. And yet he didn't always strike me as very happy. One day when I was talking to him on the telephone, he suddenly remembered a kindness someone had shown to him back when he was a graduate student struggling to find money to live on when his scholarship was held up. He broke off the story and fell into a reverie. "Everybody was very nice to me," he said quietly. "I must say they were just so nice to me. That's always been true. All my life people have been very, very nice to me and I usually didn't profit by the advantage." I thought he was going to cry.

After all his stalling and agonizing over the perils of talking to a biographer, I was startled at how easily he let down his guard. I guess he had just gotten used to me. One afternoon, reminiscing about the sad early death of his old mentor, Enrico Fermi, who may have slowly, unknowingly poisoned himself with radiation, Murray's voice trailed off into silence. "I don't know," he said after a long pause. "I wanted to write some of this myself, but you're so nice and so charming that I just tell you everything." He shook his head, a little exasperated at himself. I was starting to feel slightly guilty, sitting there in his office, the tape recorder rolling.

CHAPTER 1

A HYPHENATED AMERICAN

Scouring the old Manhattan telephone directories from the early years of the century, now relegated to decaying spools of microfilm in a dark corner of the New York Public Library on 42nd Street, one looks in vain for the curious appellation "Gell-Mann." Column after column of Gelmans and a sprinkling of Gell-mans and Gellmanns form a long, gray expanse. But year after year, the surname Gell-Mann makes not a single appearance. The name is so unusual, if not unique, that many physicists—even those who have, at one time or another, considered themselves Murray's closest friends—have long assumed it to be an affectation. Driven perhaps by his fascination with etymology and a desire to leave behind a sometimes depressing past, Murray added the hyphen (the story goes), jettisoning a rather ordinary name for one with a European flair.

After all, Murray was a master at coining amusing new appellations. This was the man who, barely in his mid-twenties, had first seized the scientific spotlight by inventing "strangeness"—an abstract quality said to cause certain cosmic ray particles to behave in a manner that, before Murray had his say, seemed to defy the laws of nature. When he and George Zweig revolutionized physics by suggesting, quite independently, that the teeming confusion of subatomic particles could be made from just three basic building blocks, Zweig, with a tin ear, called them "aces." It was Murray, thumbing his nose at the pretentiousness of scientific terminology, who picked the name that stuck: quarks. He would later remind people that he had not actually taken the name, as many believed, from James Joyce. The sound of the name, "kwork," Murray wanted the world to know, was an entirely original invention. He just wasn't sure how to spell it until he stumbled upon the now-famous line in a first edition of *Finnegans Wake* that belonged to his older brother, who goes by the unpretentious moniker Ben Gelman. "Three quarks for Muster Mark," Joyce had written with his usual opacity. Like a thousand graduate students before him, Murray liked to pore over the pages in his spare time, squinting to see beneath the surface. Was there meaning lurking in there or just an author's silly word games? It was a tantalizing text, as subtle as the quantum machinations of the subatomic world, but when it came to christening the most basic of all particles, Murray wasn't about to give Joyce any more credit than he deserved.

Making up names was not just fun; it gave one a feeling of power over the world. Quarks, the mischievous young scientist proposed, came in three different varieties, which he dubbed "up," "down," and "strange." When other scientists later found hints of three additional "flavors" of quarks in the universal ice cream shop, what else could they call them but "charm," "beauty," and "truth"? (The latter two were soon renamed, just as oddly, "bottom" and "top.") When theory demanded that all six quarks be sliced across the grain into three more categories, they were called "colors." Gell-Mann preferred "red," "white," and "blue," though it was later decided that it would be less parochial to name them after the three primary colors of light: red, green, and blue. Physics was on a roll, inventing whimsical new names and concepts as fast as they were needed. And it was Gell-Mann, more than anyone, who started it all. It was perfectly reasonable to postulate that he had also coined his own surname.

This prodigy, it seemed, was a different kind of hyphenated American. Of all the Gelmans and Gellmans who came to the United States, the only Gell-Manns who now turn up in a search of the computerized phone books that have thankfully replaced the microfilm spools are Murray and Murray—at his addresses in Aspen and Santa Fe.

Gell-Mann himself offered little help with his personal pedigree. For most of his life, he steadfastly refused to talk about his family, referring vaguely to his parents as "immigrants from Austria" or alluding to a father who had studied in Vienna. In his later years, when he was willing to be a bit more specific and forthcoming, he would simply describe his father as "the son of a forester, living in the beech woods of what was then eastern Austria, near the Russian border." It's characteristic that Murray would know the names of the trees that surrounded his father's boyhood home, just as he would, in his own childhood, lovingly and obsessively identify every tree and bird in Central Park. It is equally characteristic that he would resist simply saying that his father was, like so many proud Americans, an Eastern European Jew. With his passion for words and language, Gell-Mann was notorious for beginning a conversation with a new acquaintance by providing a detailed, on-the-spot etymology of the person's own name—including the way it should *really* be pronounced. But when it came to the origin of "Gell-Mann," he was typically cagey.

Sheldon Glashow, who won a Nobel Prize for helping untangle the deep connections between electromagnetism and the weak nuclear force, liked to tell about a party in Berkeley many years ago where a beautiful young German woman, the wife of the host, asked Murray about the notorious hyphen. As Glashow would have it, Murray, fascinated by the woman's good looks, backed her into a corner and immediately launched into a tale, apparently improvised on the fly, about the confluence in Scotland of the River Gell and the River Mann. The physicist Leon Lederman, who rarely resisted an opportunity to puncture Murray's egotistical balloons, once facetiously called himself Leon "Lede-rman" in a Fermilab biographical sketch: "His period of greatest creativity came in 1956," he wrote, mockingly referring to himself, "when he heard a lecture by Gell-Mann. . . . He made two decisions. First, he hyphenated his name. . . ." Murray got his revenge. He was famous for coining names not just for particles but for people who had crossed him—in fact, it seemed, he would become incapable of uttering an antagonist's real name. Lederman became known as "Leather Person."

Trivial as the matter was, Murray's evasiveness simply reinforced what became the common wisdom: that sometime between his birth and his first publication as a physicist, Murray Gelman had shed his dusty Lower East Side chrysalis and taken flight as Dr. Murray Gell-Mann. And the story, it turns out, is wrong. Searching through the library microfilms, one's eye is struck when suddenly in November 1920 (in those days, the phone book came out twice a year), a name pops out among the Gelmans and the Gellmans like a typographical error: "A. I. Gell-Mann, 233 E. 14th Street, Stuyvesant-7455." The affectation seems to have come not from the son but from the father, a dignified, somewhat pedantic man who, with a different roll of the dice, might have made his own mark as a scientist or scholar. It was not Murray but his brother Ben (weary of explaining the weird spelling) who had given himself a new name.

Long after he had received his Nobel Prize and was coasting into retirement in Santa Fe, Murray toyed with a popular cosmological theory in which all the different ways the universe might have unfolded—the alternate histories, he called them—can be thought of as "existing," in a purely abstract sense, simultaneously. There is a path where your father didn't meet your mother because he turned left instead of right coming out of a movie theater, or one in which the solar system congealed into sixteen planets instead of nine. Gell-Mann would puzzle at the circumstances, the long chain of accidents, that lead each of us along these ever-forking trails. In this universe of things that might have been is a path in which Arthur Isidore Gell-Mann, fresh from his philosophical and mathematical studies in Vienna, took America by storm, rising through the ranks at City College, say, or Columbia University to assume his place among New York's intelligentsia. But this was not to be. The manner in which his father's rise was so abruptly truncated, his life shuttled off onto one of history's sidings, was never far from Murray's mind. He had good reasons for not wanting to talk about the past.

In a very loose sense of the word, Isidore Gellmann (he apparently adopted both "Arthur" and the hyphen after arriving in the United States) was indeed an Austrian immigrant. He was born on June 13, circa 1886 (the exact year is a matter of dispute), in Chorostków, a small village in the province of Galicia near the Russian frontier. Seized from Poland in the late eighteenth century, Galicia lay on the northeastern fringe of the vast Austro-Hungarian Empire ruled by the Hapsburg dynasty. The volatile land was home to Roman Catholic Poles, Eastern Orthodox Ukrainians, and a community of Jews, who had been migrating into the area from Germany since the expulsions of the Middle Ages. It was an uneasy and sometimes explosive mix. As unwilling subjects of the empire, the Poles resented the Austrians. As Catholics, they were suspicious of the Jews. Those Jews who swore allegiance to the Hapsburgs—like Isidore and his parents, Moses and Celia—were considered doubly damned.

Unlike his distant cousins who populated the claustrophobic shtetls of Galicia, Moses was among the lucky few who managed to elevate themselves into the small middle class of educated Jews who helped carry on the business of the empire. At the time his son was born, Moses worked as a forester and game warden on a nobleman's private estate. As family lore had it, he was an officious man who liked to boss the peasants around as he wandered the woods with a mastiff named Hector. Though the family spoke Yiddish, they preferred German, the official language of the empire. Their family name—Gellmann—may have been a Russian rendition of the German name Hellmann. "Hell" means "bright" or "clear," and a "Hellmann" was someone with blond hair. According to some interpretations, a "Hellmann" was also someone who saw clearly, a clairvoyant. Since the biblical prophet Samuel was known as a seer, some conjecture that "Gelman" (with its variations) was another way of representing his name.

No family diaries survive, and neither Moses nor Celia seems to have passed on many tales of life in the Galician hinterlands. But an inkling of what life was like in those days comes from the novelist Karl-Emil Franzos, a German-speaking Jew who, like Murray's father, grew up in a village on the Russian-Galician frontier. Franzos's father was a doctor for the empire. In an introduction to his novel about a young man's life in Galicia, called *Der Pojaz* (The Clown), Franzos describes the netherworld the family inhabited—separate from the Christian Poles but never really part of the lower-class, Yiddish-speaking Jewish community. "I seldom went into a Jewish home," he wrote, "never into a synagogue. Neither religious rituals nor kosher cooking was practiced in my home. I grew up on an island."

I grew up on an island. Considering the isolation from the people around them that both Isidore and Murray would later feel, this last sentence could have been written by a Gell-Mann.

Sometime before the turn of the century, the edge Moses and Celia had gained over so many of their poorer neighbors began to crumble because of a devastating financial mistake. Working for the local nobility, Moses had managed to accumulate enough money to invest in a business. In those days it was not uncommon for wealthier Jews in the region to operate small distilleries and breweries, fed by the vast grain fields, or to run steam-powered mills that turned the trees of the Carpathian forests into lumber. At the recommendation of Celia's father, Moses invested the family nest egg in a lumber mill. But somehow the investment went sour, and the family was forced to leave the woodlands for the economic promise of the city. While Isidore was still a boy, they moved south to Czernowitz, capital of the neighboring Austrian crown land called Bukovina.

Going from rural Galicia to cosmopolitan Czernowitz must have been like emerging from a cave. In the late nineteenth century, Galicia was mired in the poverty and prejudices of the Middle Ages. Bukovina was known as "the Jewish Eldorado," and those years were its golden age. "Bukovina" means "the beech wood," after the forests that covered the lower slopes of the Carpathian Mountains, the geological wall separating the province from its southeastern neighbor, Transylvania. For hundreds of years, when it was still part of the kingdom of Moldavia, Bukovina had been something of a haven for Jews, buffering them from the seemingly endless local skirmishes and ethnic cleansings as Russians to the north clashed with invading Turks, or wrested away more land from the Poles. By the time Austria-Hungary took over Bukovina in 1774, a prosperous class of Jewish merchants and businessmen was already flourishing there. When the Hapsburgs began settling the area with German colonists, they welcomed the Bukovinian Jews as allies. Given their civil rights in 1867, they enthusiastically embraced the culture of the empire and were well along the road to assimilation. They were proud to be Austrians—an identity as important to many of them as their identity as Jews.

Nowhere was this spirit stronger than in Bukovina's capital city, Czernowitz. Built on the site of an old Turkish village in the Carpathian foothills, overlooking the beautiful Prut River valley, Czernowitz became the easternmost outpost of the empire, a little pocket of Austrian culture surrounded by Ukrainians, Romanians, Transylvanians, and Poles. With its baroque architecture and tree-lined boulevards studded with ornate monuments, the city earned the nickname "Little Vienna." Gathering at the coffeehouses on the Ringplatz, the town square, or strolling through Schiller Park, the Jews who formed the city's business and artistic elite thought themselves truly chosen. The mystique surrounding cosmopolitan Czernowitz was nicely captured by the novelist Franzos, who like Isidore (the similarities continue) went to school there. In *Der Pojaz* the young protagonist leaves dreary Galicia to fulfill his dream of working as an actor in Czernowitz's thriving Jewish theater. When he reaches the outskirts of the city, his traveling companion refuses to enter what is rumored to be a strange place where "the Jews speak High German and eat pork." One longtime resident had a similar view: "The bourgeoisie spoke German and were driven to synagogue in a horse-taxi on Yom Kippur. The ordinary people spoke Yiddish and walked."

The members of the elite showed off their good taste and learning at the grand concerts given by the town orchestra; they visited and gave money to the National Museum of History and Art; they read the German newspapers, which were edited by Jews. Their children studied at Franz Josef University, where even a few of the professors were Jewish. Of all the cities of the old empire, its former denizens still brag, little Czernowitz produced the highest proportion of important twentieth-century German writers, including Paul Celan, Rose Auslander, and Gregor von Rezzori.

Though Moses and Celia were not in the highest echelons of Czernowitz society—they ran an ice cream parlor—they managed to enroll Isidore in the gymnasium, the state-run secondary school, for a classical education in Latin and Greek. But this business failed too, and sometime in the first few years of the twentieth century they decided to leave the cultural mecca, following so many other Jews who told themselves that life must be better in America. It was customary in those days for the son to take the leap across the Atlantic, sending for his parents after he had scratched out an economic niche. The Gellmanns turned the usual situation on its head. Isidore, who had excelled in school, accompanied his parents partway, staying behind in Vienna to study at the university.

As it turned out, the family had seen Czernowitz at the height of its glory. Before long the brief, anomalous era when Austria was considered the protector of the Jews would come to an end. After the fall of the empire in 1918, Bukovina's golden light began to fade. When the spoils of World War I were divided, the Jewish Eldorado fell to its repressive southern neighbor, Romania. Czernowitz became Cernauti, and the Jews were slowly squeezed down and out of the upper reaches of society. Then came the Nazi horror of deportations and exterminations. Then the "liberating" Soviets, so thorough in their attempts to drive the old spirits from the city, now called Chernovtsky, that they chopped the corners off the Stars of David on the old Jewish cultural center. By the early 1990s, a Jewish population that had reached 100,000 before World War I had been cut to less than 10,000.

Isidore would later tell his children that the persecution and pogroms already occurring across the border in the Russian Pale had nothing to do with the family's dislocation. They had gotten out in time. As he immersed himself in philosophy and mathematics in Vienna, any loose grip that Jewish culture and religion may have had on him slowly faded away. Like his son later on, he was uncomfortable with being called Jewish. He preferred to consider himself a citizen of the world.

And what better place to be than cosmopolitan Vienna? Imagine this bright young man from the provinces, traveling some five hundred miles by rail through Budapest, disembarking in the very capital of the empire, the center of European intellectual life. Isidore had been mightily impressed by his teachers in Czernowitz, employees of the empire who, on special occasions, would strut about in uniforms sparkling with official insignia. Sometimes it seemed there would be nothing more wonderful than to be a gymnasium teacher, with all the prestige and regal trappings. Of course he might have to convert to Catholicism, the religion of the empire, but among many Jews in Vienna, absorbed ever more deeply into the cultural mainstream, that was not necessarily considered an obstacle.

But why just settle for teaching in a gymnasium when so many new possibilities now radiated before him? Czernowitz had given Isidore a watery taste of Austrian culture. But here was the real thing. Not just the old baroque splendor of the empire, with its Strauss waltzes and rococo style—this was the Vienna of Sigmund Freud and that extraordinary generation of not-very-observant Jews whose intellectual revolutions prefigured the fall of the Hapsburg dynasty and the dawn of the modern age: Wittgenstein in philosophy, Kokoschka in painting, Schoenberg with his twelve-tone music. Everything in the world of art and ideas was up for grabs. If the significance of Vienna in those days can be compacted into a single statement, it might be this: Wittgenstein's sister Margarete's wedding portrait was painted by Gustav Klimt, and she was psychoanalyzed by Freud. It was the perfect place for a German-speaking Jew who dreamed of becoming an intellectual, who wanted his Jewishness to be just another detail, like the color of his hair. To be in Vienna in those years was an adventure that could outshine even a journey to America. After all, the Americans were coming *here* to learn medicine and philosophy.

Isidore rented a room from a recently arrived Brazilian family and was amazed to learn that they had never seen snow. He studied mathematics, languages, and philosophy, and in his second year he followed a favorite professor to Heidelberg. Somewhere along the way, he became enchanted by the philosopher Leibniz's notion of monads, those fundamental forms from which everything in the world was supposedly constructed. Later, when Murray became the intellectual giant that Isidore had longed to be, the father would irritate the son by comparing elementary particles to Leibniz's mystical building blocks.

One can infer and infer about the effect of Vienna on this inquisitive young mind. But the details of the stay are frustratingly fuzzy. Years later, when Murray was developing his ideas on cosmology, an important consideration was the difference between fine-grained histories, where the unfolding of a cosmological event could be scrutinized with the utmost precision, and coarse-grained histories, where the story was followed from a more distant vantage point, with the finer details blurred together. All that survives of Isidore Gellmann's story is the coarse-graining. A letter he wrote while he was in Vienna suggests that he might have been a freshman at the advanced age of twenty-six. Perhaps he spent a few years tutoring before he earned enough money for tuition at the university. But with even his birth date uncertain, it is difficult to be sure. Over the years the story Murray would tell became compressed into a line or two, something about a father who had studied philosophy in Vienna and emigrated from Austria. If a listener came away with the impression that Arthur had perhaps hung out in coffeehouses with Wittgenstein when he was home for Christmas vacation, that just added to the aura.

As it turned out, Isidore's career as an intellectual was already headed for an early end. This was as good as things were going to get. Across the Atlantic in New York, on East 4th Street, Moses and Celia were having money problems again. The lore is filled with stories of lucky Eastern European Jews whose stars began to rise as soon as they left Ellis Island. For the Gellmanns, the move to America became a tale of downward mobility. Like so many of their fellow immigrants, they could find work only in garment industry sweatshops. And as for so many people working long hours in close, damp quarters, Moses's lungs mildewed with tuberculosis. Sometime after 1911, Isidore was summoned to the United States to rescue his parents from their latest failure.

Isidore's first few years in America are a blank—more lost information. Early on, he adopted "Arthur" as his first name. By 1916 his gift for languages and his dexterity with numbers helped him land a job as a teacher at the Dickensian-sounding Hebrew Orphan Asylum in Philadelphia. He hadn't yet lost his high ambitions: He also served as assistant to the superintendent and secretary of the institution. Despite his increasing disdain for religion, he was also compelled to act as secretary of the synagogue and teach Jewish history.

Three years of this was about all he could stand. On his visits home to the Lower East Side, where his parents were still living, he began looking for more suitable work. He also began courting an alluring young woman named Pauline Reichstein, the daughter of his mother's friend Clara. The Reichsteins had also emigrated from the eastern fringe of the empire—a village called Janow, in what had been part of Poland—after the death of Pauline's father, Benjamin. Like Moses, Benjamin Reichstein had worked as a minion of the empire—a court clerk—and the family proudly kept the official stamp emblazoned with his seal. Pauline, about ten years younger than Arthur, was an infant when her family made the voyage. Until 1940, when she tracked down her birth certificate, she thought she had been born in the United States. After arriving in America, Pauline's mother remarried, and Pauline's stepfather—a Yiddish-speaking Galician immigrant named Moses Fellner—brought along half a dozen children from a previous marriage. To help make ends meet, Pauline enrolled in secretarial school.

After a short courtship, Arthur proposed. The marriage certificate shows that Arthur I. Gellman (the second *n* somehow dropped away) and Pauline Reichstein were joined in matrimony by the City Clerk in Manhattan on August 12, 1919. They didn't tell their parents until New Year's Day. The reason for the secrecy has been lost, but when Murray's parents told him the story years later, he thought it added a romantic luster to those early days.

After a year of shuttling between New York and Philadelphia, Arthur quit his job teaching orphans. He was rapidly learning to speak English in the flawless way that shows the speaker is not a native. With a glowing letter of recommendation in hand from the orphanage, he landed a job on Wall Street. Raabe, Glissmann and Company of 20 Broad Street, a stock and bond trader, hired him as secretary to the president, a position that involved composing advertisements and compiling statistics. He quickly moved up to become head of the foreign correspondence department and manager of foreign securities. This was more like it. It wasn't academia, but the money was good and the titles at least sounded prestigious.

It was around this time that Arthur began trying out the hyphen, apparently pleased by the flair it added to his name. Though the marriage certificate spells his name "Gellman," on the letter of

recommendation from the orphanage it emerges as "Gell-Mann."
Over the years the name would appear in many different ways—
Gellmann, Gellman, Gelman—as bureaucrats tried to wrestle it
back to the ordinary. Who knows what Moses thought of all this?
Relatives remember him as plain old Moses Gellman, pronounced
like the more common Gelman, with a hard *G* and the accent on
the first syllable. And they might not know whom you're talking
about if you use the pronunciation Murray would later insist on,
with the stress placed almost equally on the two syllables. They talk
about "Murray Gellman" and "Arthur Gellman" as though that
hyphen had never wedged itself into the name.

When Arthur and Pauline's first son was born in 1920, they
named him Ben—short not for Benjamin (Pauline's father's name)
but for Benedict. There is no reason to suppose that he was named
after St. Benedict; maybe it was for Benedict Spinoza, the great
Dutch philosopher whose rationalist notions got him excommuni-
cated by the Jews of Amsterdam. Offered 1,000 florins a year to
keep his blasphemous notions to himself, he refused. This was a
man Arthur would have wanted his children to admire.

Enduring the mental drudgery of Wall Street, he still dreamed of
a better life and found time to take pharmacy classes miles uptown
at Columbia University. Since he couldn't afford the textbook, he
made his own, carefully copying the intricate diagrams. But when it
came time to take the qualifying exam, he couldn't afford the fee
and was too proud to borrow the money. Another avenue of possibility closed. Two years after Ben was born, the brokerage company
went bankrupt and Arthur was out of a job. He found work han-
dling correspondence for a toy company that did business with Germany, but he longed for something more dignified.

If he could no longer hope to be a professor, then perhaps he
could still make a living as a teacher. Arthur was one of those rare
people who acquire languages as easily as others learn the rules of a
card game. With the expanding acres of immigrants, he had a ready
audience and clientele. Sometime after the family settled into St.
Mark's Place, a few blocks north of Houston Street, he started the
St. Mark's School, offering lessons in English and German.

But the Gell-Manns were not at this address for long. Pauline was
a restless woman, and the family always seemed to be moving, sometimes for the most trivial reasons. (Once she insisted on relocating
because the boy next door had whooping cough.) After several
uprootings, the family migrated to the clanging thoroughfare of

14th Street, between Second and Third Avenues, occupying a floor above a restaurant in a five-story building next to the 14th Street Theatre. They shared the space with the institution that soon took a more grandiose name: the Arthur Gell-Mann School.

Arthur threw himself into his makeshift institution with all the confidence and aplomb of his old gymnasium teachers in Czernowitz. A proper school must have a syllabus, and so he began compiling a book of English lessons, cramming three looseleaf binders with pages laboriously typewritten with black and red ribbons. This was not a simple course in informal conversation. From day one, students were submerged in a numbing compilation of the dishearteningly complex rules of English pronunciation—along with all the exceptions and exceptions to the exceptions. "When a vowel is followed by 'rr,' " the syllabus advised in Rule 12, "it generally has its medial sound, except where a single 'r' has been doubled through the influence of inflectional or derivational terminations." From pronunciation, one moved on to the labyrinthine rules of English spelling, then of grammar, finally reaching the pinnacle, where, like thirty-third-degree Freemasons, the more ambitious students were exposed to the highest secrets of the craft—a catalogue of the arcane rules of rhetoric that would ensure that a point was made not only competently but with style. There were, of course, the familiar devices of simile, metaphor, onomatopoeia, and euphemism; and the somewhat more vaunted oxymoron, metonymy, and synecdoche. But an immigrant taking Arthur's classes couldn't consider himself fully versed in English until he had mastered a whole list of devices with names that sounded like rare diseases. There was prosopopoeia (the assignment of life to the inanimate, as in "Murder, though it have no tongue . . .") and paranomasia (a play on words like, "What is mind? No matter. What is matter? Never mind.") There were syllepsis, epistrophe, antistrophe, periphrasis, aposiopesis, pleonasm. It's hard to imagine an immigrant acquiring English in this top-down, unintuitive manner, but apparently that is what Arthur himself had done. Along the way, he developed a pedantic, overbearing style that would later drive young Murray up the wall—until he began half-consciously imitating it himself.

As stated in the introduction to the syllabus (and we can imagine Arthur intoning these words in the makeshift classroom as the noises of 14th Street rattled the windowblinds): "If a foreigner goes wrong in the matter of intonation, emphasizing a word that a native would pass by lightly; if he misplaces his accents; nay, even if he

occasionally substitutes a foreign speech sound for the corresponding English sound, e.g., German guttural 'r' for the English alveolar
fricative—we must bear with him. But if he stumbles over monosyl-
lables of the simplest type, mistaking 'huge' for 'hug', 'laud' for
'loud', and 'seize' for 'size'; if he articulates stressed and unstressed
vowels in the same manner, giving the same sound to 'a' in 'man'
and in 'gentleman'; if, quite unaware of the positional operations
in English orthography, he pronounces 'a' alike in 'van' and 'wan',
or in 'hard' and 'ward'; if he obliterates the difference between
sonants and surds, saying 'serf' for 'serve', 'bet' for 'bed', and
'shop' for 'job'; if, by introducing his 'glottal stop' into English, he
hacks word-groups that we are accustomed to hear in their entirety,
as an unbroken series of sounds, into a number of straggling, jar-
ring, and unmeaning vocables, thereby offending the English ear
even more than by his faulty articulation;—I say, if he goes wrong in
such matters, the point is—not that he cannot read with propriety,
but that he cannot read at all."

Here was a man who had mastered a language, flowery rhetoric
and all, a man in love with learning and with the sound of his own
voice. It's hard to imagine who he thought his audience was—other
than himself.

The nearest respite from the constant din of 14th Street was
Stuyvesant Square, a block to the north, with St. George's Episcopal
Church rising majestically on its west side. Ben came away from the
neighborhood with the typical New York memories of passing time
at the local dairy restaurant (run by someone called "Old Lady
Hammer") and of illicit games of stickball, which ended abruptly
when the cops took away the broomstick and broke it in the slot of a
manhole cover. (He decided then that marbles was a safer activity.)
For a while the language school thrived. Teachers were hired so the
school could offer the Romance languages, as well as Arthur's specialties, English and German. Life was good for a while.

On September 15, 1929, when Ben was nine years old, Murray
was born. His first memory is of being maneuvered up a dark, steep
staircase in a baby carriage. When he was two, the family and school
moved to a larger and quieter apartment on the other side of the
block: 15th Street across from the Friends Seminary. One evening, with the family sitting on the stoop of their building, Murray
looked around at the streetlights coming on and—according to
family legend—uttered his first words: "The lights of Babylon!"

It is conceivable that a child born with nimble neural circuitry and exposed to a father who regularly spouted classical allusions might have issued such an impressive utterance. But memory has a way of turning the past into a good story. Murray once recalled his mother taking him to Stuyvesant Park to watch an eclipse and feeling jealous because his brother and father had gotten to travel a little farther—to New Jersey perhaps—for a better look. But the more he thought about it, he wasn't so sure. He might have been blurring a memory of a trip to the park with a story Ben and Arthur later told of an entirely different occasion. Other memories were more vivid, like the time he pedaled his kiddie car to Gramercy Park and looked through the bars of the gated private gardens, wondering why he couldn't go inside.

Most of all, he came away from those early days with memories of being very puzzled. His parents dragged him from place to place with no explanation of what was going on. They seemed to underestimate his curiosity and ability to understand. Or maybe he needed more explanation than most children his age. Why had he been placed on this large animal, a pony, to have his photo taken? Why were they taking him to this big noisy place with lots of lights and people (Wanamaker's department store at Christmas) to sit on the lap of a bearded old man in a red suit? Nothing seemed to be connected, and he would later wonder whether his passion for pattern grew from constantly being confused.

For both Ben and Murray, the outings with Arthur were soon history. Murray was born six weeks before Black Tuesday set off the Great Depression. By the time he was three, the language school had collapsed and Arthur was left with practically nothing. He had to sell most of his books, including his cherished chess library, keeping only the volumes about mathematics and science. From then on, just about everything the family owned came from what Murray would bitterly think of as the good old days—before he was born. Later on, when his fame as a physicist brought him lucrative consulting deals and lecture fees, as well as one of the highest salaries a professor can command, he would still find it impossible not to worry about money.

In 1932, all but destitute, the family moved to cheaper quarters in the Bronx, on 188th Street, the first of three addresses they occupied over the next five years as money problems, and Pauline's neuroses, pushed them from one dismal dwelling to another. Arthur found work a few months later as the vault custodian at a bank on the Lower East Side, taking the long, rattling subway ride to work every day. Pauline thanked Franklin Roosevelt, who had just been elected, for their good fortune. Arthur was not so grateful. For the next twenty years he stoically worked at the bank, impressing Murray's friends with the gun he got to carry. He distinguished himself as a top marksman when the bank guards competed at the shooting range. But this was hardly the kind of distinction he'd had in mind. He was never really happy after his school failed. For a while he worked the swing shift at the bank, leaving home around noon, returning at 10 p.m. Sometimes he had to stay later while the clerks straightened out a discrepancy in the accounts. This happened often enough that years later Murray would sometimes wonder if his father had had a girlfriend.

When Arthur was home, he would closet himself in his room reading his science books, struggling for a layman's grasp of Einsteinian relativity. But for all his skill with numbers, he found the equations intimidating. He still liked to play with numbers, carefully filling notebooks with precise diagrams of conic sections and trigonometric relations. He compiled notebook after notebook of "Arithmetic Recreations": finding the latitude from the shadow of a pole, or the largest number that can be expressed with three digits (9 to the 9th power to the 9th power, which generated so many digits, Arthur wondrously noted, that if one recorded fifty to the line, forty lines to the page, and 500 pages to the volume, it would fill almost 370 books and take more than ten years to complete). Or he would play with perpetual calendars, filling a notebook with solutions to problems like this: "The Dutch philosopher Benedict Spinoza died on Feb. 21, 1677. What day of the week was that?" Then he would tediously calculate the same for Descartes, Pascal, Kepler, all his heroes. He constructed magic squares, where the rows and columns and diagonals added up to the same sum. Ultimately he filled eight notebooks with data about the moon—mean distance from earth, diameter, mass, mean density. In exercise after exercise, he would calculate the phase the moon was in for any given date in history. He copied out the powers of 2 all the way up to 2 to the 65th, and listed all the prime numbers up to 1,000. He must have been bored out of his mind. As time passed, he drew further and further inside himself, and Murray sometimes felt as if he didn't have a father.

Arthur seemed to hope that the notebooks, like his excruciatingly exact English syllabus, would be published as textbooks someday. They had tables of contents, appendixes, addenda—all written in his careful hand. And occasionally he would slip into the second person, addressing some imaginary reader with teacherly advice. "Before attempting to use a table of natural logs, make a careful study of its place, its range, and the size of the intervals. . . ." Or perhaps he had just copied others' books, unable to afford his own. The notes were good enough that Murray later consulted them in high school and at Yale. (He even referenced one of the books in a paper, which his instructor told him was pretentious.) But he didn't learn very much from Arthur directly. The man was just too didactic and demanding, like a Prussian schoolmaster. He wasn't even that effective a German teacher. Years later Murray would regret that his father hadn't simply spoken German to him as a child, gracing his synapses with a fluency that can never be so easily acquired later on. Instead he tried to give Murray formal lessons, grafting the German onto the English. For all his linguistic abilities, Murray never learned to get the German case endings quite right.

While Arthur grew more and more pessimistic, convinced there was little point anymore in trying to excel, Murray's mother took the opposite approach to life. She was obsessively cheerful, even when there was nothing to be very cheerful about, rarely complaining, losing herself year by year in a dream world. Murray later realized that this was the beginning of mental illness. Trying to impose some structure on what must have been a very trying childhood, he later spoke of his father as Apollonian by nature (detached, rational, analytic), while his mother was Dionysian (intuitive, romantic, involved). "I seem to have inherited a mixture of both," he once wrote, "which makes me, I suppose, an Odyssean. Being Odyssean is rather lonely, because most people seem to insist on emphasizing one style over another, even though everyone has some access to both."

With so high a wall between father and son, brother Ben (almost a decade older) became the grown-up Murray wanted to imitate. It was Ben who introduced him to science and nature. For all its dreariness, the Bronx was full of wonderful new places—the Bronx Zoo and the woodlands to the north. As they explored this world, Murray and Ben tried to imagine what it had been like when much of New York City was still a hemlock forest. They found themselves drawn to the few patches that hadn't been cut down.

There wasn't much reason to hang around the apartment. Neither Arthur nor Pauline had many real friends—just ancient relatives, Murray complained, who didn't speak English. The exception

and saving grace was their cousins the Walkers. Murray's grand-
mother had a half-sister with the same first name: Celia. The kids
called her Aunt Silly. She was married to a flamboyant gentleman
named Benjamin Walker, formerly Volkowitz, an immigrant from
Poland. He owned nightclubs and restaurants in Brooklyn, and
before running off with a showgirl, he provided his family with the
kind of solid financial foundation Arthur could only dream about.
Murray thought of them as the only normal people he knew. Aunt
Cele's warm kitchen was a haven for both the Gell-Mann and the
Walker children. It was here that Ben taught Murray to pick out
some of his first written words on a Sunshine cracker box.

Even Arthur would brighten up around his younger cousin
Israel, one of the Walkers' very accomplished sons. Before Murray
was born, when Israel was still a teenager, he and Arthur would play
chess together. Once Israel made a blunder and Arthur won the
game. He comforted his young opponent with a quote from the
Roman poet Sextus Propertius: "Quod si deficiant vires, audacia
certe / Laus erit: in magnis et voluisse sat est." "What though
strength fails? Boldness is certain to win praise. In mighty enter-
prises, it is enough to have had the determination." The motto
could have applied to both these learned men. Israel, who went on
to earn a doctorate in classics from Columbia, was studying abroad
with a fellowship in Greece when the Depression hit. Like Arthur,
years before him, he was summoned home to take care of his family.

But the Walkers had an easier time than the Gell-Manns bouncing back from the Depression. After a dreary teaching job in Hell's
Kitchen, Israel eventually became head of classics education for the
New York Public Schools. One brother thrived in business, another
as an actuary. The youngest brother, Alex, became a dentist and
treated the Gell-Mann family for free. A sister, Minna, married into
a family of coffee heirs. A tiny, doll-like woman, Minna became
known as the most generous of the Walker family. While her husband, Paul Martinson, worked as a junior partner in a prestigious
law firm, Minna took the El to teach in a school in Harlem. She felt
driven to help unfortunates—a category that included her cousins,
the Gell-Manns.

It was quickly becoming obvious that Murray was not just very
bright, that the family had someone extraordinary on its hands.
One day Israel Walker gave Murray—then about four or five—some

Roman coins he had found. "Here, Murray, is a coin from the Emperor Tiberius," he said. Murray corrected his pronunciation, and Israel felt like punching him. And after scrutinizing the inscription on the coin more closely, Murray determined that it wasn't really from Tiberius's reign but from that of a later emperor.

Before long the family was collecting "boy genius" stories. Hearing Murray expound in such detail on so staggering a range of subjects, Israel was reminded of a line from Oliver Goldsmith's eighteenth-century poem "The Deserted Village": "And still they gazed, and still the wonder grew / That one small head should carry all he knew." At three years old Murray was multiplying large numbers in his head. When he was seven, he won out over children five years older in a spelling bee. The winning word was "subpoena." Not only did he spell it correctly; he corrected the moderator's pronunciation. After all, it was from the Latin: *sub po én a.* The prize, offered by a New York radio station, was something that few young children would enjoy—an unabridged dictionary. For Murray, who was becoming fascinated with words and their histories— these crooked paths leading back in time—a dictionary must have seemed like an immense forest, as dense with flora and fauna as the hemlock groves he loved to explore.

Murray found public school—David Farragut (P.S. 44) in the Bronx—stultifying. Every time the teacher asked a question, his hand would shoot up. If she called on another child who then gave the wrong answer, Murray seemed to take it as a personal insult. "I knew that!" he would exclaim. But if some of the slower kids were put off by his intellectual showmanship, they were usually won over by his friendliness.

To alleviate boredom, he tried to start his own newspaper, but it was not enough to keep him amused. After school he took piano lessons at a nearby settlement house, where he fell head over heels for his teacher, Florence Freint. Like the Walkers, she provided an occasional escape from his parents' claustrophobic life. A striking woman in her early twenties, she was enchanted by this self-confident, immaculately dressed child. True, he was something of a know-it-all, but she was impressed by how nice he was about it. "Smarty, you think you know everything," the other kids would say. "But I *do* know everything," Murray would matter-of-factly reply. He wasn't being snotty about it, just stating a fact. He couldn't imagine why this seemed to bother people.

Sometimes Murray would drop by the home of Florence and her

husband for a peanut-butter-and-jelly sandwich. "Did you eat the sandwich?" she would ask. "Every molecule," he would reply. That self-satisfied smile and those sparkling, penetrating eyes stayed in her memory for years. "The cat that swallowed the canary"—that's how she remembered the look. And she was struck by how little he resembled the stereotypical boy genius. There was nothing physically awkward about this beautiful boy.

Despite their age difference, Florence—like Ben—became one of Murray's best friends, giving him glimpses of a world beyond the Bronx. One day the Freints introduced him to an artist friend, and Murray lectured him with what he knew about art history. Another time, hoping to inspire him to musical heights, Florence took Murray to Town Hall to hear a young Italian pianist. A few seconds into a Chopin prelude, Murray began quietly criticizing the musician's interpretation of the score: "Where are the dotted eighths?" When the concert ended, Florence figured that Murray would get a thrill from joining the crowd rushing to the stage for a closer look. No, Murray said, he was not interested. "Well, suppose it was Einstein. Would you go?" To which the eight-year-old Murray replied: "Of course. He's more my equal."

As it became clearer that P.S. 44 wasn't serving Murray's needs, Pauline made fitful attempts to find a private school that would accept him. Murray later remembered her dragging him from place to place to be tested. Stacking blocks, answering questions—he had no idea what was going on. But dreamy Pauline didn't have the wherewithal to find a school on her own. His brooding father was no help: What the family had was good enough, thank you, and any effort to do better was bound to fail. And there was in any case something about private schools that offended Arthur's populist sentiments, which were moving further and further toward socialism. In the old empire, everyone had gone to government-run schools.

The Walkers also tried to help. When Israel's daughter Ann complained about school (a Catholic teacher had scolded her for not marking a cross on her forehead for Ash Wednesday), her Aunt Minna arranged to get her tested and enrolled in Hunter Model School. She also tried to find a school for Murray.

But it was probably Florence Freint who took the strongest stand, stepping in when Pauline's scattered efforts failed. "Murray is so bored, he really ought to be in private school," she told his mother. Pauline reminded her about the family's money problems, and

Freint set out to find a school that would award him a full scholarship. She wanted to avoid the progressive schools where students called the teachers by their first names and left their desks at will. She would laugh at Murray, egging him on when he sang what struck her as the perfect spoof of what education was coming to. Sung to the tune of "School days, school days, dear old golden rule days," the parody went like this:

> *School days, school days,*
> *Dear old lack of rule days.*
> *Reading and writing and social science*
> *Taught to the tune of self-reliance.*
> *You were my uni-project pal.*
> *I was your maladjusted gal.*
> *I never did learn to read or write*
> *In these progressive schools.*

She loved the way he would purse his lips, eyes twinkling: "Taught to the too-une. . . ."

Freint consulted with her boss at the settlement house and was steered toward Columbia Grammar, a school on the Upper West Side of Manhattan. Pretending to ask advice on how to guide her remarkable student, she wrote to the head of the school. This child has to be seen to be believed, she politely insisted, requesting an interview. After meeting and testing Murray, the headmaster declared that the boy already knew as much as most college students. As he and Freint conferred, Murray was busy scrutinizing the pictures of butterflies on the office wall, reciting their Latin names.

In 1937 the Gell-Manns gave up their Bronx apartment and moved to smaller, more expensive quarters on the top floor of a brownstone on 93rd Street, the same block as the school. There was nothing fashionable then about brownstones off Central Park West, but his parents didn't want Murray to have to take the subway. On his eighth birthday he entered Columbia Grammar with a full scholarship. He had already skipped a grade and a half of public school. Columbia Grammar immediately skipped him two more, making him a very young sixth grader.

CHAPTER 2

THE WALKING ENCYCLOPEDIA

It is unlikely that anyone seeking the best private school for an aspiring young scholar would have picked Columbia Grammar as the first or even second choice. Over the years the school has produced a few famous graduates, like Steve Ross, the chairman of Time Warner, and David Wolper, the television producer. And just about every year it sends a handful of students to the Ivy League. But Murray Gell-Mann—who seems to have ended up there more or less at random—remains by far the school's main claim to fame. In 1944, when it listed its star pupils in a short historical sketch, the school seemed to be grasping at straws: There was Dr. Hamilton Holt, the president of Rollins College; the educator John Erskine; Felix Adler, the philosopher who founded the Ethical Culture Society; the diplomat Oscar Strauss. All of these were alumni to be proud of, but not really of equal magnitude to the small, bespectacled boy who would spend the next six years there—learning, he would later say, absolutely nothing.

The school Murray had descended upon was founded in 1764 to prepare students for King's College (the future Columbia University). By the time he arrived, Columbia Grammar, housed in a brick building half a block from Central Park West, had become independent and was best known not for its academic curriculum but for its basketball and swimming teams. It was certainly better than the public schools, but what it often came down to was this: If your parents were upper-middle-class and your grades were not good enough to get you into one of the more prestigious New York City prep schools, like Horace Mann, Riverdale, or Fieldston, you might settle for Columbia Grammar. This would be especially true if you came from one of the prosperous Jewish families who lived nearby in the grand apartment houses on Central Park West—the El

Dorado, the Turin—or several blocks west along Broadway, West End Avenue, or the "Gold Coast" of Riverside Drive.

In between these corridors of wealth was a no-man's-land of rundown brownstones. After an elevated train (now defunct) was erected along Columbus Avenue in 1879, the area on both sides of the tracks had filled up with working-class families. Some of the richer kids felt a little nervous walking across this terra incognita to visit one another's homes along Central Park and the Hudson River. Jeremy Bernstein, whose family lived on Central Park West and who was three years behind Murray at Columbia Grammar, was warned by his parents that the area was filled with "dangerous riffraff." Bernstein, who went on to become a physicist and science writer, was surprised to learn years later that this "riffraff" included the family of Murray Gell-Mann.

At Columbia Grammar, the few students who were not upper-middle-class Jews were likely to be swimmers and basketball players recruited on athletic scholarships. "Ringers," the other students called them, joking that some could barely sign their names. Full academic scholarships were rare. The result was a school dominated by rich underachievers and jocks. And then there was Murray—poor, brilliant, barely five feet tall, and three years younger than almost everybody else in his class.

It quickly became clear to the other students that the teachers—many of them near retirement—were no match for Murray. One of them, a doctor of divinity and defrocked Southern minister named Dow Bunyan Beene, was immortalized in a limerick that made the rounds:

> *There was an old teacher of Latin,*
> *Who taught in a school in Manhattan,*
> *He applauded the folks*
> *Who laughed long at his jokes,*
> *And promptly their marks he did fatten.*

Or there was the "old belle out of Natchez / Whose garments were always in patches. . . ."

The age of some of the teachers became a standing joke. When "the Beene" was picked to lead a patriotic school assembly, the emcee (the school paper reported) mischievously announced that the white-maned teacher would lead songs from the war he had lived through. "An envious faculty member declared: 'What was that—the Civil War?' . . . Whereupon, a bright Latin II student said: 'No, the Gallic War!' " It's easy to imagine that this anonymous voice belonged to Murray Gell-Mann.

A student named Justin Israel, whose family lived on Riverside Drive, was shocked the first time he heard his diminutive classmate correct one of the teachers. Murray would raise his hand, politely interrupting. Israel, who went by the nickname "Buzzy," was impressed by the matter-of-fact way Murray imparted his superior knowledge. He didn't seem to be flaunting his learning the way some of his classmates flaunted their wealth. He just couldn't stand to let mistakes go uncorrected. As far as Murray was concerned, his unmatchable intelligence was just one more fact of life, like the dates of the Peloponnesian Wars. It was like having an *Encyclopaedia Britannica* for a classmate. The students knew they would get the right information when Murray was in the class. When he was absent, they weren't so sure.

Though some of the older boys found Murray arrogant and a little weird, many—like Buzzy, who had himself skipped a couple of grades—took a liking to him. They called him their "pet genius." The school paper's telegraphic gossip column cryptically referred to "C.G.S.'s own small-scale model of Einstein [who] worked out a series of *intoxicating* problems last week, his final answer being fifty-nine cents a quart." Another possible hint of Murray's presence can be found in a 1943 poll by the student magazine. "What, in your opinion, is the hardest course in school?" the survey asked. Twenty-five said math, eighteen said physics, ten said Latin, ten said history, and one anonymous respondent replied that there were no hard courses at all.

But while Murray was a star in the classroom, outside it he didn't fit in. He was a handsome, almost angelic child but small with thick glasses. In the yearbook pictures, posing with the library club or the Virgil Society (for advanced Latin students), he sometimes looks like Poindexter in "Felix the Cat." Some days, after a particularly grueling physical education class in Central Park, he would come home roughed up by the older boys—they would run off with his bookbag, steal his hat. Because of the war, gym class was a little like boot camp. To prepare the boys to be soldiers, the athletic director, following the lead of the New York State War Department, stressed what he called "the combative sports." In addition to tumbling, boxing, and wrestling, there were weekly runs around the Central Park reservoir and a commando obstacle course in the gym. Signs of the war were everywhere. "Defeat the Axis" proclaimed an advertisement for war bonds in the school paper. Columbia Grammar's affluent students raised more than half a million dollars for the cause. They helped distribute ration coupons to families in the neighborhood and held scrap drives. Some seniors were pushed to graduate in the middle of the year so they would have time to attend college for a few months before joining the service.

But for Murray, serving in the military was not an immediate concern. This was just one more thing that separated him from his classmates. While they spent time at the soda shop in the afternoons, talking about the war or flirting with girls, little Murray would scurry home to his walk-up flat half a block from school. Arthur was strict about homework, and nothing Murray showed him was ever good enough. "A 98 on a math test? What happened to the other two points?" Arthur was especially critical when it came to writing. After all, this was a man who had filled three thick notebooks with every arcane rule of usage and rhetoric under the sun. No matter how good Murray's homework, he could count on Arthur to see the flaws.

Everyone, it seemed, was older, bigger, richer. There was Jack Schwadron, whose father was one of the partners of Alexander's department store. Gene Goldfarb's father owned a lucrative dress manufacturing company. When "Events Around Town," a regular column in the school paper, gave its advice for Christmas vacation fun, it recommended the Astor Roof for its food and dance band, the Wedgewood Room of the Waldorf (where Xavier Cugat was playing), the Rainbow Room, and various Broadway plays—all out of reach for the son of a bank vault custodian. He was still the boy looking through the bars at Gramercy Park, without a key to get in. Nor could Murray afford the latest "waxinations off the griddle" (Tommy Dorsey, Benny Goodman, Glenn Miller) recommended in the "Platter Chatter" column written by Robert Brustein, who later became the scathing theater critic for *The New Republic* and the founding director of the Yale Repertory and American Repertory theaters. One day when Arthur walked Murray to a friend's home to play, he insisted they must have the wrong address. The house occupied a full block on Riverside Drive—no one they knew could possibly live in such a place. Then the butler answered the door and ushered Murray inside.

With so little else to his name, the one thing Murray could use to his advantage was his intelligence. When he went bicycling in

Central Park with his friend Leonid Pratt, the son of a captain at the Russian Tea Room and the only other boy in his class with a full academic scholarship, Murray would recite the name of every bird he saw. Some afternoons, when he wasn't studying, he would go to his friend Al Davis's house to listen to recordings of classical music. Murray loved to show off the information he had collected. Whatever he was involved in, it didn't seem to occur to him that there was anything offensive about correcting other people's mistakes. Didn't they want to know when they were wrong?

While he liked to spend time at his classmates' homes, Murray rarely felt comfortable enough to bring friends to his own shabby apartment. Later the family moved a few blocks south and west to 90th Street, but the new apartment was still something of an embarrassment.

As in the Bronx, he spent as much time as he could outdoors with Ben. Of all the boys in the neighborhood, none was a better companion. Ben was also something of a prodigy, graduating from high school at fourteen. But after a year at City College, he dropped out. Everything one needed to learn, it seemed, was available within blocks of their apartment. Ben and Murray spent hours in the American Museum of Natural History on Central Park West looking at the nature dioramas, each a microcosm of a different wild kingdom. They stood at the feet of the dinosaur skeletons, gazing up at those cages of bones, so similar in structure, they learned, to those of the squirrels outside in the trees; so similar, for that matter, to their own. They walked across Central Park to the East Side and the Metropolitan Museum of Art to examine the Egyptian mummies. They even learned to read a few hieroglyphs. In later years, when Murray was a wealthy man, he would turn his own homes into museums.

But the biggest attraction of all was Central Park. From Stuyvesant Square on the Lower East Side, they had traded up to the Bronx's zoo and parks. Now they were living half a block from one of the greatest urban woodlands in the world. Murray and Ben walked along the carefully platted paths; they rode their bicycles across the lawns, identifying birds, insects, mammals, and plants; they carried specimens home in jars. Nature was so overwhelmingly diverse, yet it fell into patterns: The individual creatures belonged to species, the species to genera, and the genera to families, orders, classes, phyla, kingdoms. Each leaf on the Tree of Life might be different, but they were connected to common twigs—and the twigs to branches, the branches to limbs, all funneling into a single trunk. With just a little memorization, Murray realized, he could carry this beautiful map in his head. Life wasn't a confusing blur anymore. Outside the apartment, it made sense.

What was true for the biological world was also true for the species of languages. Sitting home on rainy days, Murray and Ben would study foreign tongues. How wonderful that Spanish *luz* and French *lumière* both came from Latin *lux*. Reaching further back, they learned that English, the Romance languages, Latin, Greek, even Sanskrit, all ultimately derived from an ancient ancestor called Indo-European. The Tower of Babel was as neatly structured as the Tree of Life. Everywhere beneath the world's confusion there seemed to be an underlying order. For now, Murray was simply memorizing preexisting taxonomies. One day he would change physics by inventing his own.

Though they were mostly interested in science and history, Murray and Ben occasionally scraped together enough money for standing room at a concert or the Metropolitan Opera. Most of the music they heard they played themselves on the piano or tuned in on the radio. They logged distant stations on the shortwave bands and mailed away for the verification cards radio hobbyists collect. One card, from Australia, made a particularly strong impression: It was decorated with a picture of a strange bird called a kookaburra. There was so much to see beyond the island of Manhattan. In 1939 precocious Ben bought that fateful first edition of *Finnegans Wake*. He and ten-year-old Murray leafed through the pages marveling at the weird language.

Ben was the first to get away. After working for a few years in a low-paying job as a photographer, he joined the Army Air Corps and became a first lieutenant, traveling to Italy as a photointelligence officer with the Air Force, briefing and debriefing the bomber crews. The snapshots he sent back of the Renaissance city of Florence and other exotic places whetted Murray's appetite to visit foreign lands. (Later on, pride of place in Murray's offices at Caltech and the Santa Fe Institute was given to a hand-tinted photograph his brother had taken of a gorilla in one of the dioramas at the Museum of Natural History. As threatening as King Kong, it was a reminder, Murray would say, not to let his mind wander.)

By senior year Murray had filled the kinds of high school niches one would expect: He was president of the debating club and manager of the chess team; he belonged to the math and history

clubs, as well as the Virgil Society. The yearbook reported that, pre-
dictably enough, he was chosen Most Studious: "In a shower of
votes, Murray Gell-Mann, the 'wonder boy,' easily won the title." He
was not entirely unathletic; he played soccer and managed the tennis team. Most surprising, considering how much he would come to
dislike journalists, he was editor in chief of the school paper, which
apparently created something of a stir. Because of the war, the
paper was published sporadically. No copies from that year survive
in the school archives. But there is a tantalizing hint in the yearbook: "The paper was printed in a solely unbiased manner and
naturally was subjected several times to severe criticism."

In fact, some of his wealthier classmates came to consider Murray
something of a rabble-rouser, a little communist. When Buzzy Israel
sat down to write the class prophecies, he invented a telling scene.
Jack Schwadron (the president of the student government) is being
sworn in as president of the United States by another classmate,
who has risen to the position of chief justice of the Supreme Court.
Then a bomb, "hurled by the Soviet emissary to America, Murray
(T.N.T.) Gell-Mann, landed next to Mr. Schwadron. However, with
his customary absentmindedness, Gell-Mann had forgotten to light
the fuse. . . ."

Murray, of course, was class valedictorian. He was fascinated by
the utopian ideas of H. G. Wells, and he had apparently mastered
some of Arthur's rhetoric lessons, as well as his politics. The result
was an address that Buzzy, who was salutatorian, remembers as daring and exciting—one that Murray would later dismiss, somewhat
embarrassed by his youthful enthusiasms, as "radical, left-wing stuff
about world government." A cartoonist who somehow got wind of
the performance captured it in a cartoon for *The New Yorker*.

Shortly after his graduation in 1944, Murray spent the night at
the Freints' home, across the Hudson in Englewood, New Jersey.
He awoke to hear the news of the Normandy invasion on the radio
in the next room. He bicycled home across the George Washington
Bridge and found in his mailbox a letter from Yale. He had won a
full scholarship from (he would later learn) the Medill-McCormick
family, newspaper magnates who owned the New York *Daily News*
and the *Chicago Tribune*. He was fourteen.

Murray wasn't in the habit of conferring with his father about
pivotal life decisions. But one day, when he was filling out a form

asking for the subject he expected to major in at Yale, he told Arthur that he wanted to be an archaeologist or a linguist. Arthur thought this was ridiculous. These were hobbies, not careers. How was he going to make money? "You'll starve," he said scornfully. After all, what had all those philosophy and mathematics classes in Vienna led to? A menial job at a bank and a family wedged into a walk-up apartment. He told Murray the story of the self-taught German archaeologist Heinrich Schliemann, who didn't start excavating ancient ruins until he had made a fortune in the indigo trade and as a Crimean War profiteer. Only then did he dip into his own pocket to pay for his expeditions, including the unearthing of ancient Troy and Mycenae. If you want to be an archaeologist, Arthur said, do it like Schliemann.

This wasn't what Murray wanted to hear. What, he asked his father, would he suggest? "Engineering," Arthur replied. Now it was Murray's turn to be exasperated. "I would *rather* starve," he said. He wanted to learn about the world, not how to construct buildings and bridges. Anything he designed, Murray said melodramatically, would probably fall down.

Then Arthur suggested physics. Murray was unenthusiastic. He had hated physics at Columbia Grammar—all that boring stuff like memorizing the different kinds of simple machines or counting the number of strings on a pulley. Part of the problem, Murray would later realize, was that each subject—optics, mechanics, electricity, acoustics, magnetism—was treated like a separate scripture, as received wisdom from on high. There was no sense of the beautiful architecture connecting all the pieces. It was nature Murray loved—birds and trees, not these dull abstractions. But Arthur knew from his reading that this rote memorization was just the rite of initiation into an exclusive club. For the student who dutifully ascended through the tiers of lesser knowledge, wonderful worlds lay ahead: relativity and quantum mechanics. Murray was dubious, but since the discussion was going nowhere, he decided to humor the old man. He could always change his major after he arrived in New Haven.

Because of the war, Yale was on an accelerated schedule, with terms beginning in July, November, and March. Many high school graduates were cutting short their vacations to enroll in the summer session, rushing to get as far along as possible before being snapped up by the draft. But at fourteen, Murray still didn't need to worry about the military. Since he had signed up to begin in

November, a long summer vacation lay ahead. To fill his time, he did volunteer work for Franklin Roosevelt, who was running for a fourth term. Most of the job, like opening envelopes and counting up contributions, was boringly routine. But he also helped arrange for guest speakers, giving him a chance to meet some famous FDR supporters. There was Frank Sinatra—not exactly Murray's style— as well as semi-intellectuals like Dorothy Parker and Helen Keller. Listening to their presentations, he would think about how a more just world order might be built after the war was won.

When Murray finally arrived in New Haven late that fall, Yale was not exactly the idyllic refuge it strove to be, with its Oxbridgian courtyards and pseudo-Gothic architecture. The whole school had been retooled to train people for the war. Murray had already glimpsed the uncomfortable transformation a year earlier when he went to visit Ben, who was in military training there. The old campus, where freshmen normally boarded before entering one of the prestigious residential colleges, had been converted into an aviation cadet school. All but three of the ten residential colleges— Jonathan Edwards, John Calhoun, and Timothy Dwight—had been commandeered for Army and Navy training. There were only 440 students in the entering freshman class, about a third the normal size. This handful of civilians was pushed to work straight through the year—spring, summer, and fall—until their deferments ran out. Living side by side with these anxious youngsters were the older students, returning veterans resuming or just beginning their college careers. People you started with weren't there when you graduated; people you graduated with hadn't been there when you arrived. With such a constant flux, there was little sense of cohesion, of belonging to the class of 1946, 1947, or 1948. And once again, Murray was out of the mainstream, one of the few people, along with those given 4-F medical deferments, who would stay straight through for four years.

And, unlike at Columbia Grammar, Murray had another count against him. Even in normal times, Yale was not the easiest place for a Jew from New York. The campus was slowly changing from an exclusive finishing school for New England WASPs to a more cosmopolitan university. But the pretensions and prejudices were still there. With its residential colleges, each with a headmaster and faculty fellows, Yale was trying hard for that image of dignified antiquity enjoyed by Oxford or Harvard. But the history was a thin veneer. The colleges had been around only for a decade, built in the early 1930s with a donation from a wealthy alumnus.

Any way you looked at it, there was something incongruous about a Murray Gell-Mann taking up residence at Jonathan Edwards, named for the Puritan teacher whose jeremiads had inspired the Great Awakening of New England Protestantism in the eighteenth century. There was something incongruous, in those class-conscious days, about a Murray Gell-Mann graduating (as he would later learn) with George Herbert Walker Bush. On one of his first mornings at Jonathan Edwards, Murray sat down at breakfast with a student named John Knowles. Knowles, who had just come from Phillips Exeter Academy in New Hampshire, would be famous one day for his novel *A Separate Peace*, about a tragic friendship at a toney New England prep school. He asked Murray which school he had gone to. Andover? Choate? . . . Columbia Grammar? You went *where?*

Murray was not the only oddity. Kenny Wolf, another alarmingly young prodigy, was at Yale to study music, and he and Murray would hurl good-natured insults at each other across the lunch table. Not everyone there was a preppie—Murray's precocious high school friend Buzzy Israel was also at Yale. In fact, of the 440 entering freshmen, forty-four were Jewish. Not forty-three, not forty-five, but precisely forty-four. The quota had been dutifully fulfilled. One day before Rosh Hashanah, the entering Jewish students (Murray had yet to arrive) were called into the office of the freshman dean, Norman Buck. Buck, along with two other colleagues, was the coauthor of an elementary economics text—Fairchild, Furniss, and Buck. Some students hated the dean so much that they called the book "the three F's." "This is a Christian school run on a Christian calendar," Buck told the nervous freshmen, "and we want you to be sure to know that we expect you to be in class on Jewish holidays." After some of the students protested, the school's president issued a statement assuring them they would be excused for the High Holidays after all.

Murray himself didn't find the idea of Jewish quotas particularly disturbing. After all, he reasoned, the school also had quotas for people from the western United States. It was an advantage to be from Montana. Life was difficult enough for him. He didn't feel any compulsion to align himself with an oppressed minority.

He and Buzzy did not stay friends for long. Israel was embarrassed to be seen with this small kid with thick glasses. Though Buzzy was still only sixteen, he had shot up six inches in high school. Unlike Murray, he looked old enough to blend in. The last thing he needed was to be associated with this junior Einstein. They had lunch together a few times, and that was it. Making new friends was going to take a while. Murray didn't really get along with his roommates. When one of them was drafted after the first term and the other one quit, he asked for one of the college's few private rooms. Located near the entrance, with a window facing a wall, it wasn't the best of accommodations. Some nights he would awaken to drunken students stumbling home late or singing loudly.

If he felt lonely or intimidated by this strange new world, he showed little sign of it. Just a few days into freshman term, he was sitting in one of the suites at Jonathan Edwards with a group of students listening to the Dewey-Roosevelt election returns on the radio. Dressed in a brown suit with a vest, providing a running commentary, he seemed perfectly in his element.

If Murray had been serious back then about wanting to be a physicist, there were much better places he could have gone to college. It would have made far more sense for him to travel up Broadway from his family's apartment to Columbia University. The distinguished physicist I. I. Rabi was teaching there, and Enrico Fermi had just departed for the University of Chicago, which would have been another good choice. Or he might have gone an hour southeast to Princeton, where Eugene Wigner was teaching (with Einstein himself just down the road at the Institute for Advanced Study), or upstate to Cornell, the domain of Hans Bethe, or to Berkeley or Caltech, the two schools where J. Robert Oppenheimer held a joint appointment. In the years before World War II, all these universities had moved to the forefront of physics. In 1944, Murray's freshman year, these star professors were busy with war research, but the departments they had helped create were undeniably among the best. Yale's main claim to fame in physics was still Josiah Willard Gibbs, who had done pioneering work in thermodynamics in the 1870s.

Though far from the cutting edge, Yale was still a decent place to study physics—or just about anything—and for the first time Murray actually admitted to learning something in class. Though some professors had been recruited by the military, there remained a solid core of teachers who at least knew what physics used to be about. One of the first courses Murray took was an introduction to modern physics for undergraduates taught by Louis Kovarik, one of the pioneers of radiation research. Kovarik, then in his mid-sixties, was hard of hearing; he would ask questions but couldn't understand the answers. Sometimes he seemed to be living back in the glory days. He would show slides of himself posing with great experimenters like the Curies and Ernest Rutherford, or with Sir James Chadwick, the discoverer of the neutron. He literally steeped himself in radiation: His laboratory office was full of uranium and pitchblende. When he retired, the radiation safety inspectors came in with their Geiger counters (invented by Hans Geiger, another old Kovarik cohort) and found that the room where he had spent so much of his life was hot enough to drive the needle dangerously up the scale.

The other term of modern physics was taught by the person Murray came to consider the best physics teacher at the university: Henry Margenau. He and Murray didn't get along very well. But under the guise of a course on the philosophical foundations of physics, he introduced his students to the wonders of relativity and quantum mechanics. True, the man's philosophical bent was sometimes annoying. Murray had tried to read the great philosophers in high school, but had found their ideas about the world frustratingly speculative and maddeningly imprecise. But through Margenau, Murray learned that Arthur had been right about one thing: Quantum theory and relativity were absolutely fascinating, and not nearly so hard as his father had led him to believe. It was not only the birds and trees and other creatures of the earth that were tied together into one grand system. Hidden orders pervaded everything, and with mathematics you could sometimes tease them out.

Whether he was taking courses in electricity and magnetism or classical optics, Murray was rapidly pulling ahead of everybody else. In one class meant for graduate students, the professor insisted on strict attendance because there was no text, just the lecture notes. Murray, who was accustomed by now to learning mostly on his own, found he could master the material from books, some of them in German, in the university library. He cut the lectures and easily passed the final exam.

Some of his fellow students were crestfallen at how effortlessly the most difficult subject matter came to him. In sophomore year, another resident of Jonathan Edwards, Richard Brandt, was sitting in a math class, feeling he was in way over his head, when he noticed Murray following along with no trouble whatsoever. Brandt had switched his major from chemistry to physics because he kept breaking glassware. He wanted something more theoretical and abstract, where you didn't have to worry about cutting your fingers. But he didn't have the prerequisites for this particular class. Suddenly this young kid shows up and seems to know all the answers before the professor asks the questions.

One evening Brandt knocked on Murray's door at Jonathan Edwards and asked him how to solve a problem that had stumped him in class. Murray gave it a little thought and told Brandt the answer. Brandt said, "Murray, that's not what I'm asking. I'm asking *how* you get from here, step by step, to the answer." Murray, growing a little impatient, repeated the answer. Brandt gave up in frustration. The deductions seemed to go through Murray's mind so quickly that he was barely aware of them. For Brandt this was a turning point. He decided that if this was typical of physics majors at Yale, then there was no hope for him. He switched to mathematics, took a lot of philosophy, and after graduation went into the family business, running a chain of movie theaters. Of course Murray was anything but typical, and Brandt would later regret giving up so easily.

Bud Rosenbaum, who lived in the room next to Murray's—the connecting door was always open, making them almost roommates—noticed that his neighbor never seemed to study. He would pore over the *New York Times* and the *Herald Tribune*, easily unscrambling the Cryptoquote, doing the crossword in ink. In one class, taught by an irascible Russian-born theorist named Gregory Breit, who came to Yale after working on the Manhattan Project, Murray was the only undergraduate. Breit didn't want to be bothered with calculating grades—you either made the cut or you didn't. But Murray insisted that he needed a grade for his average. Breit, famous for his short fuse, flew into a rage and said Murray was persecuting him. He didn't want to put together a special exam just for one precocious student. "Why don't you just write up your notes?" he said. Murray made a halfhearted attempt. Breit flipped through the pages and said, "How about a 90?" Murray agreed, and that was his grade in the class.

Before long, Murray was taking Introduction to Theoretical Physics, the grueling course that was the rite of passage for first-year graduate students who thought they wanted to become physicists. It was presided over by Leigh Page, a shy, introverted man who had graduated from Yale in 1904, when Josiah Gibbs was still there. As a young man Page had come up with something he called the emission theory of radiation. Then quantum mechanics came along and eclipsed him. Sometimes he talked as though he wished the quantum revolution would just go away. His course, concentrating on classical physics and including archaic techniques invented by Gibbs himself, met five days a week, beginning first thing in the morning. On Saturdays there was a question session, and every fourth or fifth Saturday a four-hour examination. Abner Shimony, a classmate of Murray's, found it amusing that the Saturday sessions were held in the afternoon, preventing the graduate students from going to football games. It seemed entirely deliberate, as though they were being asked to join a monastic order. Nor were the acolytes required to think very creatively. Tests consisted of regurgitating the lecture notes.

Shimony found the ease with which Murray mastered the material discouraging. He had come from his high school thinking pretty well of himself, but Murray was showing him up again and again. While Shimony labored through an electrodynamics class taught by Page, Murray solved the problems so rapidly that it almost seemed his knowledge was inborn. When Murray saw that Shimony had marked up and underlined page after page of his textbook, he was appalled. "You should never mark up a book like that! You should have respect for a book." That was fine for him to say, Shimony thought. It all came so easy. In a topology course he took, he struggled as Murray breezed right through.

Those who could tolerate Murray's know-it-all attitude enjoyed his company. Since he didn't study, he was always ready to go out for a beer or a late-night snack and the inevitable political bull sessions. Walking across campus with his friends, he would play Twenty Questions or show them how many birds he could identify. He was not just brilliant but funny. "Whither away?" Shimony asked Murray one day when he saw him on campus. "*You* wither away," Murray bitingly replied. He was so self-conscious about his age that when he came down with whooping cough, he refused to admit he had a childhood illness and continued to go to classes, sometimes coughing uncontrollably. When asked what was wrong, he insisted he had a rare tropical disease he called lazamerchi, acquired from the bite of the Malaysian fruit bat.

And sometimes the rebellious streak Murray "T.N.T." Gell-Mann had displayed as a budding communist at Columbia Grammar would show through. He joined a group of students who climbed East Rock, a nearby landmark. They were met on top by policemen who told them that next time they would be arrested. Sometimes at night Murray and an older student named Gerald Haggerty would explore the network of steam tunnels that underlay the campus, emerging in dark, deserted buildings. Haggerty, an aspiring lawyer and antiestablishmentarian, was obsessed by the Yale senior societies, which met in secret chambers called "tombs." Once he and a group of students mocked the self-conscious solemnity of the societies by dressing in hooded robes and marching behind a group returning from the tombs to the dormitories. Murray joined an expedition led by Haggerty that broke into the rooms of the most secret of the societies, Skull and Bones, whose members included George Bush.

With his utopian ideals, Murray was a natural to join the Political Union, a Yale debating society divided like the British Parliament, with the Laborites on the left, the Liberals in the center, and the Conservatives on the right. Murray, along with Haggerty and a classmate named Harold Morowitz, joined the Labor party. If there had been a Communist party, Murray might well have signed up. The Reds, he believed, were the good guys—until February 1948, when the coup in Czechoslovakia shocked him into realizing that there was nothing idealistic about Stalin's plans for the world.

Several years later, when a young alumnus named William F. Buckley wrote his famous tract *God and Man at Yale*, lamenting the undermining of conservative values, he described the Political Union's Laborite wing as "a small, militant group of doctrinaire leftists and socialists." Murray and his cohorts (who included the future political scientist Samuel P. Huntington) liked to think of themselves as raging radicals, sometimes grandiosely calling themselves the FAR, Fathers of the American Revolution. One year when Haggerty ran for president of the Political Union, the campus was littered with flyers announcing the support of something called the Nonpartisan Committee for Haggerty. The group was nothing other than the Labor party, gathered one day around a mimeograph machine, with Haggerty turning the crank.

As much as he enjoyed being part of Haggerty's whimsical revolution, Murray was finding himself drawn to some of the rich kids he and the FAR liked to mock. Later on, when Murray reminisced about his days at Yale, the classmate he remembered most fondly was a boy named Carteret Lawrence. Groomed for success at the Hotchkiss School in Connecticut, Lawrence had recently left the Army Air Corps with a serious kidney infection. He came to Yale to study medicine. Like Murray, he had fallen in love with science and mathematics, and the two spent hours talking together. Lawrence, who died in the 1970s, must have been impressed by the depth of Murray's knowledge. And Murray was fascinated by his family's wealth. Lawrence was one of the lucky ones, born to a life of privilege. He was, as Murray would later put it, his bridge to the Social Register crowd. For all his accomplishments and brilliance, Murray had been instilled by his father with a dim view of life's promise. No matter how smart he was, in the back of his mind like a constant presence was a dark foreboding—a sense that all his efforts were doomed, like his father's language school. Lawrence was just the opposite. He believed anything was possible. Some of this bravado began to rub off on his smart young friend.

Before long, Murray found that he could hold his own with Lawrence's friends. It was a little amazing. Here he was socializing with the kind of rich kids who had so often spurned him at Columbia Grammar. They went skiing together, camping in the snow. When it came to issues like preserving the environment and nuclear disarmament, Murray would never lose his liberal sentiments. But he was starting to feel that there was nothing wrong with having money.

Little by little, he was putting his past behind him. In his four years at Yale, his friends rarely heard him talk about his family. Morowitz realized one day that he had no idea what Murray's parents were like or where they lived in New York. Nor did Murray ever allude to being Jewish. Shimony found this puzzling. He was proud to be a Jew and wondered why anyone wouldn't feel the same.

Murray continued to visit his parents on holidays. A snapshot from around that time shows him standing in Central Park gallantly holding his mother's arm. Murray—much taller now and dressed in a tie, a long overcoat, and a snappy fedora—smiles confidently, almost beaming. Self-assured, relaxed, he was beginning to look like the consummate Ivy Leaguer (though he quickly abandoned the hat as hopelessly out of style). On one of his visits home for Thanksgiving or Christmas, his younger cousin Ann Walker was walking on Central Park West with her mother and sister when they encountered this handsome young man. Murray picked up Ann, put her on the back of his bicycle, and rode through the park, pointing out the trees by name.

He finished his course work in January but didn't want to graduate midyear. Extending his scholarship for an extra semester, he took what he thought of as gut courses for football players—arts and humanities. In one class the students had to critique a play. This was not Murray's kind of thing, and for the only time in his life his grades suffered. The last semester was really a vacation, and because of it Murray graduated second in science. In 1948, when most people his age were finishing high school, he was awarded a degree from Yale. It was time to start working on a doctorate.

By all rights the Yale physics department should have been waiting for him with open arms. But it decided to pass. He was accepted instead by the mathematics graduate school, but that was no good. He wanted to study nature, not abstractions. The problem may have been with his senior thesis adviser, Henry Margenau. Stricken with a bad case of writer's block, Murray never finished his thesis— he barely started it. All he had to do was study the way charged particles were pulled one way or the other by negative and positive electrodes and then write something up. It didn't have to be earth-shaking. But as good as he was at solving puzzles, the problem of writing an original paper seemed insurmountable. He didn't know where to begin. He had no idea you were supposed to research the literature to see what others had done, to find approximate solutions to the problem and zero in from there. He wasn't about to admit defeat and ask Margenau for help. And most oppressive of all was the image of his father reading over his shoulder. Nothing he wrote would ever be good enough.

Murray suspected that Margenau was not giving him good recommendations. Something was going wrong. One by one, every Ivy League school he applied to sent him a disappointing reply. He was turned down by Princeton. Harvard admitted him, but the school didn't immediately follow up with the necessary scholarship and Murray couldn't motivate himself to pursue the matter. Then, in the midst of his depression, he received a letter from Victor Weisskopf, as accomplished a physicist as one could hope to find. Murray had never heard of him. Weisskopf offered to hire him as his assistant while he worked on his Ph.D. The only problem was the location: the Massachusetts Institute of Technology, a dreary campus that Murray considered hopelessly inferior to the Ivy League.

"How could I go to that grubby place?" he kept asking himself. Years later he liked to say that ultimately he was faced with a choice between going to MIT or committing suicide. "A little reflection convinced me that I could try MIT and then commit suicide later if I wanted to, but not the other way around." The process, a mathematician would say, was not commutative: $A \times B$ did not equal $B \times A$. That summer, after spending a month working as a shipping clerk in a sportswear store in Manhattan, he took a road trip across the West with a group of friends. In the fall he arrived on the campus in Cambridge looking, as one classmate remembers, like a Yalie, right down to his saddle shoes.

CHAPTER 3

A FEELING FOR
THE MECHANISM

After four years enveloped in Yale's carefully crafted antiquity, a newcomer might indeed have found MIT a little grubby. In going from blueprint to construction, the ever-so-functional, deliberately unalluring design championed around 1910 by the institution's original architect had been softened somewhat. The workmanlike concrete exterior was replaced with stereotypical limestone, and a fake Greek temple was added to form a grand entrance that looked out across the Charles River toward Boston. But despite these small concessions to the popular notion of what an institution of higher learning is supposed to look like, MIT stood as an affront to the Ivy League. Perched on a landfill at the river's edge near the opposite end of Cambridge from Harvard, the Institute almost seemed as though it wanted to be as far away as possible from its neighbor's lofty airs. MIT's buildings had numbers instead of names. They were connected not with a sweeping tree-filled quad and meandering landscaped pathways but by a matrix of walkways and tunnels that one writer has aptly compared to "the network of supply canals and millraces that fed the hydropower mills of the nineteenth century."

Coursing through these pipelines were the students—practical sorts who had always been good with engines or electricity. MIT would mill them into engineers. The backbone of this carefully constructed ant farm was the Infinite Corridor, a seemingly endless hallway beginning at the top of the front steps on Massachusetts Avenue and running, as one of MIT's aspiring engineers might have told you (pointing with his yellow slide rule), precisely 762 feet through five buildings. Twice a year, the earth would line up just so, allowing the setting sun to shoot straight down this dingy tunnel—a reminder that there was a world out there waiting to be tamed.

By the time Murray arrived, MIT was just beginning to shed its image as a glorified trade school, a place to train people to build bridges and buildings, to string power lines, to design and operate factories—exactly the kind of stuff he had told Arthur he would not do. In 1940 Lee DuBridge, an energetic physicist from the University of Rochester, had arrived at the school to head a government-funded inquiry into a mysterious new military technology called radar. It was with the Rad Lab, as it came to be called, that the physicists got their foot in the door, and the school began developing a taste for more theoretical pursuits, though always with an eye toward practical applications. It wasn't just architecture and the long stretch of Massachusetts Avenue that separated MIT from Harvard. The two schools had different attitudes toward education.

Victor Weisskopf, the man who recruited Murray, had come aboard just after the war. Following the success of the Manhattan Project and the bombing of Hiroshima, MIT was determined to become a major player in nuclear science. Weisskopf, who had been hired by Oppenheimer to work at Los Alamos, was one of the science's bright young stars. He jumped at the chance to move to Cambridge. After the cultural desolation of Los Alamos and his previous post at the University of Rochester, he was eager to live in the environs of Boston. With its museums and symphony orchestra, this was a place where he could resume the kind of life he had known as a boy in Vienna.

Though still smarting over being snubbed by the Ivy League, Murray quickly came to appreciate that there was nothing grubby about Victor Weisskopf. This was no gung-ho builder of dams recently arrived from a land-grant college in the Midwest. Here was a man who really was an Austrian immigrant—everything that Arthur had once aspired to be. Viki, as his friends and students called him, had grown up in a wealthy Viennese family, in a well-appointed home near the finely manicured park across from the Rathaus, Vienna's stately city hall. The family was attended by maids, a cook, and a nanny. Like so many Viennese Jews, the Weisskopfs had, in those more innocent days, proudly considered themselves Austrians who also happened to be Jewish. The empire had treated them well. Viki's maternal great-grandmother had made a fortune by taking over a failing bank, and there had always been money for the opera, concerts, the theater, piano lessons—not at a Bronx settlement house but with a private instructor. The children were taught to revere Goethe and Schiller and the great German composers. It was the kind of life Arthur had once imagined for himself.

After World War I, Vienna suffered from food shortages. But Viki's family had enough money to send the children to live with a family of peasants, who were paid to keep them well fed. When the Weisskopfs briefly had to give up their trip to Villa Charlotte—their summer house in the Austrian Alps, named after Viki's wealthy great-grandmother—they vacationed instead at a nearby farm with enough to feed them. They always seemed to land on their feet. Viki's upbringing had left him with the polish and self-assurance of someone who, as he liked to say, had led a charmed life. He wasn't being smug, just grateful.

When Viki showed an interest in astronomy, his father didn't try to steer him into more practical pursuits, as Arthur had with Murray; He went out and bought him a telescope. One summer night when the Perseid meteor shower was pouring forth full blast, fifteen-year-old Viki and a friend left their homes in Vienna and hiked high into the mountains. Sitting back to back, they trained their gaze toward the constellation Perseus and watched as one shooting star after another crashed and burned against the sky. Back down on terra firma, they wrote up their results and were astonished when—with a nudge from the local astronomy club, Friends of the Stars—the paper was accepted by a national publication called *Astronomical News*.

Sure, as only a teenager can be, that he had found his calling, Viki devoured popular accounts of the great new scientific theories. Einstein, in his wonderfully strange theory of relativity, had overturned the most ingrained assumptions, showing that time and space do not provide a rigid and absolute frame—they appear to squeeze and stretch, conspiring to ensure that nothing can exceed the speed of light. And the faster a particle flew, the more massive it became, approaching infinity as it neared the speed of light. Indeed, that was why nothing could go any faster.

But wait, Viki mentally objected. He had also read about the other great revolution, the new quantum theory, built on Max Planck's realization that energy must come in packets. Light itself was now believed to be made of particles—the quanta that came to be called photons. Photons, by definition, are moving at the speed of light, so why isn't *their* mass infinite? They should not be able to move at all! A less curious child would have let the matter slide, assuming that he had missed something. Viki dashed off a letter to Planck. The boy was vacationing with his family at Villa Charlotte when he got a postcard. "Dear Sir!" it began and went on to cordially and confusingly explain that though light was indeed packaged as photons (as Einstein himself had demonstrated with his photoelectric effect), it still traveled through space in the form of waves. Thus, unlike electrons and other particles, photons were not subject to the Einsteinian speed limit.

Never mind that Planck had got it wrong. The solution to the seeming paradox still lay in the future. In the coming years, physicists would learn that it was not just light that had this dual nature—acting sometimes like a stream of particles, other times like a wave. An electron could take on wavelike qualities; a beam of these particles could be refracted and diffracted like a shaft of light. No, the answer to the puzzle, Weisskopf would later learn, was that, unlike an electron, a photon at rest has no mass. Since there is no heft to begin with, there is nothing to grow heavier as the particle accelerates. Viki had no way of knowing that the physicists of his youth had not yet caught up with his question. The important thing was that he had been able to fool the great Planck into thinking he was worth writing to. That, he would say later, more than anything, convinced him to become a physicist.

Not even Murray had been this precocious. Arthur's sour outlook on life had kept his son from developing such aplomb. When it came to knowing things, Murray hardly suffered from an inferiority complex. His uncanny memory ensured that he would have more facts in his head than just about anyone else around. His mind was a kind of intellectual vacuum sweeper, sucking clean books and museums. But it took a special kind of assurance to sift through and combine these leavings into new ideas for others to discuss and learn. Murray hadn't even been able to muster the discipline to write his senior thesis. He had a lot more than facts and techniques to learn from Weisskopf. He could use some of his intellectual courage.

Murray's snobbish attitude toward MIT soon gave way to the excitement of being at a place where people were really doing physics, not just lecturing out of books. At Yale, the closest thing to a name-brand physicist, one still working at the forefront, was Gregory Breit, whose mad rages frightened off potential acolytes.

Now here was the kindly, if somewhat distant, Weisskopf, who had held an assistantship in Berlin with none other than Erwin Schrödinger, the passionate physicist-philosopher whose puzzling equation for the electron blurred the distinction between particle and wave. And before that, Viki had been in Göttingen with Max Born, who showed that these particle-waves—these wavicles, as some physicists called them—were not physical waves, like those lapping at the shore. Schrödinger's equations, Born had realized, described waves of probability, mathematical undulations expressing the likelihood of where an electron would be at a certain time. There were inherent uncertainties on this tiny scale, as Viki had learned from the masters. The electron was a haze of possibilities, all hovering together in simultaneous superposition. It was only when you measured an electron that this probability wave collapsed and the particle assumed a precise location in space and time, but the outcome of the experiment was inherently impossible to predict. Viki had also studied at Leipzig with Werner Heisenberg, whose famous uncertainty principle laid bare the curiously complementary nature of the subatomic world. The more closely you pinned down an electron's position, the fuzzier its momentum became, and vice versa. You had to decide which quality—position or momentum—you wanted to know. Measuring one destroyed the possibility of knowing the other. Like enrolling at MIT and committing suicide, the two quantities did not commute.

Imagine working under the people who had confronted these puzzles when they were still shiny and new. Viki had even been in Copenhagen with Niels Bohr, the Socratic pipe-smoking Dane whose enigmatic comments prodded his disciples to seek out truths so wispy and subtle that it was never entirely clear whether there was any substance to them at all. It was Bohr who had quantized the atom, showing that electrons could only take on certain energies, moving from level to level in quantum jumps. How can it be that an electron can leap from one orbital to another without traversing the space in between? Why, for that matter, can a particle's position and momentum not be known at the same time? "When it comes to atoms," Bohr had told Heisenberg, "language can be used only as in poetry." It was a different world down there in subatomic-land. There was no reason to believe that the concepts that worked so well in the macroscopic world could be exported to that foreign realm.

Lest he be influenced to wander too far in such mystical directions, Viki had served time in Zurich under the icy scrutiny of Wolfgang Pauli, the corpulent, mercurial thinker whose pointed letters, berating his colleagues to be more careful in separating sense from nonsense, were sometimes signed "The Wrath of God." (Here is everyone's favorite story about Pauli: Young Wolfgang is sitting in class listening to a visiting German professor. At the end of the lecture, the boy rises to his feet: "What Professor Einstein has just said is not so stupid.") Viki had spent time in Cambridge, England ("the real Cambridge," Murray called it), with Paul Dirac, who had confronted head-on the seeming clash between relativity and quantum theory that had puzzled Viki as a child. The result, Dirac's celebrated "relativistic wave equation" for the electron, allowed physicists to treat the particle in a way that took both these great theories into account. (As a by-product, the equation predicted the existence of the positron, identical to the electron but positively charged. No one was more surprised than Dirac himself when this mathematical figment of "antimatter" was found in a stream of cosmic rays. "The equation was smarter than I was," he said.) Finally, working on the Manhattan Project, Viki had learned the chilling reality behind all these abstractions, crouching in a bunker in the southern New Mexico desert with Enrico Fermi to witness the first nuclear blast.

For Murray these people had just been names in textbooks, more nomenclature to learn. In the stories Weisskopf told his students, the men who had made quantum physics came alive. But Viki, who was barely forty when Murray came to MIT, wasn't some old warhorse, like Kovarik at Yale, mired in a glorious past. Though he never rose as high as his illustrious teachers, Weisskopf was working on knowledge's crumbling edge.

The quantum theory that had begun with Planck and been honed and polished by Schrödinger, Heisenberg, Born, Bohr, Pauli, Dirac, and others had left a wonderfully precise, if maddeningly strange, picture of how subatomic particles behave. But that was just the beginning. While Viki was learning physics, his teachers were combining quantum mechanics and Einstein's special relativity into a magnificent new creation called quantum field theory. Just as photons were the particles—the quanta—associated with the electromagnetic field, so the electron was associated with an electron field and the proton with a proton field. Every kind of particle was intimately intertwined with a field, and every kind of field with a particle. Since there were gravitational fields, there must be particles (still undiscovered) called gravitons. In fact, in a seeming reversal of figure and ground, the fields themselves, rather than the particles, came to be seen as the bedrock of nature. Electrons, protons, neutrons, photons, and so forth were just secondary manifestations—epiphenomena.

For a long time, physicists had been bothered by the spookiness of phenomena like gravity, magnetism, and electrostatic attraction and repulsion. The sun tugging at the earth, a magnet tugging at a piece of metal—what could account for movement without mechanism, action-at-a-distance? Quantum field theory removed the mystery. When two particles interact—an electron repelling another electron or being pulled by an oppositely charged proton—the carriers of the invisible force are none other than photons. In the picture provided by quantum field theory, the particles influence each other by bouncing photons back and forth.

Not all the spookiness had been exorcised. Where did the photons come from to mediate the repulsions and attractions? They were created out of thin air. This might seem to violate the law of conservation of energy, which holds that you can't get something for nothing. But Heisenberg's uncertainty principle provided an exception to the rule. Quantum uncertainty applies not just to position and momentum but to other complementary quantities, like time and energy. The result is that energy can spontaneously arise from nothingness as long as it doesn't stay around too long. Quantum uncertainty provides the loophole through which these apparitions—these "virtual" particles—make their brief appearance.

This weird picture, so out of joint with the world of everyday experience, led to such precise predictions that most physicists willingly swallowed such oddities as Heisenberg uncertainty, virtual particles, and wave-particle duality. But other implications of these finely wrought equations were much harder to accept.

Consider the version of quantum field theory—called quantum electrodynamics, or QED—that dealt with electromagnetism. One of the most vexing problems that came to preoccupy Weisskopf in his early days as a physicist was called the self-energy of the electron. When two of these particles approach each other, their identical charges cause them to push each other away (bouncing photons back and forth between them). But each electron also interacts with its own field, generating a kind of feedback loop. The result was an infinite regress—the snake swallowing its own tail—in which an electron's charge and mass became infinite. On paper anyway.

Everywhere physicists looked in their theories, these infinities seemed to grow. Just as troubling as self-energy was a phenomenon called "the polarization of the vacuum." In the mind's eye, we conceive of particles surrounded by empty space. But another of the surprising ideas of quantum field theory is that the vacuum isn't really empty. It seethes with virtual particles that flit in and out of existence so quickly that they are barely there. Included among them are positrons, the electron's positively charged antimatter equivalent. Take the notion of a vacuum buzzing with virtual positrons and combine it with the age-old realization that opposite charges attract and you get this disturbing prediction: Every electron must surround itself with a cloud of virtual positrons. And the closer you get to the electron, the thicker this cloak becomes. If the electron is considered a dimensionless point, as it was in the physicists' equations, the strength of the positron haze becomes essentially infinite. And that, it seemed, should cancel out the electron's own charge altogether.

So here you had one theory implying that the electron's charge was infinite and another saying it was zero. Obviously, neither result is what is observed in the real world. When Murray first heard about these problems at Yale, he worried that theoretical physics, this brotherhood he longed to join, was in a state of disgrace. The maps weren't describing the real world.

Weisskopf first made his mark with a couple of papers in the mid-1930s that eventually helped point the way out of the mess. Perhaps these two countervailing infinities, caused by self-energy and vacuum polarization, could be made to offset one another, to cancel each other out. By balancing the two effects just so, the "real" charge of the electron (the one observed in the laboratory) might shine through. Weisskopf never completely solved the problem—it lingered for years—and he was not the only one confronting it. But he laid things out in such a straightforward, clear-headed style, devoid of mathematical filigree, that people began to take notice.

Before long, physicists were developing all kinds of crude techniques for canceling out these and the other infinities that seemed to infect QED. These ploys allowed them to calculate and make sensible predictions. But what was one to make of this trickery? If one looked at QED simply as a tool for solving problems, the situation wasn't so disturbing. If the mathematical manipulations worked, then who cared what they might mean? But those physicists who wanted to believe their theories really mirrored the subatomic world worried that the infinities they were blithely crossing out were

warning signs that something was seriously amiss in their understanding. Maybe if they had a better theory they wouldn't have to go
through this questionable rigamarole.

In 1947, when Murray was still exploring the tunnels at Yale,
Weisskopf was invited, as one of physics' rising stars, to ponder
these problems at an old inn on Shelter Island, tucked between the
North and South Forks of Long Island, New York. Gathered with
him under one quaintly shingled roof were some of the great luminaries of the day. There was I. I. Rabi, the short, scrappy physicist
who grew up in Brooklyn in an Orthodox home so devout that he
still remembered his surprise at learning that the earth was just one
of the many planets circling the sun. It was Rabi who had turned
Columbia University into a powerhouse of physics; three years
before Shelter Island he had been awarded a Nobel Prize. There
was Hans Bethe, a kindly, lumbering German with an encyclopedic
knowledge of nuclear physics second to none. It was Bethe who had
theorized about how thermonuclear fusion made the stars shine.
Joining Rabi and Bethe at the meeting were the Hungarian wonders John von Neumann and Edward Teller. Von Neumann helped
develop the mathematical techniques for modeling the first nuclear
explosions and was one of the inventors of the digital computer.
The ever-hawkish Teller became the leading proponent of the
hydrogen fusion bomb. (Along with Eugene Wigner and Leo Szilard, who also hailed from Budapest, they inspired whimsical speculations that Hungarians were actually advanced beings from Mars,
their strange language providing a perfect camouflage.) There was
John Archibald Wheeler, the disciple of Niels Bohr whose mysterious musings always reminded his colleagues of how truly weird
physics could be. There were younger theorists like Abraham Pais, a
Sephardic Jew from the Netherlands who had spent his first years
out of school hiding from the Nazis, and the two prodigies, Richard
Feynman and Julian Schwinger, each in his twenty-ninth year and
already competing for the title of smartest kid on the block. Murray's nemesis at Yale, Gregory Breit, was also there. And presiding
over the occasion was J. Robert Oppenheimer, Viki's boss from his
Manhattan Project days. This was truly one of those proverbial occasions when a strategically placed bomb would have snuffed out an
entire field.

For all the great theorists at Shelter Island, it was an experimenter who made the biggest splash, puncturing quantum field
theory once and for all—and pointing the way toward the necessary

patch. QED predicted just which orbits an electron can occupy as it hovers around the nucleus of a hydrogen atom. But recent measurements had suggested that two of these energy levels, which the equations said should be identical, were actually separated by a tiny gap. The first morning at Shelter Island, Willis Lamb, an Oppenheimer protégé, explained how he and a colleague had measured this shift with unprecedented precision. The discrepancy was unmistakable. Quantum field theory was giving wrong answers. After mulling over the problem, the physicists came to suspect that the so-called Lamb shift was probably being caused by none other than the self-energy of the electron. This feedback effect hadn't been taken into account. Self-energy was not just a mathematical curiosity but a phenomenon, real and physical, that left its shadow on the world. What's more, it was a finite effect, something that could be dealt with in the calculations. Now that they had a precise measurement of the Lamb shift, the equations could be rejiggered. The whole theory didn't have to be thrown out.

Weisskopf left the conference with a sick feeling that he had missed the boat. He immediately realized that he might have predicted something like this way back in the mid-thirties, when he first wrote about self-energy. Maybe he would have had to wait decades for experiment to catch up with theory, but so what? When the Lamb shift was announced, he could have gone back into the files, pulled out his paper, and perhaps laid claim to a piece of the Nobel Prize Lamb would later get. Even worse, just months before the conference, Weisskopf and one of his MIT students had begun doing this very calculation. But they had become bogged down in the details and hadn't seen any reason to hurry.

On the train ride back to Cornell from Shelter Island, Hans Bethe did a quick-and-dirty calculation (ignoring the effects of relativity) and showed that self-energy could indeed be used to explain most of the Lamb shift. Weisskopf, rushing to make up for lost time, computed the effect more exactly. Ever cautious, he traded results with the boy wonders Feynman and Schwinger. Their answers were identical to each other and different from Weisskopf's. Viki knew he was a weak mathematician—because of a stupid error, his first paper on self-energy had resulted in an embarrassing correction—so he decided not to publish. As it turned out, Weisskopf was right, and Feynman and Schwinger had made the same mistake. Feynman's apology has become enshrined in the folklore: He recognized Weisskopf's priority and bad luck in a paper, playfully manipulating the citations so the acknowledgment came in footnote 13. Schwinger, magisterial and aloof, apparently saw no need to apologize.

Schwinger, Feynman, and a Japanese physicist, Sin-itiro Tomonaga, went on, in different ways, to completely reformulate QED, banishing the infinities for good. In a kind of mathematical shell game, one could manipulate the equations, using one set of infinities to cancel out another. Finally those that remained were sequestered into two values: the electron mass and electron charge. Since it was known that these numbers were not boundless, the infinities could be crossed out and the real values, determined from experiment, plugged in. The theory, as physicists put it, was "renormalized." Not everyone was impressed. Dirac objected strongly: "Sensible mathematics involves neglecting a quantity when it turns out to be small—not neglecting it because it is infinitely great and you do not want it!" But resistance was quickly overcome when renormalized QED went on to achieve one experimental triumph after another, predicting with great precision how electrons and photons interacted. Years later, the three scientists were honored for this work with a Nobel Prize—another one Weisskopf might conceivably have shared if he had been more persistent. And luckier. His personal life may have been charmed, but his career had been deflected more than once by a missed opportunity.

The problem, Weisskopf later decided, was that he never concentrated long on the same problem, jumping instead from one question to another, wherever his curiosity pulled him. He tried to be stoic. He once said he wouldn't take a Nobel in exchange for the broad general knowledge his fitfulness had given him. "I prefer to know nothing about everything rather than everything about nothing," he said in his typically ironic, self-deprecating style. He lacked a quality Gell-Mann would later call the killer instinct. Weisskopf found solace in believing that his low-key manner made him a good teacher, that he never felt he really understood a subject until he could convey it to others. But sometimes his pronouncements seemed a little halfhearted. In some ways, the mentor Murray had stumbled upon was a disappointed man.

For all the inspiration Viki provided with his tales of life on the quantum frontier, Murray soon found that he was not the greatest lecturer. To simplify his calculations, Viki used what his students called "Weisskopf units." He would arbitrarily set 4 pi equal to 1, or 2 equal to 1, or −1 equal to 1—anything to get through those damned equations and on to what he really loved to talk about: how the subatomic machinery worked. Sometimes Weisskopf's calculations were so riddled with errors that they canceled one another out, like those quantum infinities, serendipitously yielding the right answer. One day Murray and some other students watched with amusement as Viki stood at the blackboard trying to derive a formula about the nucleus, one that he himself had discovered with Eugene Wigner. On the first attempt, the formula came out upside down, with the numerator and denominator reversed. Viki took another stab at it. This time the formula was right side up, but one of the terms had the wrong sign (positive instead of negative, or vice versa). Eventually, he got the right answer.

Weisskopf continued to rationalize these weaknesses away, seeing himself as someone with an unusually developed "physical intuition," a feeling for the reality the mathematical formalisms strained to describe. He never tired of telling students how the great physicist Paul Ehrenfest had urged him not to be overly impressed with fancy mathematics. "Physics is simple but subtle," Ehrenfest had said. It was dangerously easy to get carried away playing with numbers and forget that the object of the game was to talk about the real world. And he liked to quote Niels Bohr, who said expertise comes from making all possible mistakes.

Sometimes Viki's suspicion of higher mathematics seemed like a defensive reaction to his own carelessness: He didn't always grasp the loftier mathematical formalisms, therefore formalism was not very important. But after all the erasures and broken chalk, the ideas he was trying to express finally came through. That was what really mattered. And the joy Viki found in his haphazard presentations made his students love him. They would join him for weekly dinners at a Chinese restaurant or a French place in Boston that sometimes served, of all things, bear meat. They would excitedly discuss theoretical physics, nuclear disarmament (one of Viki's favorite causes), and just about anything else on their minds. Viki could be warm, understanding—sometimes maybe a little too understanding. He always put his name last on a paper, even if he was the principal contributor. This kind of generosity left some of the MIT graduates unprepared for the cutthroat competition they would encounter later in their careers.

Murray always looked forward to the sessions with Weisskopf. But the two never really shared the easy camaraderie Viki developed with some of his other students. Murray was a little too brash for his teacher's refined Viennese sensibilities, and a little too enamored of equations. And the professor may have been just a little jealous of the ease with which his student glided through mathematical manipulations that left him stumbling and confused.

Viki liked to ask students to give talks, to make sure they understood a subject. Murray, he thought, was a bit of a show-off. Sometimes it almost seemed that he was being purposefully elliptical, more interested in impressing the audience than in conveying knowledge. Outside the classroom, when Murray reeled off the names of birds or obscure facts about history, Viki assumed at first that he was faking—leveraging a small investment in trivia into the appearance of a great wealth of knowledge. But, Viki came to admit, a little prodding usually revealed that the foundation was really there.

Who was this energetic know-it-all who never talked about his family? And what was the story with the hyphen? There had been plenty of Gelmans back in Vienna, but no one with the highfalutin name Gell-Mann. The name, Viki thought, seemed a deliberate attempt to obscure a Jewish ancestry. To some extent, he could sympathize. Jewishness was an accident of birth, so why should it be part of one's identity? But for all his secular upbringing, Viki had learned that he could never wander too deeply into the abstract wonderlands of physics without being snapped back into a real world where anti-Semitism seemed as rooted as natural law. At the University of Berlin, Weisskopf had watched in horror from his office window as Nazi thugs beat up Jewish students in the courtyard. (Sometimes he would pull one of the victims inside and help him sneak out a back door.) Despite their influence, Viki's family lived in fear of the Nazis and barely managed to get out of Europe in time. Jewishness was not just a detail, like the color of your hair, and it was hard to understand someone who could so blithely ignore his ancestry.

Years later, when Weisskopf was reflecting on what kind of influence he might have had on Murray's meteoric rise—a phenomenon he never foresaw—the teacher liked to think that perhaps some of his straightforward style, his feeling for the mechanism, had indeed rubbed off on his student. He had taught Murray that nature came first, not numbers. And Murray would say that Weisskopf was one of the few people he had learned something from. Physics was more than just conjuring with numbers. Because of Viki, Murray would say, he had seen the importance of "avoiding cant and pomposity," of "prizing agreement with the evidence above mathematical sophistication" of always searching, "if at all possible, for simplicity." But at the time, Murray's extraordinary talent wasn't taken all that seriously. As at Yale, he just didn't seem able to connect, to convince his teachers that there was depth to the dazzle. When Weisskopf wrote his autobiography, he made just a passing mention of his most famous student, blandly describing him as winning a Nobel Prize "for his ideas about the substructure of the proton." He found the name "quark" ugly.

If Murray was hurt by his teacher's inattention, he didn't let it faze him. He was used to learning mostly on his own. Walking around campus with his friends, animatedly discussing physics, he was coursing with energy, dancing up and down stairways. He was no longer on the outside looking in but right at the center of the action. Norbert Wiener, the eccentric mathematician, would ride his unicycle in the hallways, smoking cigars and eating peanuts. Sometimes Murray would find him asleep on a stairway, blocking traffic with his elephantine body.

Marvin Goldberger, a postdoc who had come to MIT from the University of Chicago, where he had studied under Fermi, was startled one day when this small child—that is how Murray appeared to him—walked up and announced (the cat who ate the canary) that he had read every one of his papers. Goldberger thought, "There, there, little boy," and mentally patted him on the head. But it turned out that Murray wasn't kidding—he really *had* read the papers. And from then on, they were friends. Murray and Murph, as Goldberger's friends called him, would gather around a blackboard with a group of students, arguing loudly about the latest theory in the *Physical Review* or some perplexing new experimental result, writing and erasing and crossing out until everyone was covered with chalk dust.

Murray thought Murph was absolutely amazing. He always seemed to know about every important experiment, every important theory—where they meshed and where they clashed. Though just a postdoc, Murph seemed tuned in to the latest developments as acutely as Weisskopf himself. If you wanted to know about the latest calculational techniques—which ones worked and which ones didn't—you asked Murph Goldberger. And, unlike Viki, Murph had an incredible ability to focus and march through a chain of calculations without a false step. Weisskopf would stand at the blackboard looking befuddled at his own markings, the stereotypical absentminded professor. Murph would start at the top of the board, work his way line by line to the lower right-hand corner, and with a few clicks of the chalk draw a box around the answer—the Fermi style, Gell-Mann would later learn.

Two or three times a week, Murray and a group of aspiring physicists would pile into Goldberger's car and drive up Massachusetts Avenue to Harvard to sit in on a seminar by Julian Schwinger. Schwinger's students were in awe of their nattily dressed, somewhat imperious young master and the way he enveloped his theories in an elegant, rigorous, and ultimately arcane personal notation that nobody but he could really understand. The lectures—seamless, polished to a sheen—were already legendary. One day when Goldberger—this interloper from the trade school down the road—had the temerity to ask a question, every head in the room turned around and stared.

Driving up the street to Harvard was like entering another world. Once, at a theoretical seminar held jointly by the two schools, a Harvard graduate student was asked to describe his Ph.D. work. It was one of those typical microproblems that aspiring physicists are given as a rite of passage: Calculate the characteristics of the lowest energy state that can be assumed by the element boron.

Particles are described by their quantum numbers, including charge, mass, and a quantity called spin, which captures how a particle rotates on its own axis. The student at the seminar had laboriously calculated that the spin of the boron nucleus should have a value of 1. Murray, imagining himself up there, was wondering what the teachers thought of the performance when a little man who hadn't shaved for days (Murray thought he looked like he had crawled out of the basement of MIT) piped up in a thick accent: "Hey, da spin ain't one. It's t'ree. Dey measured it!" Murray was impressed. This was not like college, where you imbibed and expelled the received wisdom. People were actually talking about unsolved problems, teetering on the cutting edge. The game was not just to make the highest possible grades and impress the teacher. A theory, no matter how beautiful, was worthless if it didn't predict true phenomena. Physics, this life he had stumbled into, was going to be fun.

Murray had little trouble making friends with the other residents at the MIT graduate house, at least with those who could overlook his sometimes prickly personality. As at Yale, if someone wanted to go out for a beer, Murray would lead the way, holding forth on science and politics, trying to win his friends over to his leftist views. Though he still wouldn't talk about his family, the scars of the Depression showed through. If people wanted to be protected from the blind economic forces that had crushed his father, they were going to have to support a more progressive political agenda.

Once a week, a group would drive to a restaurant across the river in Boston's Chinatown. The upper level was for Americans, the lower level for Chinese. Murray and his friends sat, of course, on the lower level, where he would try to order for everyone. It was a trick he would hone to perfection: quickly picking up just enough of a language to fool people into thinking he had been speaking it all his life. It was not all just show; Murray mastered a number of languages with stunning fluency. That just made it all the harder to tell when he was putting people on. One day he surprised Jay Meili, a Swiss-American chemistry major who lived in the same suite at the graduate house, by explaining jokes to him in Schwyzertütsch, the dialect of German spoken in Switzerland.

One advantage of socializing with chemists like Meili was that they had access to a limitless supply of pure ethanol. When it was mixed straight with grape juice, the result was an occasional party that was over by nine p.m. with everyone nauseated or passed out on the floor. At one of the tamer parties, Murray's friends were impressed when he walked through the door squiring Weisskopf's beautiful secretary. The woman, with her slight Germanic accent, was so alluring that young men would walk by her office terrified, longingly stealing a glance inside. Murray had actually worked up the nerve to ask her out. It did wonders for his reputation, but he was still uneasy around women. One evening Meili and his girl-friend from Wellesley arranged a double date with Murray. He was so awkward that it was uncomfortable for everyone.

Though barely twenty, Murray began to take on an aura of legend. How, his friends wondered, did he have time not only to learn nuclear physics but to forage for all the obscure facts he loved? He still didn't seem to spend much time studying. His roommates decided he must have really crammed when he was at Yale, storing up the mental fuel he was burning now. Then one of them met a former college classmate who told them Murray had seemed just as casual back in New Haven. It was almost as though he had been born with an innate knowledge of theoretical physics, and a lot of other subjects as well.

Murray's education wasn't really as effortless as it appeared. He spent hours poring over MIT textbooks, which he found so much better than the ones at Yale. Toward the end of the first year, he announced that he was ready for his "majors," the oral exams most students don't take until later on. Viki was supportive, but some faculty members were offended by this upstart and determined to cut him down to size. During the exam, one professor—a practical sort less interested in theory than in issues like the health effects of radiation—asked Murray a question he was sure would stump him, something about hole theory, where the absence of an electron is treated like a positively charged particle. Murray didn't really understand the question, but he had read enough in related areas to snow his questioners and pass the exam.

After his first year, Murray, Meili, and a couple of friends decided to leave Cambridge behind for a few weeks and drive as far as they could into Mexico. They pooled their money and bought a beat-up 1936 Nash Rambler for a couple hundred dollars and headed south. By shifting the car into overdrive, they could pass anything on the road. But there was no front bumper, and the body was rusting out. Halfway through the trip a lock fell out, so they had to open the door with a screwdriver.

Once they crossed the border, Murray took on the job of interpreter. With their tight budget, the boys camped out when they could, staying the rest of the time in cheap hotels. The first night on the way to Monterrey, they were so afraid of bandits that they hid the car in a cactus grove. The next morning they had four flat tires. Eventually they got to Lake Pitscuaro, where they camped on land owned by an old Spanish Civil War veteran. Murray would speak French to him and he would answer back in Spanish. Since the car's headlights didn't work, the boys tried to drive only in daylight. But one evening the setting sun left them on a dark, winding mountain road, feeling their way by tailgating a truck through the night. On the way back, they were startled to see the perilous terrain they had driven through: steep cliffs dropping off from the side of the road.

They made it as far south as the Isthmus of Tehuantepec, camping on the beach and looking up at the stars.

Back in Cambridge the next semester, Murray was starting to worry about his dissertation. Never mind that most of the graduate students were just getting their feet wet. Murray was determined to earn his Ph.D. in two years. After the fiasco with his thesis at Yale, he was going to do things right this time. He told Meili that he wanted his dissertation to be as important as one of Einstein's big three papers—special relativity, general relativity, and the photoelectric effect. He tried out one problem after another, exploring each and finding it just wasn't as interesting as he had hoped. Finally, with time running out, he had to settle on a problem Viki gave him, something he considered rather minor but tractable: a formal treatment of the theory of the nucleus, and what happens when it is bombarded by a neutron.

It was a problem only a nuclear physicist could love. According to a model developed by Niels Bohr, the protons and neutrons in a nucleus were tightly bound together like the molecules in a drop of water. Fission was like a drop stretching and splitting into two droplets; in fusion, two droplets merged together to form a larger whole. By the early 1950s, however, experimenters were finding that this simplification didn't always hold. Some nuclei acted as though the protons and neutrons were less like water molecules forming droplets than like freely orbiting electrons in an atomic shell. Which metaphor was better? The problem facing Murray was to reconcile this "weakly coupled" shell model of the nucleus with the "strongly coupled" Bohr model—to craft a mathematical compromise that would show more clearly what was going on inside the cores of atoms.

But he just couldn't get into it. The writer's block that had sabotaged him at Yale was now deeply ingrained. He was afraid that no matter how hard he tried, he would fail. He would work a few days, then get bored and depressed and spend the afternoon at the library studying Chinese or reading the Tibetan Book of the Dead. To make matters worse, his bank account was running dry. The Atomic Energy Commission had agreed to pay his way through MIT, but they still hadn't come through with the money. This was the McCarthy era, and AEC officials had been mortified when one of their earlier scholarships went to a member of the Communist party. The Senate responded with an amendment requiring that future scholarships go only to students who could pass a security clearance. In his interview, Murray had bluntly declared that he used to be a Red. He hadn't recruited anyone for a revolution, he told them, or belonged to any Communist organizations. He had simply believed for a while that Communism was a good thing. But the AEC apparently wasn't satisfied with this explanation.

As Murray waited and waited for his scholarship money, he wondered what was going wrong. Maybe the investigators had learned that the family of his old piano teacher, Florence Freint, had been Communist sympathizers. Maybe the AEC was suspicious because one of Ben's friends had put Murray on the mailing list for bulletins from the Soviet embassy. Couldn't they tell it had just been a lark? In those red-baiting days, it can't have helped matters that Arthur—avidly anti-Stalinist all along—belonged to the American Labor party. Anyway, the money had been frozen. To help Murray through the final weeks of graduate school, which dragged on into months, his friends set up a fund for him.

According to his self-imposed draconian schedule, he was supposed to be done with the dissertation in June 1950, when Weisskopf was leaving to teach in Paris for a year. For a while Viki had talked about taking Murray along—the nice thing about the AEC fellowship was that it was portable. Murray, whose only contact with Europe had been through the photographs Ben sent back from the war, thought living in France would be a dream fulfilled. Then one day Viki said he had changed his mind: Taking along a student didn't seem appropriate. Murray thought the excuse sounded a little flimsy. He was hurt that his mentor didn't want him to come along.

Instead, Weisskopf tried to find him a position somewhere else. Maybe Murray could go to Harvard as a member of the Society of Fellows. Viki arranged for him to give a seminar there, but it went badly. Intimidated by his august listeners, Murray assumed they would already know about everything he said. He glided over important details, nervously tossing a piece of chalk in the air and catching it. An interview with a faculty member, a humanities professor, also bombed. For the second time since he had left Yale, Harvard disappointed him. Then Viki called up his old boss Robert Oppenheimer, who was now the head of the Institute for Advanced Study in Princeton. Oppie agreed to give Murray a visiting position beginning that fall.

But September passed and Murray was still trying to get into his albatross of a dissertation. Since he was no longer officially enrolled at MIT, he had to give up his bed and camp out on a couch in the living room of his old suite. Oppenheimer would call, impatiently asking just how late Murray planned to be. An appointment at the Institute was quite an honor; people generally showed up on time.

Finally on January 1, 1951, hung over from a New Year's Eve party, Murray started seriously writing. In a few days he was finished and got his degree. He was depressed over the way the dissertation came out, with the problem left half done. Later he would learn that the work was nothing to be ashamed of, that it contained some useful tidbits. But, still mired in the gritty details of the subject, he had no idea what was useful and what was not. A few years later, Wigner solved the problem more thoroughly. Murray, like Weisskopf, hadn't focused hard enough, and he paid the price with a missed opportunity. It wasn't something he intended to let happen again.

CHAPTER **4**

VILLAGE OF
THE DEMIGODS

After the Institute for Advanced Study opened its doors in 1933, Albert Einstein occasionally found the place so oppressive that he would write, as a return address on his envelopes, "Concentration Camp, Princeton." His description of the town, in a letter to his friend the Queen of Belgium, has become legendary: a "quaint, ceremonious village of puny demigods on stilts."

It was a melancholy time for Einstein. If it hadn't been for the Nazis, he would have preferred the vibrancy of Berlin, where he had just resigned his professorship. By the time he was forced to emigrate, all his important work was far behind him, and he was mired in an unsuccessful effort to weave together his beautifully crafted theory of gravity—general relativity—with the equally elegant picture of electromagnetism drawn in the previous century by James Clerk Maxwell. The result, Einstein dreamed, would be the ultimate theory, explaining everything from the tiny subatomic particles to the motion of the galaxies. Never mind that Maxwell's equations had been retooled into quantum field theory for use inside the tiny spaces of atoms. In his grand unification, Einstein fully intended to rid physics of the "evil quanta" that defied commonsense notions of causality, to show once and for all that God was not a crapshooter. Hardly anyone thought he had a prayer of success. "I have locked myself into quite hopeless scientific problems," Einstein confided to a friend, "the more so since, as an elderly man, I have remained estranged from society here."

In a way, this estrangement was exactly what the Institute's founder, Abraham Flexner, had in mind for his stable of luminaries. The Institute was to be, as he grandiosely put it, "a haven where scholars and scientists could regard the world and its phenomena as their laboratory without being carried off in the maelstrom of the immediate." All that was missing was white robes, laurel wreaths, and the Acropolis looming in the background. Like Plato's academy, Flexner's institute was to be a place where a few of the chosen could live the life of the mind. In Flexner's haven, no one need worry about classes or curricula. No diplomas were awarded. When it came to research, the faculty and visitors were accountable to no one but themselves. Compensation was so generous that the place became known as the Institute of Higher Salaries. But in return for the largesse, Flexner exacted a price. He controlled the place with the humorless intensity of a dictator. When Einstein accepted an invitation from the Roosevelts for dinner at the White House, an indignant Flexner tried to cancel the event: He wasn't going to have his scholars distracted with such pettiness. Einstein went to the dinner anyway, and later threatened to resign if Flexner didn't stop meddling in his life.

The money for Flexner's unusual think tank came from a wealthy family named the Bambergers. Louis Bamberger and his sister, Caroline Bamberger Fuld, were heirs to a fortune created by their family's New Jersey department store. The family had the good luck to sell out to Macy's in 1929, right before the stock market crashed. Wishing to recycle some of the nickels, dimes, and dollars the citizens of New Jersey had spent buying socks, underwear, shirts, shoes, and blouses, the siblings wanted to build a medical or dentistry school somewhere in or around Newark. To help realize their vision, they contacted Flexner, an irascible individual who had recently become famous for his Flexner Report, which denounced most of the medical schools in the United States as third-rate diploma mills. How, the Bambergers wanted to know, could they use their money to build a school great enough to win even Flexner's praise?

Flexner, they quickly learned, thought the last thing the country needed was another medical school. And ugly Newark seemed like a particularly bad location for one. He persuaded the Bambergers to use their money to help him realize his own dream: a platonic academy where the great thinkers of the day would be free to stroll among the trees and gaze at idyllic views from their office windows, advancing the frontiers of knowledge. The Bambergers, wide-eyed and willing, quickly agreed that such a place—the New Jersey Institute of Higher Learning, or something like that—would indeed be a credit to their beloved Garden State. But Flexner wasn't finished. He wasn't about to have his academy in Newark, and he saw no reason for "New Jersey" to muck up the name. Before they knew it, the

Bambergers were agreeing to bankroll the Institute for Advanced Study in the picturesque town of Princeton. If Flexner was going to have to put his dream institute in New Jersey, then at least it would be in its Athens. By the time the Institute had moved from temporary quarters on the Princeton University campus to a permanent location on an old farm on the outskirts of town, with five hundred acres of woods for strolling, Flexner had recruited an all-star cast that included not only Einstein but the mathematician and bomb theorist John von Neumann, the economist Oswald Veblen, and the logician Kurt Gödel, famous for his incompleteness theorems. Younger scientists like Gell-Mann would be invited to linger for a year or two to soak up the ambience, then be sent on their way.

By the time Murray made his way, late as usual, to the Institute, Flexner had been defanged, sidelined into an innocuous advisory role, and the atmosphere had markedly improved. Einstein was living out the last years of his life, and Murray would sometimes see him walking the tree-lined streets with Gödel, a man so diminutive that he made even Einstein look tall. Murray thought of them as the cartoon characters Mutt and Jeff, and wondered if they were puzzling over important scientific matters or just passing the time of day, two old men no longer in the loop. In Murray's eyes, Einstein had become a little pathetic. He would later find it sad that the image of Einstein that became enshrined in popular culture was that of the disheveled old man who had foolishly rejected quantum theory. Why not the dashing twenty-six-year-old who, all in the same year, completed his Ph.D., wrote his earthshaking papers on special relativity and the photoelectric effect (through which photons striking metal spark off electrons), and showed how Brownian motion, the random vibration of particles suspended in a liquid, was caused not by some mysterious life force or the spontaneous generation of energy but by the bombardment of invisible molecules?

Murray later regretted not taking the time to get to know Einstein, but he couldn't imagine what to say to someone so hopelessly mired in the past. It disgusted him to see some of the other young Institute fellows trying to ingratiate themselves with the old man. Murray was determined not to be among the sycophants. He said hello to Einstein when he saw him walking to the Institute (the old man refused to take the shuttle), and Einstein would say *"Guten Morgen."* That was the extent of the contact between the man who was the most revered physicist of the first half of the century and the boy who would do so much to revolutionize the second half.

Murray found the Institute, and especially the surrounding forests, a pleasant enough place to linger. Oppenheimer, who had recently left Los Alamos to become its director, was now imposing his own style. Gaunt, brooding, chain-smoking, the brilliant under-achiever, Oppenheimer had made some incisive early contributions to theoretical physics. But his talents lay more in orchestrating others, prodding them with his intense questions, charming them with his wit. For managing the bomb project, he had been anointed a national hero. Landing him as director of the Institute was a coup. Full of his usual self-importance, he would walk the grounds, agonizing over what he and his colleagues had done (he liked to allude to the myth of Prometheus, stealer of fire) while proudly keeping his classified documents in an office safe protected round the clock by armed guards. The parties at his Princeton house, which came to be called "Bourbon Manor," were legendary, the drinks so strong and the chili so hot that guests would stumble home in a daze.

It was said that Oppenheimer was fluent in eight languages, that he wrote poetry and read Plato in Greek. Years after the unleashing of nuclear energy had changed the world, General Leslie Groves, the military leader of the Manhattan Project, asked Oppenheimer why he had chosen the name Trinity for the site in southern New Mexico where the first bomb was detonated. Oppie replied that he had been inspired by a John Donne poem. While he was watching the explosion, his cultivated brain had extruded the now-famous line from the Bhagavad-Gita: the god Vishnu saying, "I am become Death, the destroyer of worlds." There was later some dispute over whether Oppenheimer had actually studied the Hindu epic in the original Sanskrit, as legend had it, or in English translation. Oppie basked in the aura he projected—of someone who was a little better than those whose erudition was confined to quantum field theory. It was an image—the multilingual expert on everything—that Gell-Mann himself was already perfecting.

Oppenheimer was also a good talent scout. The cast of promising young physicists he had gathered around him included T. D. Lee and C. N. (Frank) Yang, two Chinese theorists who had come to the Institute after completing fellowships together at the University of Chicago under Enrico Fermi and Edward Teller. He recruited Abraham Pais, who had been one of the younger theorists at Shelter Island, and Freeman Dyson. While still a graduate student at Cornell, Dyson had polished up calculations made by his teacher Hans Bethe on the Lamb shift—the crucial experiment establishing that

something was rotten with the original version of quantum electrodynamics. Then, in intense conversations with Feynman, he joined
in the effort to "renormalize" QED, smothering those annoying
infinities. Feynman and Schwinger had come up with competing
approaches; but which one was right? Dyson was determined to find
out. Schwinger's theory was cast, as usual, in a baroque private language, as prim and proper as Schwinger himself. His rival had taken
a scrappy, more intuitive approach—solving problems using the
simple drawings of arrows and squiggles that became known as
"Feynman diagrams." Claiming that he couldn't understand quantum mechanics the way it was taught in school, Feynman had rein-
vented it himself. In his "sum over histories" approach, one would
calculate the trajectory of an electron by imagining that it took
every possible path from A to B—straight paths, curvy paths, zigzag
paths, even paths where it went backward in time. All of these would
then be combined, according to Feynman's rules, adding and canceling like overlapping waves to yield the answer to the problem.
Dyson thought something so beautiful had to be right. In the
course of a cross-country bus ride, somewhere on the Nebraska
plains, he was struck by the answer: If you looked beneath the surface, Schwinger's mathematical filigree and Feynman's cartoons
were really saying the same thing. By combining the two methods,
one could have both precision and deep understanding.

The Institute was becoming a magnet for such young prodigies.
In 1951, Murray's first year there, Dyson was twenty-seven. Yang was
twenty-eight. Lee was twenty-four. Pais, born the same year as both
Feynman and Schwinger, was a ripe old thirty-two. And Murray,
youngest of the lot, was twenty-one.

Shortly after Murray's arrival, Oppenheimer brought the new
recruit to afternoon tea at Fuld Hall, the Institute's elegant main
building, and introduced him to another young theorist, Francis
Low, with whom he would be sharing an office. Low had just
received his Ph.D. from Columbia, and had come to the Institute in
September, when Murray was supposed to have begun. "I'd like you
to meet Murray Gell-Mann," Oppenheimer said. Puzzled by this
august-sounding surname—Low thought it must be European—he
asked Murray if he had just arrived from abroad. "Only if you call
Cambridge, Massachusetts, abroad," Murray said.

Because of the war, Low was four years behind in his career, and
Murray, of course, was four years ahead—age twenty-one to Low's
thirty. "They put this child in my office," Low complained. He

quickly found that Murray wasn't self-conscious about his boyish looks. Or if he was, he compensated by pushing his ideas with a fierceness that Low—reserved and sardonic—found startling. With Gell-Mann, he quickly realized, you had to be tough and fight to hold your own, because Murray could be just as forceful when he was wrong. Fortunately, and sometimes maddeningly, that seldom seemed to be the case.

Low was in need of a smart collaborator. He had been struggling with one of the remaining puzzles of quantum electrodynamics. With the infinities vanquished, QED, the version of quantum field theory that dealt with electromagnetic interactions, was in great shape. This theory of electrons and photons—the very basis of chemistry and biology, of what we experience most directly about the world—was and remains the most solid and successful accomplishment of particle physics. As Feynman later put it, the theory matches experiment as closely as if one tried to predict the distance from New York to Los Angeles and was off by no more than the thickness of a human hair. But there were still a few loose ends. Low had been working on one of them—something called the relativistic two-body problem—and getting nowhere.

If you wanted to describe a one-body system—a single electron, say—using the language of quantum theory, you could pull out Schrödinger's famous wave equation, in which all the possible states of the particle (its positions or its momenta, whichever you wanted to know about) hovered in superposition. If you wanted to take special relativity into account—the electron might be whizzing by at near the speed of light, causing those strange Einsteinian distortions—then you could use the equation derived by Paul Dirac (the one that had surprised its inventor by predicting the existence of the positron). But calculations involving two-body systems—a proton and an electron tugging at each other or an electron and a positron—were a stickier matter. The Dirac equation could be used if you made certain approximations. But if you wanted to precisely describe these "bound states," better tools were needed. In 1929, Murray's teacher at Yale, the mad Gregory Breit, had cobbled together a way to describe bound states using the language of quantum field theory. But Breit's method didn't work very well when relativity was a major concern. Even worse, the Breit equation (like a bad piece of software) was a little buggy: inconsistent and easy to misuse, resulting in papers that spouted mathematical nonsense. Something better was needed. If two particles were not moving fast compared to each other, like the proton and electron clasping to form a hydrogen atom, then you could get by without the relativistic correction. But in nuclear physics, particles bounced off each other so quickly that time seemed to dilate and length contract, and there was no trustworthy way to combine the Einsteinian effects with all the quantum oddities and make a precise prediction.

More than one theorist had fallen flat trying to find an answer. When a physicist came up with a new discovery, it was customary to make a pilgrimage to the Institute and try out the new idea on Oppenheimer and his young geniuses. If your brainchild passed their hard scrutiny, you could feel that much more confident about releasing it to the world. If the geniuses shot it down, it was back to the office or laboratory. One hopeful physicist, passing through Princeton on a lecture tour, drew a crowd of thirty-five or forty—big for the Institute—when he announced that he had a solution to the two-body problem. When it became clear, as the lecture dragged on, that he wasn't living up to his billing, Oppenheimer was merciless, berating the man for wasting everyone's time. Some young scientists, whose enthusiasm crumbled under his interrogation, left the lecture room in tears.

Gell-Mann and Low thought they could do better. They had already begun talking about the problem when a young postdoc named Edwin Salpeter, another Bethe prodigy from Cornell University, arrived to present his own solution. As Gell-Mann and Low listened, Salpeter described the rough outlines of an equation he and Bethe were working on. They seemed to be on the right track, but Salpeter and his mentor hadn't really spent much time on the theory, basing it loosely on some earlier work by Feynman. For the equation to be accepted as something solid—a reliable, precision tool, unlike the rickety Breit equation—it would have to be rigorously derived from quantum field theory, showing that it meshed with the rest of physics' tight weave. Salpeter knew the work was a little loose and premature; he and Bethe had just given what is known in the trade as a plausibility argument, showing that the equation intuitively seemed true. Anyway, Salpeter was already drifting away from field theory to make his mark as an astrophysicist. He was startled when a bespectacled youngster with dark curly hair interrupted the talk, pointing out the holes and suggesting how they might be filled. Low soon chimed in. Everybody was arguing back and forth, and before the afternoon was over, Low and Gell-Mann had sketched out the steps that would lead to an airtight derivation. Salpeter had heard of Low, whose reputation was already rippling through the field, but who was this child physicist, Murray Gell-Mann? He went back to Cornell feeling ignorant and, at age twenty-six, like an old-timer. Surely Murray, despite his youthful appearance, had been doing physics for years. How had he missed him?

In the following days, as the two young scientists put the finishing touches on their derivation, Low was struck by the clarity with which Murray could see to the heart of a problem, taking a hazy idea that somehow felt right and, as the hours sped by, crystallizing it into a structure, an architecture, something you could turn over in your mind. He had an incredible capacity to tease out and organize hidden patterns. It was the beginning of a long friendship. Sometimes with a single comment Gell-Mann would illuminate an idea, and for the rest of his life Low would have a deeper understanding. Murray had an instinct for cutting away the clutter and exposing the deeper truths. "Why don't we do this?" Low would say, proposing a trick to get around some immediate obstacle. "No," Murray would interject. "There is a much more general way of doing it."

Just six months after Gell-Mann's arrival, he and Low submitted what was to be Murray's first paper: "Bound States in Quantum Field Theory" was published four months later in the *Physical Review*. It wasn't exactly a major breakthrough, more a tidying up around the edges of QED. The tool for solving two-body problems would become known as the Bethe-Salpeter (not the Gell-Mann–Low) equation. The two scientists felt compelled to acknowledge, in a footnote, that the indefatigable Schwinger had apparently done a similar derivation in his Harvard lectures. Years later, when winnowing his list of some 100 publications down to 27 principal works, Gell-Mann passed this one by.

He should have been more excited. At twenty-one, he had published in the premier journal of the field and demonstrated full command of the tools of his profession. His pin had been added to the map of physics. But he didn't feel very happy. The paper seemed overly formal to him—something Viki wouldn't like. It was more about playing with numbers than explaining how nature operated. Oppenheimer loved the work, but Murray thought that just reflected badly on Oppenheimer, showing that he was too impressed with fancy footwork. Sometimes it seemed to Gell-Mann that half the scientists Oppie brought to the Institute had made their mark with some vacuous mathematical technique. For some reason, Murray complained, Oppenheimer was a sucker for that stuff. He seemed to practically worship Dyson. "I'm sure Dyson will sum the series for quantum electrodynamics within a few months," Oppenheimer would say. Murray was tired of hearing it. What had Dyson ever done that was so great? As far as Gell-Mann was concerned, he had just shown that the Feynman version of QED and the Schwinger version were formally equivalent. But Murray thought that Feynman had done that himself, just not as smoothly. (Not everyone shared this view. Though Dyson was modest about his own accomplishment, some felt that his original insights into the nature of field theory illuminated Feynman's work and deserved a share of the Nobel Prize.) When Gell-Mann and Low were putting the finish on their derivation of the Bethe-Salpeter equation, they were stumped by the need to somehow define the vacuum mathematically. Oppenheimer suggested they use a quantum wave packet—one of those expressions in which every possible state hovered together in superposition. Murray thought this was bizarre; the vacuum, by definition, had just one state, the lowest possible one. What could Oppie have been thinking? Sometimes Gell-Mann and Low laughed behind his back. For all his brilliance, he sometimes seemed the perfect example of someone who kept forgetting that the mathematics had to reflect an underlying world.

Gell-Mann generally liked Oppenheimer, but he couldn't help wondering if life wouldn't have been better if Viki had taken him to Paris. Princeton was nice when it wasn't raining and he could walk from his boardinghouse near the university campus to the Institute, bird-watching along the way. He made friends with some other young bachelors, including David Bohm, a Princeton University physicist who was struggling to reconcile his Marxist beliefs with the maddening indeterminism of quantum theory, and David Pines, who was collaborating with Bohm on something called plasmon theory. They would all have dinner together or walk the streets and talk. Most of the time he felt depressed. The maelstrom of the immediate was very much with him. His mother's mental health seemed to be deteriorating, and many nights he would stay up until 2 a.m. reading and brooding, showing up at the Institute in the early afternoon.

But his work was making an impression. In a letter home to Arthur and Pauline, Murray wrote that Oppie had told him that he could certainly stay on for a full year, maybe even a year and a half. He complained of his security clearance problems—he was trying to line up a "draft-proof" job at the government laboratory at Oak Ridge, Tennessee, just in case he needed it. It would help, he wrote, if his parents didn't belong to left-leaning organizations like the American Labor party. Like Einstein, Murray seemed to find Princeton a little boring. He told his parents he had made a few friends but that any kind of real social life required trips to Philadelphia or New York. He wished he had his own car. He promised to try to hitchhike into the city that weekend for a visit. Murray brightened a bit when the money from the AEC scholarship finally came through. Suddenly, with that and his $3,000 salary from the Institute, he was feeling flush. He paid off all his debts to his former roommates at MIT and bought an old Chevy to knock around in. And, with his credentials secure, he was invited to work during the summer on a classified radar project in Urbana, Illinois, theorizing about how to make computers out of unreliable parts like vacuum tubes. It was the beginning of a lucrative sideline as a military consultant.

As it turned out, Murray didn't have to worry about the draft board snatching him up. His friend Murph Goldberger, now back at the University of Chicago as an assistant professor, was talking up Murray's qualifications to Enrico Fermi. It wasn't easy. Murray's dissertation had been less than stunning. The paper with Low hadn't yet appeared, so Murray didn't have a single publication to his name. But Murph was so enthusiastic that he secured his friend an instructorship with Fermi's prestigious Institute for Nuclear Studies. At first, Murray didn't like the idea of a mere instructorship, but with the end of the Manhattan Project a glut of young physicists had filled up most of the university positions. The more coveted assistant-professorships were scarce.

Murray talked to Frank Yang about what it had been like at Chicago. "Are there dozens of instructors competing with one another?" Murray asked. "No," Yang assured him. "Is there a very small chance of being promoted to assistant professor?" "No. A very large chance," Yang replied. It was sounding better and better. At $4,300 a year, the pay seemed lavish, and though he would be called an instructor, he would actually have to teach only one course. Anyway, this was Chicago, the lair of Fermi. Even joining Fermi as the most junior of faculty members was something of a coup. Murray relished telling Oppenheimer that he would be leaving early. He was especially pleased that he had never had to ask him for help in landing a job. He was already in demand.

CHAPTER 5

THE MAGIC
MEMORY

That someone of Fermi's caliber could be found working in Chicago, Illinois, instead of Rome, Italy, showed that World War II had shifted the center of nuclear physics from Europe to the United States. Around the turn of the century, it had been Europeans—Wilhelm Roentgen in Germany, Henri Becquerel and Marie and Pierre Curie in France—who had discovered radioactivity, the surprising ability of exotic substances like uranium and radium to spontaneously emit powerfully invisible, penetrating rays. Carrying on her parents' work, Irène Curie and her husband, Frédéric Joliot, performed a landmark experiment early in the 1930s, showing that ordinary substances could also be made radioactive. They bombarded plain old aluminum with alpha particles—helium nuclei consisting of two protons and two neutrons. The result, they were startled to find, was that the seemingly innocuous metal briefly became radioactive, setting off the staccato clicking of their Geiger counter. Scientists had assumed a clear and immutable distinction between nonradioactive elements, like aluminum, and radioactive elements, like uranium. The transmutation of one into the other caused quite a stir.

Fermi had made his mark at the University of Rome in 1932, when he decided to see if he could intensify the Curie-Joliot effect by using neutrons instead of alpha particles as his ammunition. The protons in the alpha bullets, he reasoned, were being deflected by the identically charged protons in the nuclei of the aluminum. With chargeless neutrons, which had only just been discovered, he hoped to slice more deeply into the very core of the atom. The results were dramatic enough to win Fermi a Nobel Prize. Shooting a neutron at a nucleus, as I. I. Rabi later put it, was "about as catastrophic as if the moon struck the earth." Systematically striking some sixty different elements with penetrating neutrons, Fermi and his lab found they could make forty of them radioactive. And though he didn't realize it at the time, he was the first to use a neutron bullet to split apart a uranium atom. He had barely missed discovering nuclear fission. After fleeing Fascist Italy with his family, Fermi moved to Columbia University, and from there to the University of Chicago to supervise the building of the first fission reactor in a campus squash court beneath the football stadium. Inside the brilliantly engineered pile of graphite bricks and uranium, neutrons split uranium nuclei, yielding more neutrons—shrapnel that became the bullets for further fissioning. It was the first controlled nuclear chain reaction.

The tedium of carrying out experiments, even monumental ones, was not really Gell-Mann's forte. But he knew his future boss was as adept at fashioning new theories as he was at fabricating and fine-tuning laboratory equipment. About the time of his famous experiments in Rome, Fermi had been the first to explain a perplexing form of radioactivity called beta decay, in which disintegrating nuclei emitted electrons. It was weird enough that nuclei, made of protons and neutrons, were spewing out these negatively charged particles. Where did the electrons come from? Even worse, the decaying nucleus seemed to lose more energy than the electrons carried away—a violation of the principle of conservation. Wolfgang Pauli halfheartedly suggested a way to set the ledgers straight. All one needed to do was suppose (or pretend) that beta decay somehow emitted a second, still-undiscovered particle with just the right energy to make up the difference. Fermi latched onto the notion and, in an imaginative leap, saw beta decay as a new kind of reaction in which a neutron actually changed into a proton, emitting along the way an electron and a particle he called the neutrino—the "little neutral thing."

The manner in which Fermi puzzled through this problem would become second nature to the generation of young physicists, Murray among them, who went on to draw the new maps of the subatomic world. The new particle had to be neutral, Fermi reasoned, or it would create more problems than it solved. Charge, like energy, had to be conserved. Add up the charges going in and coming out of a reaction and they should be exactly the same. So, a neutron (charge zero) decays into a proton ($+1$) and yields an electron

(−1) as a by-product. Minus 1 added to plus 1 already equals zero, so any additional particle produced in the exchange, such as the hypothetical neutrino, must have a zero charge as well.

Fermi knew that beta decay also had to obey yet another law, the conservation of the quantum number called spin. Two electrons in the same orbit around an atomic nucleus can be thought of as "spinning" on their axes in opposite directions. Other particles have spin as well. In the units physicists used to measure this quantum analogue to rotation, the neutron and its decay products—the proton and the electron—each had a spin of ½. The figure was written +½ if the particle was spinning counterclockwise and −½ if spinning clockwise. In beta decay, a neutron of spin ½ gives rise to two half-spin particles, the proton and the electron. Depending on which ways these new particles were spinning, the total might add up to 0, 1, or −1. This was no good. Any answer other than the original ½ was unacceptable; it violated the conservation of spin. But if a third particle, the neutrino, was also produced in the reaction, and its spin was also ½ or −½, then the books could be balanced again. The neutrino was a very useful accounting device.

Since no one had seen one of these convenient little particles (they were not discovered until the late 1950s), Fermi supposed that the neutrino's rest mass must be very tiny, or even nonexistent. That would help explain why it had slipped through the sieves of the laboratory detectors. Chargeless and with a mass equaling or approaching zero, the neutrino would be a particle that barely existed, zipping through the earth almost as easily as a light beam through space. It was impressive enough in those days that Fermi had the audacity to invent a new particle—a practice that in Murray's time would become commonplace. Fermi's theory also implied the existence of a new force of nature: The cause of beta decay—the disintegration of a neutron into a proton, an electron, and a neutrino—came to be called the weak nuclear force (to distinguish it from the strong nuclear force, which holds the protons and neutrons together into a nucleus).

Whether he was shooting neutrons at elements or putting the polish on a new set of equations, the intensity with which Fermi tore into one project after another never seemed to flag. The man was a dynamo—a "physics machine," Goldberger called him. Fermi had streamlined his life for problem-solving. He would arrive at the University of Chicago at 9 a.m. having already spent hours working at home. He could walk into a classroom and lecture on any aspect of physics—astrophysics, particle physics, classical thermodynamics, even geophysics. Stop him in the hall to ask a question and his attention was all yours. Fermi was not like most people—pretending to listen, nodding absently while thinking about where he would rather be. He would focus his mind on *your* problem—the highest flattery.

Like Weisskopf, Gell-Mann was quick to notice, Fermi shied away from fancy mathematics. The formalisms were always just a tool for describing nature, never an end in themselves. He hated philosophy. Pauli scathingly called him a "quantum engineer." His experiments shooting neutrons into everything he could lay his hands on were reminiscent of Edison trying one filament after another until he had a lightbulb. Genius, Edison had famously said, was 1 percent inspiration and 99 percent perspiration. Fermi would not have disagreed. Give him a problem and he would pull out every device in his mathematical toolbox. If calculation wasn't enough, he would head for the laboratory. If he couldn't find the right piece of equipment on the shelf, he might turn it himself on a lathe. The best piece of equipment was in his head, what he liked to call "the magic memory." Most of us are never sure if what we remember is quite right, or even if we have recorded it correctly in our notebooks. Fermi had a rule: If it got into his magic memory, then it could be trusted.

Chicago was clearly one of the best places an ambitious young physicist could be. The faculty also included the Hungarian physicists Edward Teller and Leo Szilard; in chemistry there were Harold Urey and Willard Libby. But the main attraction was Fermi. As he interacted with the bright new stars around him—experimenters like Richard Garwin, Valentine Telegdi, and Roger Hildebrand, theorists like Goldberger and Gell-Mann—the result was a chain reaction of ideas as magical and intense as the ricocheting particles in an atomic pile. Telegdi, an acerbic chain-smoking Hungarian who arrived at the university shortly before Murray began working there in January 1952, liked to compare Chicago to Copenhagen under Bohr, or the Cavendish laboratory under the great Ernest Rutherford. "It was a time," he would say, "when you could be proud to be the dumbest one."

Telegdi, who had studied under Wolfgang ("Wrath of God") Pauli, seemed to have something bad to say about almost everybody. Murray, along with Fermi, was quickly admitted into the tiny group of people Telegdi admired. He had just moved with his wife, Lia (the former secretary of Szilard), into a sparsely furnished apartment in Hyde Park, the southside academic ghetto that surrounds the school. The Telegdis invited Murray over as their first dinner guest, the beginning of what became practically a weekly ritual. Telegdi was amazed that Murray had a magic memory that rivaled Fermi's. But unlike Fermi, who seemed to have no interest in anything but physics, Murray would regale the Telegdis with his knowledge of classical history, ornithology, and restaurant Chinese, along with a hodgepodge of other unpredictable obsessions. Though he had not yet been to Europe, Murray spoke French with what struck the cosmopolitan Telegdi as a creditable accent. "The only thing wrong with this young man," he said to his wife, "is that on Saturday nights he is always free."

There was still no romance in Gell-Mann's life. He was sharing an apartment with an old suitemate from MIT named David. When it came to family problems, his were even worse than Murray's. He would tell Gell-Mann about his conflicts with his father, a military man, and about the girlfriend he had just broken up with. But Murray didn't realize until it was too late how depressed his friend really was. Lately he had even seemed upbeat and optimistic, talking excitedly, almost euphorically, about a trip out west, how he would go mountain-climbing there with a friend. The more he talked about it, the more excited he became. Maybe that was the tipoff that it was all a fantasy and a deception. As it turned out, David's real plan was to drive to Florida and kill himself in a motel room.

Murray never discussed the suicide with his friends. Suddenly needing a new place to stay, he moved into a house owned by Willard Libby, the chemistry professor, who rented a couple of rooms to bachelor scientists. Murray decided he hated Chicago. When he wasn't working, he walked the streets of Hyde Park, thinking how much nicer it had been in Princeton. He wanted to be someplace with wilderness nearby. Or, if he had to live in a city, why couldn't it be an interesting one like New York or Paris, with beautiful neighborhoods to explore? The area around the University of Chicago seemed like an endless expanse of slums. He tried to find something good about the place. But for someone who had grown up on the Upper West Side, Chicago in the 1950s seemed like a cultural wasteland. As far as he could tell, the only play in town was something called *Made in the Ozarks,* which had been running so long that it threatened to never close.

For lack of other entertainment, Murray and his friends hung out in restaurants. There was a place not far from campus where he could polish his Chinese. Another young physicist, Arthur Rosenfeld, often joined him there for conversations about arms control and the cold war. Murray would borrow Rosenfeld's *New York Times,* scan it for ten minutes, then give his friend his analysis of the news—the world according to Gell-Mann. When the United States began exploding thermonuclear bombs in the Marshall Islands, Rosenfeld and some other physicists set up a monitoring station on top of the physics building and issued their own fallout reports. Rosenfeld was glad to have Gell-Mann as an intellectual ally. But arms control was just one of Murray's interests. One night on the way to the Chinese restaurant, they stopped off at Rosenfeld's apartment to pick up a coat. Rosenfeld's roommate, a theology student, was working on a paper on medieval religious thought. Murray asked him if he would like to come along to dinner. "No, I can't," he replied: the paper was due the following day, and he expected to be up all night. "Oh, come on," Murray said. "I know all that stuff." And so they all went to the restaurant, where Murray gave the aspiring theologian an impromptu seminar.

The one escape from the urban surroundings was Lake Michigan, on Hyde Park's eastern border. Gell-Mann and Robert Gomer, a young chemist who played billiards with Murph Goldberger at the faculty club, bought a sailboat together, a fifteen-foot sloop. Murray quickly became an expert on nautical terminology, mastering it as easily as he did any exotic language. He was about as interested in the physical act of sailing as he was in doing nuclear experiments. But he loved to talk about it. Goldberger thought of him as a theoretical sailor. The boat, too, was a little small for the gusty Chicago winds. As it cut across the water, Murray would nervously think that if it capsized he would be swimming in Chicago's sewage.

The most effective way to shake off depression was work. One day Telegdi walked into Murray's office, a couple of doors down from his own, and found him deep in concentration. Telegdi interrupted: "You know, there is a chapter in Dirac's book which I have never understood, and I have asked several theoreticians to explain it to me. Do you understand it?" Murray said, "Of course," and walked to the blackboard. "Dirac uses a certain language," he said a little snidely, "and other people maybe use a different language, but the essence is this." Clicking the chalk against the slate, he cut away the distracting details and gave Telegdi the clearest explanation he had ever heard. Particle physics, Gell-Mann would say, was the highest of intellectual callings. In fact, he could be rather snobbish about it. Everything else in physics, he would declare, was second-rate: You were just applying the received wisdom to new situations. In particle physics you were seeking the most basic laws, the foundation on which all others were built.

What he needed was a good problem to work on. In Princeton, Gell-Mann and Low had helped tie up one of the loose ends of QED, the field theory explaining electromagnetism. Far more troubling than electromagnetism, which was essentially a done deal, were attempts to understand the strong nuclear force, which held atomic nuclei together. It was this very phenomenon that scientists had tapped to create the Trinity explosion, but they still didn't have a good theory of how it worked. The situation was something of an embarrassment. Fermi had recently fired up his new atom smasher, the synchrocyclotron. In the data that came pouring out were clues to how the strong force operated, if only they could be deciphered. Murray had barely settled into Hyde Park when, first with Telegdi and then with Goldberger, he began puzzling over these results.

The mystery had arisen decades earlier with a seemingly innocent question: What holds the nucleus of the atom together? What keeps all those positively charged protons from pushing each other apart? Since opposites attract, it was natural to think, in those early days, that the negatively charged electrons might act as some kind of electrical glue. Wedged between the protons, they might hold the structure together. But then the protons would also have to keep the electrons from repelling each other. It would be a delicate balancing act, something like trying to get a handful of little magnets to clump together into a ball. Put one particle in slightly the wrong position and the whole thing would fly apart. And yet, even radioactive elements aren't that unstable. The discovery in the early 1930s that nuclei contain neutrons as well as protons only complicated matters. How were neutral particles and charged particles made to adhere? There seemed to be something entirely novel involved, a binding force powerful enough to overcome the electromagnetic repulsions between protons while somehow causing them to stick to the chargeless neutrons.

But what was the mechanism for this hypothetical force? Physicists first tried to make do with the particles already on hand. One can often get a handle on a strange problem by reasoning through analogy, looking for some familiar process that might be adapted to the new situation. In the early 1930s, Werner Heisenberg, groping in the dark, had thought there might be a clue in the chemical bonds connecting atoms into molecules: Electrons bounce back and forth between the two atoms' outer shells, holding them together. Maybe something similar happens inside the nucleus: Protons and neutrons also bounce electrons back and forth, forming some kind of resonating link. More specifically, Heisenberg suggested, the neutron briefly emits an electron and is converted into a proton (losing its neutralizing scrap of negative charge); a proton absorbs the electron, turning itself into a neutron. These electrons wouldn't actually be part of the nucleus but rather virtual particles, those wispy theoretical conveniences that appear and disappear on cue.

One of the strongest signs that a theory is on the right track is that it obeys the various conservation laws—the guarantors that you can't get something for nothing. Heisenberg's theory successfully conserved electrical charge. A positively charged proton absorbing a negatively charged electron would yield a neutrally charged neutron: $+1 + -1 = 0$. But when it came to the conservation of spin, he couldn't get the arithmetic to work. Start with a neutron (spin $\frac{1}{2}$) and you get an electron ($\frac{1}{2}$) and a proton ($\frac{1}{2}$). Whether the spins are clockwise or counterclockwise, plus or minus, there is no way for them to add up to $\frac{1}{2}$ again. You get an answer of either 1 or 0. Either spin is magically created—as though a tornado has split into two tornadoes rotating at the same speed—or else it disappears without a trace. For the theory to work, Heisenberg realized, he needed some kind of spinless version of the electron—something that didn't seem to exist.

His work was not a complete loss; in the course of his theorizing, he invented a powerful new concept that years later would become crucial to understanding the strong force. Again reasoning by analogy, he started with the increasingly familiar concept of atomic spin and abstracted it into something deeper. Since a spinning top (or electron) will reverse direction if flipped upside down, physicists talked about clockwise as "down-spin" and counterclockwise as "up-spin." (They remembered this arbitrary convention with the right-hand rule: Hold your right arm in the most natural position. When your fingers curl counterclockwise, your thumb points up.)

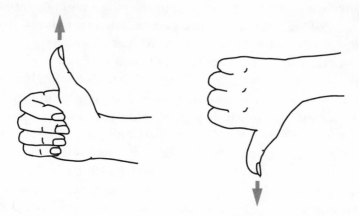

The right-hand rule. Curl the fingers of your right hand in the direction the particle is rotating. For counterclockwise, the thumb points up and the particle is said to spin up. For clockwise, the thumb points down and the particle has a down spin.

Heisenberg proposed that the neutron and proton (identical except for their charge and the tiniest difference in mass) be considered shadows of a single particle, called the nucleon, which was "spinning" (in a very abstract sense) in two different directions. Turn the nucleon "counterclockwise" (giving it an up-spin) and it would manifest itself as a proton; give the nucleon a "clockwise" down-spin and it would become a neutron. The quotation marks are a warning that this "rotation" is not taking place in the ordinary three-dimensional world, but in an imaginary mathematical space. (This is not as mysterious as it sounds. A simple chart showing the performance of a stock over the last few months has two dimensions: One axis measures price, the other measures time. The peaks and valleys in this artificial space are meaningful, but have no more physical reality than the kind of spinning Heisenberg was talking about.) This so-called isotopic spin (isospin, for short) was not a literal description—nothing was really spinning—but an abstract analogy for understanding the relationship between protons and neutrons.

In trying to make sense of the nucleus, Heisenberg had done nothing less than invent a new physical quality, another quantum number. A particle could be described not only by its charge, mass, and spin, but also by its isospin. The idea was fleshed out in 1936 with the publication of three papers (one coauthored by the ubiquitous Gregory Breit). But for the moment, the theoretical gambit did not pay off. Isospin fell out of favor, only to be revived decades later as an important factor in explaining the subatomic world. In the meantime, it was left to other scientists to try different approaches to the strong force.

In Japan, a physicist named Hideki Yukawa was puzzling over Heisenberg's failed theory when he decided, like Pauli and Fermi with the neutrino, to execute a bold stroke. Since it couldn't be electrons holding the protons and neutrons together, as Heisenberg had suggested, maybe there was some undiscovered particle acting as the binding agent. To avoid the problems that had sunk the Heisenberg scheme, Yukawa would have to assign this hypothetical carrier of the nuclear force a spin of zero; then protons and neutrons could convert from one to the other without violating the conservation law.

Yukawa also figured that, unlike an electron, the carrier of the strong force would have to be very massive. His reasoning went like this: Particles that act over small distances must be quite short-lived, disappearing before they can reach very far. Since the strong force was confined inside the tiny space of the nucleus, the lifetime of its carriers must be very short indeed. That was step one of the argument. Next, by combining Heisenberg uncertainty and Einsteinian relativity, he could get the mass. According to Heisenberg, energy and time (like position and momentum) are reciprocally related: The briefer a particle's life span, the higher its energy. And energy, of course, is equivalent to mass, $E = mc^2$. Taking all this into account, Yukawa estimated the mass of his new particle as 200 times that of the electron. Its brief lifetime would also help explain why so important a particle had escaped notice for so long.

This was sounding better all the time. Mulling over his equations, Yukawa decided that his particle must be not only spinless and very massive, it must also come with both positive and negative charges. A proton emitting a positive Yukawa particle gives up its charge and becomes a neutron. A neutron emitting a negative Yukawa particle becomes a proton. (A third, neutral Yukawa particle was later added to the picture: It would interact with a proton or neutron without changing the charge.) Yukawa found the analogy with electromagnetism compelling. Electrons and protons bind to each other by bouncing photons back and forth. Protons and neutrons bind together by shuttling these new particles. Here Yukawa was relying on another of physics' laws of aesthetics: the conservation of style. For all the invention that went into his theory, it retained the same general shape as QED.

He had barely finished polishing the equations—it was now the mid-1930s—when he learned that signs of a new particle, much like the one he predicted, had been found by Carl Anderson, an experimenter at Caltech. Anderson used a device called a cloud chamber—a vessel filled with a rarefied haze of water vapor—to study the tracks left by cosmic rays. These streams of particles, which rain constantly from the sky, leave a distinctive mark that depends on their mass and charge. It was in just such a manner that Anderson had discovered Dirac's positron—a particle that left a track like that of an electron but, when bent by a magnetic field, curving the other way. (It was possible that these mirror-image particles were simply electrons shooting up from the earth instead of down from the sky. But even when the bottom of the detector was shielded, the inverse signatures appeared.) More recently, Anderson had found tracks that seemed to be caused by particles midway in mass between the light electrons and the heavy protons. These middleweights—later dubbed mesotrons, or mesons—seemed to come in both positive and negative varieties. It wasn't long before physicists came to believe that Anderson had found Yukawa's carriers of the strong force.

But the excitement gradually gave way to despair as the new-found particles refused to behave as they were supposed to. For particles meant to carry the strong force, these mesons didn't seem to react with matter very strongly. Indeed it seemed that if they really were Yukawa's particles, they would never reach the surface of the earth, having eagerly interacted along the way with the atoms in the atmosphere. (The failure of the meson to behave as the theorists had predicted was one more reason Gell-Mann had decided at Yale that theoretical physics was in a state of disgrace.)

It wasn't until the Shelter Island conference in 1947 that a way was found to clear up the confusion. A young physicist named Robert Marshak proposed that perhaps there were two kinds of mesons. The first (the true carrier of the strong force) would be very unstable and decay rapidly into the second kind of meson—the kind that had been snared in Anderson's trap. This all seemed a bit ad hoc. But as it turned out, experimenters lofting a detector high above the earth in a balloon had already recorded what looked like the decay of one of Marshak's particles, which came to be called pi-mesons, or pions. The particle Yukawa had predicted had finally been found. The other particle, the mu-meson, or muon, turned out to be a fat version of the electron. A disgusted Oppenheimer called it a "ten-year joke."

Now that physicists had finally found the pion, they needed to develop Yukawa's ideas about the strong force into a full-blown field theory—the equivalent of the QED apparatus, with pi-mesons playing the role of photons. Meson theory was particularly hot in Chicago, where experimenters were using the synchrocyclotron to shoot positive and negative pions at atomic nuclei, studying the ricochets for clues. Earlier experimenters had found that the strong force seemed to be "charge-independent." Whether you scattered protons off protons, protons off neutrons, or neutrons off neutrons, the strength of the interaction was pretty much the same. Fermi's team quickly discovered that pion-nucleon scattering followed the same rules. Whether pions collided with protons or neutrons seemed irrelevant, having no more effect than the color of a billiard ball on its trajectory across the green felt of the table. The theorists, instinctively anthropomorphizing, described the situation like this: From the point of view of electromagnetism, the proton and neutron obviously look like different particles, one positively charged, one chargeless. But in keeping all those protons from repelling one another, the strong force apparently ignores electrical charge. From its point of view, neutrons and protons are identical. Suppose you take two electromagnets, one turned on and one turned off, and drop them from the Leaning Tower of Pisa. They will hit the ground at the same time. Gravity doesn't care whether something is magnetized. Similarly, the strong force doesn't care whether a particle is charged.

As it turned out, all of this was related to Heisenberg's ingenious notion of isotopic spin: that protons and neutrons are a single particle, the nucleon, spinning in opposite directions in an artificial mathematical space. A sphere of uniform color is symmetrical under rotation in physical space. Turn it in any direction, and its appearance does not change. And so it was with charge-independence: The nucleon was symmetrical under rotation in isospin space. Whether this particle turned clockwise, making it a proton, or counterclockwise, making it a neutron, the strong force treated it the same. In the case of electromagnetism, physicists would later say, the symmetry is broken: Isospin is not conserved, and the direction of rotation makes all the difference in the world. To put it another way: The strong force conserves isospin, while electromagnetism does not.

This is where things stood at the time Gell-Mann was casting around for a problem. First, he and Telegdi picked a tiny piece of the puzzle to concentrate on, working out more of the details of how electromagnetism and the strong force intertwined. To study the strong force, experimenters bounced pions off nuclei; to study electromagnetism, they used photon bullets instead. Gell-Mann and Telegdi pondered the difference between the two interactions. For one thing, when a pion bounces off a nucleon, the resulting strong interaction can be described by a simple rule: You must always end up with the same amount of isospin that you started with. The quantity must be conserved. But when a photon is scattered off a nucleus, the situation is different. Here the electromagnetic force, not the strong force, comes into play. Isospin need not be conserved. Other rules had to be used to describe this kind of scattering.

As they played with these ideas, trying to interpret some recent experiments, Telegdi found his young colleague's intensity almost thrilling. Every possible explanation was torn to pieces, then reassembled to see which ones were sturdy enough to survive. It didn't seem to matter whether the idea under attack had been put forth by a rival, by Telegdi, or by Murray himself. Gell-Mann was as hard on his own ideas as he was on others'.

The result of this early work was hardly monumental, just a microdevelopment in what would turn out to be a long haul toward understanding the strong force. But it was respectable physics, and Telegdi was exasperated that Murray kept procrastinating in writing it up. Telegdi, who couldn't help comparing people to Fermi, noted that for all Murray's strengths, he lacked Fermi's tenacity. He was easily deflected from problem to problem, working on too many things at once. And there was always that writer's block, the shadow of his father reading over his shoulder, never satisfied. It took months to get the paper ready for submission to *Physical Review,* and by then Murray was immersed in something deeper.

Telegdi was competing for Gell-Mann's attention with Goldberger. In the spring of 1952, Murph and Murray gave a series of lectures on special topics in quantum mechanics. As they argued at the blackboard or strolled afterward through the streets of Hyde Park, they mulled over what at first must have seemed an obvious question: What exactly is meant by "scattering"? The word was used in the broadest sense to describe what happens when two particles collide. But it struck them that no one had really sat down and, starting from the very basics of quantum mechanics, developed a theory so general and powerful that it could be used to encompass scattering of all kinds. This was the kind of question Murray could really get his teeth into.

Part of the problem was, again, how to make sense of the strong force. But a good scattering theory would have to go even further, explaining not only strong interactions like pions glancing off nuclei but also electromagnetic reactions—the scattering of high-energy photons off a proton, for instance, or an electron off another electron. Surely there was a way to unite all these phenomena into a single framework. Gell-Mann, with an instinct for seeing how special cases are related by general laws, seemed like the perfect collaborator for this ambitious project, which went by the unpretentious name of dispersion theory.

In the classical world, scattering is a fairly straightforward affair. Two balls on a frictionless surface collide and recoil, their trajectories depending on the energy and angle of the impact. But in the quantum realm, nothing is ever so simple. Consider the seemingly trivial matter of an electron repelling another electron by trading a photon back and forth. This can be represented by a simple Feynman diagram:

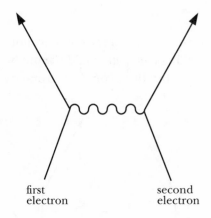

first
electron

second
electron

The two electrons approach each other from the lower corners of the picture and exchange a photon (the wavy line), which causes them to veer off in opposite directions. Sounds simple enough. But to be as precise as possible, one also had to consider the possibility that, during its journey, this photon might spontaneously disintegrate into an electron and its antimatter twin, the positron. (The reason? When particles of matter and antimatter come together, they annihilate each other in a flash of light, creating photons. According to the symmetries of physics, the mirror image of this reaction must occur as well: A photon can spontaneously turn into an electron and a positron or some other matter-antimatter pair.) And suppose that immediately after the photon turns into an electron and a positron, these particles instantaneously reunite to form a photon again. This possibility can be represented by a more complicated Feynman diagram:

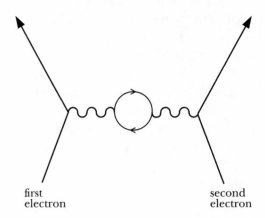

But that is just the beginning of the complications. There is no reason why the two repelling electrons couldn't emit two photons, either or both of which might briefly become electron-positron pairs:

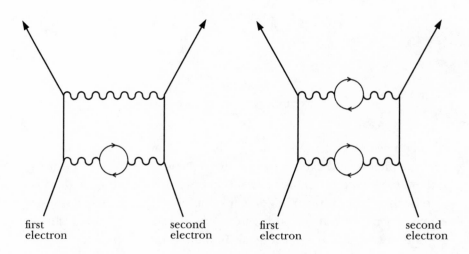

In any of these situations, each electron might further complicate matters by emitting and then quickly reabsorbing one or more photons, which might also split into electron-positron pairs, resulting in tangles like this:

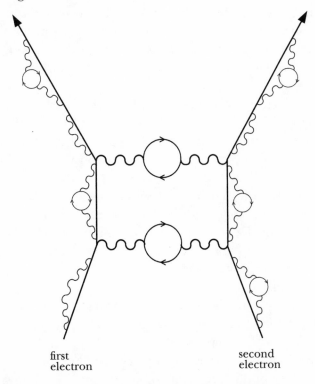

first
electron

second
electron

There are, in theory, an infinite number of possibilities that can occur during this deceptively simple case of one electron repelling another. The result of this bizarre situation is that the equations of QED cannot be solved exactly; the best one can get is a good approximation. Every electromagnetic reaction is represented by an infinitely long series of calculations, each of which can be described by its own Feynman diagram. The first term represents the simple situations in which the two repelling electrons exchange a single photon. The second term deals with the two-photon situations, the third term with the three-photon situations . . . and so forth, on to infinity, each diagram becoming ever more complex. This is called a perturbation expansion: Each term can be thought of as a small correction, or perturbation, to the simplest case in which the electrons interchange a single photon.

The reason it is possible to calculate how two electrons interact, without taking into account the endless number of terms in the perturbation expansion, lies in the nature of the electromagnetic force. As you proceed down the length of the equation, the possibilities become increasingly byzantine and less likely to occur. The chance of the scattering electrons emitting two photons is very tiny, three photons tinier still. In fact, you can get a very good approximation of how two electrons will behave by ignoring all but the first few terms. Even a so-called first-order approximation, involving just the single-photon situation, is precise enough for many applications.

The reason this convenience works is that electromagnetism is fairly weak. The strength of a force is represented by its coupling constant. For electromagnetism, the number is approximately $\frac{1}{137}$. The way the perturbation equation is constructed, each term is $\frac{1}{137}$th weaker than the one before it, quickly fading to near-nothingness. Such a well-behaved series makes good approximations possible. But Goldberger and Gell-Mann became convinced that to really understand the phenomenon of particle scattering, one needed a way to make exact predictions. They hoped dispersion theory would provide this new tool—an alternative to the perturbation technique.

When it came to the strong force, the need for a new method was even more acute. Here not even approximations were possible. With the notions of isospin, charge-independence, and Yukawa's pi-mesons, physicists were starting to piece together an aesthetically pleasing picture of the force that bound protons and neutrons into nuclei. But a good theory must not only explain, it must predict. With the strong force, not even the perturbation technique could be used to churn out the forecasts experimenters needed to test the theorists' idealizations against reality.

The problem is that the strong force is so strong. The coupling constant was not a tiny fraction like $\frac{1}{137}$ but more like 10. Suppose you're trying to calculate what happens when a pion is scattered off a proton, or a proton and a neutron are bound inside a nucleus, trading virtual pi-mesons as electrons trade virtual photons. Again you end up with an infinitely long equation, describing the endless possibilities as clouds of virtual particles interact in every imaginable way. But in this case, each succeeding term in the perturbation expansion is ten times *more likely* to occur than the one before it. Unlike with electromagnetism, these increasingly complicated diagrams cannot be ignored. It is because of this buzzing cacophony of virtual interactions that the strong force is so intense. The terms get bigger, not smaller, and approximations are impossible. Perturbation theory was useless to explain the results of the pion-nucleon scattering experiments going on while Gell-Mann was in Chicago. The theorists needed another way.

It was discouraging that this apparatus, quantum field theory— finally honed to perfection, the infinities banished, for use with electromagnetism—failed when it came to understanding the meson physics of the strong force. Just before Murray's arrival in Chicago, Fermi lamented this sad state of affairs. "Of course, it may be that someone will come up soon with a solution to the problem . . . and that experimental results will confirm so many detailed features of the theory that it will be clear to everybody that it is the correct one," he halfheartedly noted. "However, I do not believe that we can count on it, and I believe that we must be prepared for a long hard pull."

What Goldberger and Gell-Mann had in mind was a tool that would do no less than replace approximations with exact answers. But there was even more: Through these new theoretical spectacles, the scattering caused by the strong force and the scattering caused by electromagnetism could be seen as special cases of a more general phenomenon.

They knew this route had been tried before, back when it seemed that the pernicious infinities might sink QED along with the whole notion of quantum field theory. In the late thirties and early forties, Heisenberg had even tried to replace the whole concept of the field with something called the scattering matrix, or "S-matrix." Because of the virtual particles, quantities would show up in the field equations that seemed to represent situations that were absurd—negative probabilities, for example. What could it mean for something to have less than a zero chance of occurring? Mathematical "particles" would show up that had negative energies or momenta represented by imaginary numbers, those involving the square root of -1. In the jargon, these pseudophenomena were opaquely referred to as being "off the mass shell." Goldberger had a better term: Borrowing from *Alice in Wonderland,* he called this "physics in the never-never land."

Heisenberg wondered if some of this craziness could be banished by replacing the field with the S-matrix. Instead of all the rigmarole of field theory, with its endlessly multiplying virtual particles; instead of swearing allegiance to an idealized abstraction called a field that purported to describe what happens at *every point* in space and time—as though it were possible to really know such a thing—he called for a more concrete approach. Just concentrate on what can be directly observed and base your calculations on that. In any experiment, a set of particles with certain characteristics—certain quantum numbers—is converted into a set with different quantum numbers. There are many possible ways the conversion might unfold. The likelihood of each can be assigned a probability, and these numbers can then be arrayed into a grid, the S-matrix, describing all the possibilities—a kind of quantum version of the before-and-after photograph. There would be no need for virtual particles and the other questionable abstractions of field theory. The S-matrix would replace the field as the fundamental metaphor for particle interactions, and all the problems would go away.

That was the hope anyway. Before the technique could be perfected into something one could use to calculate—to derive predictions for the experimenters to test—QED was renormalized. Unphysical quantities still showed up in the equations, but they seemed a small matter now that the infinities were gone. Quantum field theory, at least for use with electromagnetism, was vindicated, and the enthusiasm for S-matrices waned. It remained, as one historian put it, "little more than a vessel containing well-arranged experimental data."

But now the problems with the strong force were once again undermining confidence that all reality could be explained in terms of virtual particles and fields. Maybe dispersion theory would provide a way out of the never-never land. Like the early proponents of the S-matrix, Gell-Mann and Goldberger decided to start from scratch, to forget about the details of field theory and make only very simple assumptions—things, as Murray put it, that "hardly anyone would challenge"—and then build from there. This platform would include the notion of causality (if A causes B, then A obviously must precede B) and relativistic invariance—time and space appear to stretch and contract to preserve the absoluteness of the speed of light. They would assume the truth of the various conservation laws—energy, charge, and spin, along with the conservation of probability: If the likelihood of either A or B occurring is 50 percent, then the likelihood of both events occurring simultaneously had better be smaller, not larger: .50 × .50, or 25 percent.

By now Goldberger was familiar with the way Murray's mind operated. Still, as they began working on the project, he was impressed by the way Gell-Mann seemed to surround the problem from all sides, hitting from one direction, then quickly regrouping to attack from a different angle. (Francis Low once told Goldberger that Murray's mind seemed like a searchlight sweeping across a landscape, illuminating dark mathematical spaces.) Like Fermi, Murray wasn't a purist: He would use any tool at his disposal, whatever it took to get the job done. The collaboration paid off. As he and Goldberger struggled with the mathematics, they developed dispersion theory into a precision technique for predicting the scattering of massless particles like photons.

During the next few years, Goldberger and others expanded the idea to include pion-nucleon scattering. (Murray was off working on other things.) For a while it seemed that the strong force was starting to yield its secrets. Those unruly perturbation expansions were not quite such an obstacle. Strong interactions could be scrutinized instead through this fine new lens. And, as Goldberger had hoped, the numbers that emerged from the equations weren't just approximations, like those of perturbation theory; they were exact predictions. So exact, he proclaimed, that the experimenters would have to fine-tune their equipment before they could put the theory to the test. For years physics had been driven by the experimenters confronting the theorists with new phenomena—muons, pions, and so forth—that needed to be explained. Now the theorists, no longer in disgrace, were pulling ahead. "This is an unusual and gratifying position for the theoreticians to be in after all these years," Goldberger gloated before a gathering of physics' inner circle. "I'm sure every red-blooded experimentalist will want to rush back to his laboratory in an effort to produce data to contradict the equations." Dispersion theory, it seemed, was succeeding where the old S-matrix had failed. Heisenberg's empty vessel seemed on its way to being filled.

But the celebration turned out to be premature. No one could have known that, for all the promise some saw in dispersion theory, a decade later the strong force would still be largely a mystery, that scientists would still be debating whether field theory or some version of the S-matrix was the proper framework for understanding the subatomic world. And for all their power, the ideas involved with dispersion relations were so abstract that they didn't make much of a splash outside a small subset of particle physicists. Goldberger was once asked to write a popular article on the subject for *Scientific American*. He gave up in despair. It just didn't seem possible to translate such airy mathematics into words.

At this point, Gell-Mann wasn't sure whether his and Goldberger's invention should be expanded into a complete replacement for field theory, a whole new way of doing physics. After all, QED was a grand success. Maybe a field theory could eventually be found for the other forces, explaining them too in terms of virtual particles bouncing back and forth. Meanwhile dispersion relations were at least a helpful tool. They were a way of doing field theory "on the mass shell," as Murray liked to put it, avoiding some of the pitfalls of the never-never land. Ultimately, it seemed, field theory would have to be either overhauled or overthrown. But for now he was keeping his options open.

And he had other things on his mind. For some time he had been savoring a different mystery—how to explain certain strange cosmic rays that were not behaving as physicists predicted. This, it turned out, would be his first solo flight as a theorist. He had cut his teeth on the work with Low, Telegdi, and Goldberger. Now he had finally found a problem of his own, one that would lead to his first big discovery. As it happened, this was science that could be explained to a popular audience, and so his reputation would grow.

CHAPTER 6

"NO EXCELLENT BEAUTY"

Few people are lucky enough to name a mountain, a river, a planet, or a star. Rarer still are those who find themselves in the dizzying position of naming a new physical phenomenon, a previously undiscovered essence of nature. Adding to the list of quantum numbers that included mass, momentum, charge, and spin, Heisenberg had invented isospin, helping to explain protons and neutrons and the nuclear glue, the pions, that seemed to bind them together. But none of these qualities—these symmetries, as they increasingly came to be called—was adequate when it came to explaining the curious new particles that first left their mark in a cloud chamber in 1946, six years before Gell-Mann's arrival in Chicago.

Scientists from the University of Manchester were studying cosmic rays in a laboratory high in the Pyrenees Mountains when they first spotted these V particles, so-called because of the sharp angular mark they left on a photographic plate. The vertex of the V represented some unseen spark of matter disintegrating into positive and negative particles (the arms of the V), each pulled to one side or the other depending on its charge. Conservation laws implied that the mother particle must be chargeless (positive plus negative equals zero). As they scrutinized the tracks, physicists also concluded that the V particles were extremely massive. Similar particles with electrical charges were also soon found. Before long, physicists were using their accelerators to produce all kinds of strange new particles, smashing together nucleons and pions and examining the shards. Lambdas, kaons, sigmas, xis—there seemed to be no end to the stream of interlopers messing up the neat picture physicists had drawn of the world. "If you run out of Greek letters, you can always use the names of Pullman cars," Von Neumann said at

the time. Oppenheimer provided the name that stuck: the subnuclear zoo.

Some longed for the good old days when everything could be
explained with a few basic particles: Nucleons clasped to form
atomic nuclei by trading pions back and forth. The nuclei and electrons formed atoms by bouncing photons back and forth. Throw in
the neutrino to balance the books of beta decay, and you had a
simple world of five basic particles—ten counting their antimatter
shadows. That, it seemed, was all the particles that nature should
require. Stumbling across the muons, those fat electrons from space,
had been unsettling enough. "Who ordered that?" Rabi famously
said when the particle arrived like some unwanted concoction
slapped onto the linoleum table of a Chinese restaurant in Morn-
ingside Heights. The V particles were even more maddening. Not
only did they seem to fill no role in the cosmic scheme, they apparently violated one of the cherished symmetries of physics. The particles were produced so abundantly that they must be spawned by the
powerful strong force. But the strong force should then break down
its creations just as readily as it created them. Easy come, easy go.
The V particles and their cousins lingered far longer than theory
predicted. Something was seriously wrong.

When Murray finally came up with the new essence, the new
quantum number that would resolve this dilemma, he called it
"strangeness." It was the need to conserve something called strangeness that prevented the strong force from tearing down the particles it so effortlessly made. Then eventually, a hundred-millionth of
a second later—an eternity on the subatomic scale—the weak force
would wear the particles away. The weak force could do this because
it did not respect strangeness—another broken symmetry. Strangeness would be the first step in Gell-Mann's scheme to bring order to
the subatomic world.

Years later Murray sat down with an editor at *Scientific American* named Ted Rosenbaum (the brother of his old Yale roommate, Bud) to describe the intellectual adventure that had led to this
surprising invention. At the beginning of the article, Murray put a
favorite quote from Francis Bacon: "There is no excellent beauty
that hath not some strangeness in the proportion." Science with a
literary flair. You might say that Murray had out-Oppenheimered
Oppenheimer.

. . .

Spending time worrying about strange cosmic rays was not an obviously swift career move for a young theorist in the early 1950s. Pion physics—the attempt to understand the strong force—was where the action was. Many theorists considered the weird behavior of the strange particles "some kind of dirt effect"—noise in the particle detectors that was best ignored or wished away.

When Gell-Mann, maybe a little bored with meson theory, first started thinking about the strange particles early in 1952, he didn't yet see the need for devising a new quantum number. He was thinking about charge, mass, spin, isospin, trying to sort the new particles into the familiar categories, arrange them into patterns that were elegant and consistent with the known laws of nature.

In trying to put a new particle in the proper bin, a physicist might first consider the value of its spin. A particle like an electron, proton, or neutron, with a spin of ½, can rotate either counterclockwise, with its axis pointing up (a spin of +½) or clockwise with its axis pointing down (a spin of −½). (Again, this can be remembered with the right-hand rule.) These particles, with two degrees of freedom, are called fermions and are sometimes described, to oversimplify a bit, as the building blocks of matter. Fermions also obey the famous Pauli exclusion principle taught in every high school chemistry class: No two fermions can have the same quantum numbers. If there are two electrons in an atom's lowest shell, something must distinguish one from the other. If one is spinning clockwise, then the other must be spinning counterclockwise. And once this orbit is filled, additional electrons will have to occupy new, higher shells. Ruled by the exclusion principle, electrons and nuclei are built up into the atoms that form the material world.

Two degrees of freedom. Fermions can spin either up (counterclockwise) or down (clockwise).

But there is another kind of particle, the bosons, with very different qualities (different "statistics," the physicists say). Their spins are stated as whole numbers. Examples include the pion with spin 0 and the photon and certain mesons with spin 1. Compared with half-spin particles, spin-1 bosons can have as many as three degrees of freedom—three options instead of the fermions' two. Such particles can spin up, down, or sideways.

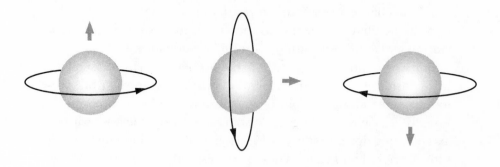

Three degrees of freedom. Some bosons can spin up, down, or sideways.

Unlike fermions, bosons are immune to the Pauli exclusion principle. Any number of them can crowd into the same state, as happens when photons are lined up in lockstep and sent through space in a needle-thin laser beam. Most significantly of all, bosons can serve as the carriers of the forces—electromagnetism and the nuclear forces—through which particles interact.

Having sorted particles into fermions and bosons, one then might focus on the more abstract quality called isospin. Since isospin was modeled after ordinary spin, similar patterns arise. A nucleon with an isospin of ½ has two states—the positively charged proton (isospinning counterclockwise) and the chargeless neutron (isospinning clockwise). Murray liked to think of it like this: If you could somehow flick a switch and turn off charge, neutrons and protons would be identical, collapsing into a single particle called the nucleon. A particle like the pion with a whole-number isospin has three degrees of freedom: It can have a charge of $+1$, 0, or -1. Isospin is intimately related to charge. Again, if you could turn it off, the three kinds of pions would be indistinguishable.

The first step to understanding the strange particles was to try to arrange them, along with their more familiar counterparts, according to charge and isospin. Start with the nucleon and the pion:

CHARGE

−1	−½	0	+½	+1	
		n		p	nucleon (isospin = ½)
π^-		π^0		π^+	pion (isospin = 1)

n = neutron   p = proton π = pion

As a physicist would say, the nucleon is a doublet (it can be positive or neutral) and the pion is a triplet (positive, negative, or neutral). It is also useful to speak of a particle's "charge center." Look at the rows on the chart representing the nucleon or pion and find the midpoint. For the nucleon, the charge center is $+\frac{1}{2}$; for the pion, it is 0.

But how did the V particles and their cousins fit into the scheme? In some ways, the strange particles seemed like the ordinary components of the nucleus, where there are heavy fermions (the proton and neutron) and middleweight bosons (the pions and other mesons). Likewise, the strange particles seemed to consist of heavy fermions, called "hyperons" (the lambda particle was one example) and middleweight bosons, called K particles, or kaons.

Anyone who had thought about the subject assumed that the heavy hyperons belonged on the chart with the nucleons—an exact overlay. Like the proton and neutron, the hyperons would be doublets with an isospin of ½ and a charge center at ½. All this means is that there would be two types of hyperons, distinguished only by the presence or absence of charge.

CHARGE

−1	−½	0	+½	+1	
		n		p	nucleon (isospin = ½)
		H^0		H^+	hyperon

n = neutron   p = proton   H = hyperon

By this logic, the K particles would belong with the pion: triplets with an isospin of 1 and a charge center at 0.

CHARGE

-1	$-\frac{1}{2}$	0	$+\frac{1}{2}$	$+1$	
π^-		π^0		π^+	pion (isospin = 1)
K^-		K^0		K^+	K particle

π = pion   K = kaon

This was the kind of mental map Gell-Mann would have had in his head when he was trying to make sense of the strange particles. Some were heavy, with isospins of $\frac{1}{2}$ (allowing two degrees of freedom); some were lighter, with isospins of 1 (three degrees of freedom). He also assumed that the strange particles, like the ordinary particles, would obey the isospin conservation law: The total going into a reaction must equal the total coming out. Maybe the strong force is barred from breaking down the strange particles because to do so would somehow violate this law. If the total isospin entering and leaving the decay didn't match, then nature would never allow the reaction to proceed. Immune to breakdown by the strong force, the strange particles would have to wait around for the weak force to come into play. But Gell-Mann just couldn't get the scheme to work. Mulling over the idea with Goldberger and another theorist, Ed Adams, Murray realized that he had forgotten about electromagnetism: This force, in between the strong and weak, would mess up the plan. It would cause the particles to decay before the weak force got to them. And this would make their lifetimes far shorter than was actually the case.

Isospin conservation alone wouldn't solve the problem. Murray put the idea on hold; there was no sense in writing it up. He was annoyed a few weeks later when a young Columbia University physicist named David Peaslee (he had preceded Murray as Viki's assistant at MIT) published a letter in the *Physical Review* describing a similar scheme—along with an explanation of why it wouldn't work. "Why bother publishing a wrong idea?" Murray grumbled to himself. It was painful enough writing papers that were right. He couldn't help noticing, a little smugly, that Peaslee's paper had come out on April Fool's Day.

A month later, on a visit to the Institute for Advanced Study, Murray was asked to give an informal talk on Peaslee's paper, explaining why isospin conservation wasn't enough to solve the mystery of the strange particles. Partway through the talk, he made a dumb mistake. Referring to the heavy strange particles, the ones that seemed to be related to the nucleons, he said they had an isospin of 1. What could he have been thinking? The particles were, after all, fermions, and fermions were doublets, not triplets. The isospin would be measured in halves. Then he was quiet for a few minutes, the gears silently turning. Staring at the equations he had been writing on the blackboard, he walked a few steps backward and sat down. Then he stood up again and, now on autopilot, finished going through the steps of Peaslee's argument. But the mistake kept sticking in his mind. "By the way," he said when he was done, "a few minutes ago I got what I think is the right answer."

Maybe, against all expectations, the isospin of these weird particles really was 1, not ½, and maybe this was the key to their perplexing behavior. They would be fermions that came in three varieties instead of two. There was, after all, something undeniably strange about these particles, so readily produced and so reluctant to break down. Maybe the strangeness had something to do with their having a different isospin from normal fermions. The idea was only partly formed. The details would develop over the coming months. For now Murray could dimly see that if one dropped the prejudice that the isospin must be ½, then maybe—just maybe—you could keep the strange particles from quickly disintegrating, allowing them to live long enough for the weak force to come into play.

Murray would later wonder if the idea had been in his head all along, the result of unconscious machinations that hadn't yet surfaced. Maybe the random misfiring of a neuron had caused him to misspeak, jogging his brain out of a rut and down this new avenue.

For now, he gave a quick explanation. But the audience, which included Francis Low and T. D. Lee, was unenthusiastic. After the talk, Abraham Pais threw more cold water on the embers, telling Murray that his idea was not really very important—merely a subcase of a theory Pais himself was working on.

Pais had first described his own ideas back in January at the second of a series of physics conferences at the University of Rochester, held, perversely, in the dead of winter in the miserable cold of upstate New York. At the meeting, Oppenheimer had coined the word "megalomorphs" to describe the strange new particles. Picking up the cue, Pais gave a talk which appeared in the proceedings as "An Ordering Principle for a Megalomorphian Zoology." (The precious title wasn't his fault; it had been imposed by Oppenheimer.) Attempting to explain the strange particles, Pais made up a quantum number called N. N would be set at 0 for regular particles and at 1 for the new ill-behaved strange particles. Then he supposed that nature had decreed this rule: If you added up the amount of N going in and coming out of a strong nuclear reaction, the evenness or oddness of the number had to be conserved.

This was the first step in a rather convoluted scheme. First of all, Pais's conservation law meant that the strange particles could only be produced in pairs—a phenomenon called associated production. Suppose you started with a group of protons and neutrons. Since these were ordinary particles, their N was zero, and the total N would also be 0, an even number. If the particles collided and gave rise to a strange particle—with N, by definition, equal to 1—the result would be an odd number, violating Pais's rule. Hence there must be a second strange particle created to preserve the evenness. Strange particles could only come in pairs. And here was the clincher—the process was irreversible. The strong force could produce strange particles in pairs, but it couldn't break them back down again. An individual strange particle (with an odd-numbered N) disintegrating back into normal particles (with even N) would violate his rule. The process would be prohibited. Only the weak force could cause the individual particles to decay. Why? The weak force, Pais proposed, did not respect even-odd conservation.

Murray found the theory bizarre. *Conservation of evenness and oddness?* It all seemed so ad hoc and unphysical, just playing with numbers. There was no reason to believe there was any such thing as N, or that nature had reserved some special place in its heart for even and odd numbers. Pais seemed to be pulling new symmetries and quantum numbers out of thin air.

For now, Murray bit his tongue. He found Pais's tone dismissive and patronizing. But he didn't feel confident enough to push the idea, and he couldn't explain it very clearly. Having sat through his father's dinner-table lectures all those years, Murray hated sounding didactic. And he was terrified of being wrong. Arthur had taught him that mistakes were inexcusable. Anyway, who was he to take on Pais? At this point, Murray's work with Telegdi and Goldberger had yet to be published. He had only one paper to his name. Pais, on the other hand, had been made a permanent member of the Institute, a rare honor, at the young age of thirty. Now, at thirty-four, he had been a full professor for two years. Murray didn't really care for the man. He saw him as one of the sycophants always buddying up to Einstein, walking to lunch with him and Gödel, and as one of Oppenheimer's pets. (He wasn't being fair. In fact, Pais's affection for Robert—he refused to call him Oppie—was mixed with pity and sometimes anger at the way he mistreated others.)

A few weeks later, in June, Murray left from Princeton for his first tour of Europe—Britain, France, Switzerland, Italy, and Spain. In Paris he dropped in at the Ecole Polytechnique, where Bernard d'Espagnat, a French physicist he had met in Chicago, introduced him to the great cosmic-ray researcher Louis Leprince-Ringuet. Then in his early fifties, Leprince-Ringuet was something of a legend. Spending weeks at a time in the high, cold air of his observatory near Mont Blanc, he was one of the swashbuckling physicists from the days, just beginning to pass, when particles were captured from the sky instead of manufactured on demand in accelerators. A true romantic, he rejoiced in the "splendid isolation" of the mountain observatory, where "one can do a year's work in a month," and in the exhilaration of emerging from the cramped confines of the scientific station to ski above the clouds.

Quite a different life from that of a theorist sitting in an office with a blackboard—and a window, if you were lucky. Murray was delighted when Leprince-Ringuet invited him to his country home in Burgundy, where his researchers were meeting. With some of France's great experimenters seated around him, he tried again to explain his notion that the very strangeness of the particles might lie in their weird isotopic spins. Suppose that particles like the hyperons—which were made from the collisions of ordinary nucleons, which decayed (ever so slowly) back into ordinary nucleons, which by all rights should, like the nucleons, have an isospin of $\frac{1}{2}$—suppose instead that their isospin was 1. Murray still wasn't ready to press the idea. He was leery of getting too deeply into abstractions. His listeners, after all, were experimenters more comfortable with talk of particle detectors. He let the idea stew a while longer, content to enjoy the good food and wine.

Back in Chicago that fall, he struggled again to get someone besides himself to understand what was in the back of his mind. Every week the theorists held an informal seminar, a kind of "Quaker meeting," as Gell-Mann called it, to air whatever physics had been occupying them. All the big guns were usually there: Fermi, Teller, Urey. Gregor Wentzel, a very proper Herr Professor who had spent most of his career in Zurich, would act as emcee. Wentzel found the informality of the Chicago physics department a little unnerving.

Shortly after his arrival, Fermi had made the mistake of asking Wentzel to join him for lunch with the graduate students at the cafeteria. After nervously making his way through the food line, Wentzel sat down only to realize that he was expected to carry his own silverware to the table. He returned with knife, fork, and spoon to find that an overeager busboy had already dumped his tray. After that he ate at the faculty lounge. Wentzel was the department's senior theorist (Fermi was currently in experimental mode). As he presided over the Thursday meetings, he would point one by one at each of the young scientists. "What have you been working on?" Then the budding theorist would go to the blackboard for show-and-tell.

When it came Murray's turn, no one could understand what he was talking about. A young experimenter named Roger Hildebrand found the ideas utterly opaque, and Wentzel kept saying, "What about this? What about that? This can't be right." At the end of the talk, another experimenter, Richard Garwin, belittled Murray's idea: "I don't see what use it could possibly be."

Murray knew he was on to something, but this universally negative reaction was too much. He didn't bother trying to write up the idea. Instead he continued his work on dispersion theory with Goldberger, and in the early summer of 1953 he went to the University of Illinois at Urbana, where Francis Low and T. D. Lee were now working. David Pines, whom he had met at Princeton, was also there. Murray's mind always caught fire around Low. As they had back in Princeton, the two scientists decided to take on another loose end of QED. Why not try stretching this enormously successful theory to the breaking point, to see if it would work at ridiculously high energies, at distances so short they were barely imaginable? Energy and distance are related in a very direct way. The closer together you push two identically charged particles, the more energy you need to overcome the repulsion. Low and Gell-Mann wondered how electromagnetic reactions would behave at distances vastly smaller than a single proton—far smaller in relation to the proton than the proton was to the visible universe. Pushing two particles this close together would require thousands of times more energy than you could get if you put the entire mass of the earth into a nuclear furnace, converting every milligram into energy with $E = mc^2$.

The Illinois summer was hot and sticky, one of the worst on record. Gell-Mann and Low had one of the few air-conditioned offices, but it still was so uncomfortable that they would join Lee and Pines in the basement of the student union, one of the coolest places on campus. For hours the physicists would sit there trying to do physics. The jukebox drove Murray crazy. Once he went over and pulled out the plug. It seemed to him that you should be able to feed it with coins and choose silence as an option, purchasing time to think.

In a few weeks he and Low had hammered out a theory. The key to making QED work at these extremes, they found, lay in the coupling constant for electromagnetism, the number that tells you how strong a field is. For electromagnetism, the number the physicists plugged into their formulas was the familiar $\frac{1}{137}$. As they played with their equations, Gell-Mann and Low discovered that the constant wasn't really so constant. It was a running constant, varying according to distance and energy. They described it using a mathematical device that later came to be called the renormalization group. Many years later, at a sixtieth-birthday celebration for Low—these Festschrifts are one of the more hidebound traditions in the physics world—the theorist Steven Weinberg called the paper, which was all but ignored when it appeared, "one of the most important ever published in quantum field theory." When he first read it, Weinberg said, he was thunderstruck at the far-reaching conclusions Gell-Mann and Low had managed to wrest from a few artful assumptions and heroic calculations. So little went in, and so much came out. The paper seemed to violate what Weinberg called "the First Law of Progress in Theoretical Physics, the Conservation of Information." Or, as he explained it, "You will get nowhere by churning equations."

In what was becoming a bad habit, it took Gell-Mann more than a year to finish the paper with Low and get it published. Part of the problem was simple laziness. He loved to work on the problems, arguing and joking with his physicist friends, but committing the results to paper was sheer torture. And the strangeness problem was still eating away at him.

Returning to Chicago in July, Murray found no letup in the hot, muggy weather. As at Urbana, air-conditioning was a luxury, usually limited to rooms containing delicate laboratory equipment. Murray couldn't believe how cheap the university seemed. If electronics needed cooling, then why not brains? He toyed with the idea of finding a wax pencil to write his equations and then claiming he needed air-conditioning to keep his "equipment" from melting.

To make matters worse, he found a draft notice sitting in his pile of mail: The secretary had forgotten to send in his yearly deferment request for scientists working on important research. Compared with the possibility of guard duty in Korea, writing up his ideas on strange particles didn't seem quite so bad. But it was impossible to concentrate very long in the heat. Then Val Telegdi came to the rescue—there were, it seemed, advantages to being an experimenter instead of a theorist. Telegdi had access to an air-conditioned room where the department kept an analytical balance, and he let Murray use a desk there.

While Murray sat in Chicago trying to describe his ideas, Pais was busy promoting his own scheme. Earlier that year, he had delivered a paper at a conference in his homeland, the Netherlands, honoring the hundredth birthday of two Dutch physicists, Hendrik Lorentz (famous for the Lorentz transformation in relativity theory) and Heike Onnes (discoverer of superconductivity). It was a prestigious affair. Bohr, Dirac, Heisenberg, and Pauli were there. In his talk Pais laid out his solution for the strange particles—the even-odd rigmarole that Murray found so dissatisfying. Then, going further, he presented the scheme as an example of what, he presciently predicted, would soon become the dominant theme of particle physics: the invention of new quantum numbers to bring order to the ever-expanding subnuclear zoo.

Pais had hinted at this program at the very end of his Rochester talk on "megalomorphian zoology." Perhaps a new physics was unfolding "in which one talks of families of elementary particles rather than of elementary particles themselves." It was not the proton and the neutron as individual entities that were important, he suggested, but the group that they formed—the pair of almost identical particles with opposite values of the quantum number called isospin. Seeing the world in terms of isospin was like putting on a pair of corrective lenses through which protons and neutrons were symmetrical, two sides of the same coin. Isospin was a higher vantage point from which the superficial details—the different charges of the two particles—blurred away. But that, he proposed, was just the beginning. Stand back far enough, and the different stitchings on two baseballs become irrelevant. They look the same. Stand back even farther and a marble, a baseball, and a basketball can all be seen as manifestations of something even more general: the sphere. Pais proposed that physicists seek ever-higher lookouts from which the particles in their bestiary would fall into symmetrical groups as neatly as the elements in the periodic table lined up into columns and rows. Each new group would be united by a symmetry involving a new, yet-to-be-discovered quantum number.

The Netherlands talk was published in the European journal *Physica* (along with favorable comments by Heisenberg and Pauli) just as Gell-Mann was getting down to work. He didn't read it. The slight he had felt from Pais back in Princeton still hurt. When he heard that Pais's theory was all the rage that summer at an international physics conference in Japan, Murray was livid. *Time* magazine had even called the traditional Japanese *ryokan* in Kyoto where Pais was staying with Feynman to ask that a copy of the paper on strange particles be dispatched posthaste to the magazine's Tokyo bureau. Murray found the situation intolerable. This annoying little man was getting all the attention for an idea that had to be incorrect. Murray simply had to break through his writer's block and get his own theory in print.

By the end of the summer, with the draft board staved off once again, he completed his first solo paper, "Isotopic Spin and Curious Particles," and sent it off to *Physical Review*. The editors hated the title, so he amended it to "Strange Particles." They wouldn't go for that either—never mind that almost everybody used the term— suggesting instead "Isotopic Spin and New Unstable Particles." "New Unstable Particles" was what they had been called in the original paper reporting their discovery, and was the only phrase, Murray complained, "sufficiently pompous" for the editors of the august publication. In the end, he yielded. After all the procrastination, he needed to quickly stake a claim. Now that he had finally begun writing, the ideas came pouring forth. He also dashed off an unpublished working paper (a "preprint"), "On the Classification of Particles," and a third paper that was never circulated, much less published, and was eventually lost. Scattered among these fitful efforts, the details were now falling into place.

He was sure now that he had been right when he'd made that slip of the tongue—more than a year earlier—back in Princeton: The heavy strange particles, the hyperons, really did have an isospin of 1 instead of ½; they really were triplets instead of doublets, having positive, negative, and neutral charges. Strange for a fermion, but that was the point. Unlike the nucleon doublet, with its charge center at ½, the hyperon triplet would have its charge center at zero, shifted half a unit to the left. Like this:

CHARGE

−1	−½	0	+½	+1	
		n		p	nucleon (isospin = ½)
H⁻		H°		H⁺	hyperon (isospin = 1)
			←		(charge center displaced to the left)
					strangeness = −1

n = neutron   p = proton   H = hyperon

The hyperon's defining characteristic, that which makes it strange, would be this displacement. Murray would later call it "strangeness," but not quite yet. For now, he labeled the misalignment "y." For convenience in calculating, it can be defined as *twice* the shift in position. The hyperons, then, had what soon would be called a strangeness number of −1.

Now consider the inverse situation. The second kind of strange particles, the middleweight K particles, are cousins to the pions. But instead of being triplets, like other good bosons, they are doublets. Their isospin is ½ instead of 1. On the table, their charge center is displaced to the right of that of the pions. Thus their strangeness would be +1.

CHARGE

−1	−½	0	+½	+1	
π⁻		π°		π⁺	pion (isospin = 1)
		K°		K⁺	K particle (isospin = ½)
			→		(charge center displaced to the right)
					strangeness = +1

π = pion   K = kaon

There was one slight problem. A negative version of the kaon, the K-minus, had been discovered in 1947, along with the positive and neutral kaons. Wouldn't this wreck the whole scheme, making kaons triplets, not doublets—just like normal pions—with no strange shift at all? Murray saw an ingenious way around this obstacle. Maybe the K-minus was instead part of a second doublet. What would its partner be? Well, the K-minus could be thought of as the antimatter twin of the K-plus. And the K-plus's partner was the K-zero. So the other half of the K-minus doublet would be the antimatter döppelgänger of the K-zero. Like this:

CHARGE

-1	$-\frac{1}{2}$	0	$+\frac{1}{2}$	$+1$	
π^-		π^0		π^+	pion
		K⁰		K⁺	K particle
		$\longrightarrow$ (charge center displaced to the right) strangeness $= +1$			
K⁻		anti-K⁰			anti–K particle
		$\longleftarrow$ (charge center displaced to the left) strangeness $= -1$			

π = pion   K = kaon

The editors of *Physical Review* didn't like this at all. What was really the difference between the K-zero and the anti–K-zero? Charged particles could be distinguished from their antiparticles because one was negative and the other positive. But neutral particles and their antiparticles both had to be of zero charge. How could you ever tell the difference? The photon is described as its own antiparticle—the two are indistinguishable, a mere bookkeeping device. Murray pointed out that this didn't have to be the case for all neutral particles. Neutrons and antineutrons had different spins, causing slight variations in their behavior. Anyway, the K-zero and the anti–K-zero had opposite strangeness numbers. The referees were not convinced. Since the K-zero was considered to be a boson, like the photon, the editors thought there should be no difference between the particle and the antiparticle.

"It's all right. They can be like that," Murray kept insisting. These were, after all, strange particles. This time it was *Physical Review*'s turn to acquiesce. The theory was in place, and the strange particles were starting to make a weird kind of sense. Ordinary particles, like the proton and neutron, had strangenesses of zero. They were the gold standard against which isospin displacement was gauged. Strange particles had strangenesses of -1 or $+1$, depending on whether they were shifted to the left or right. Murray was also able to account for the behavior of yet another recently discovered hyperon called a cascade particle (soon to be named the xi). In the Gell-Mann scheme, it had a strangeness number of 2. (It was, as he later put it, "doubly strange.") The example that had been found in the detectors was charged, and Murray predicted that a neutral version must exist as well.

But the acid test was whether this new idea of strangeness would explain why these particles were immune to decay by the strong force, living so much longer than they should. So here was the grand finale: Suppose that strangeness—like charge, energy, spin, isospin—is conserved by the strong force. Nucleons and pions collide to form strange particles. Since the strangeness of these "ordinary" parent particles must be 0 by definition, then the total strangeness of their progeny must also be zero. If a particle with a strangeness of +1 is produced in the collision, there must also be a partner with strangeness −1. More elegantly than Pais's even-odd rule, Murray's scheme explained associated production—why strange particles must be created in pairs. For either of these two particles to decay back into ordinary particles would violate strangeness conservation: You can't go from +1 or −1 back to 0. Murray went on to show why the particles would also resist decay by electromagnetism. Finally the weak force, which does not conserve strangeness, would break the particles down.

It was a virtuoso performance, made all the more impressive when, about that time, associated production was finally observed in accelerator experiments. Gell-Mann called up Brookhaven Laboratory on Long Island, New York, where some of the experiments were going on, to ask about the details. If his theory was right, certain pairs of particles were more likely to arise than others. You should get, for various reasons, something later called a sigma-minus along with a K-plus but not a sigma-plus and a K-minus. Courtney Wright, an experimenter from Chicago, answered the phone.

"What possible difference does it make?" Wright said. "Who cares?"

"I care," Murray said. "Please find out."

Wright put down the phone and went to check. It was indeed sigma-minus and K-plus. Wright didn't know why Murray sounded so happy.

But it would take more than a few experiments to convince anybody that a boson could really have an isospin of ½, or that the K-zero could be different from its antiparticle. Frank Yang, who was at Brookhaven that summer, was telling people that Murray had to be wrong.

. . .

Enrico Fermi classified and quantified everything and everybody, not just subatomic particles. People were numerically rated according to their looks, their wit, their sex appeal. He put scientists into three categories: those not worth listening to, those worth listening to, and those he could learn something from. Telegdi noticed that Murray definitely fell into Fermi's tiny third category. But that didn't mean he bought the strangeness scheme. In fact, Murray thought he sounded dubious in the extreme.

When Fermi objected to an idea or didn't understand something, he wasn't shy about interrupting the speaker. Whether it was a seminar or a class, everything would be stuck on hold for the hours or even days it took to satisfy him. First of all, Fermi wasn't so sure that Murray was treating electromagnetism correctly in his new theory. One day he walked up to the blackboard and wrote down an elaborate reaction that, if it really occurred in nature, would throw a wrench into the strangeness scheme. Murray was sure that this time Fermi was wrong. The jury-rigged reaction might appear to occur on paper, but nature would never put up with something so arbitrary and belabored. The master himself was forgetting his own rule: You have to look past the mathematics to the mechanism— what is really going on in the world. "You're violating a fundamental principle of electromagnetism," Murray bitingly told Fermi. "Namely, that it doesn't do dirty little jobs for people." The electromagnetic currents, he insisted, must emerge directly from the properties of matter, not be imposed, top down, by theorists trying to make a clever argument. He grandiosely called this "the principle of minimal electromagnetic interaction," the physicists' equivalent of *Fiat lux*, "Let there be light." Telegdi took some of the wind out of Murray's sails when he showed him that he had merely restated one of the most basic truths of physics, Ampere's law: Moving charges give rise to currents.

But that wasn't the full extent of Fermi's objections. Sitting in on one of the courses Murray was teaching, Fermi piped up and said he didn't like the notion that the K-zero and the anti–K-zero were different particles. Murray felt a sense of déja vu. Had Fermi been the referee who put this silly prejudice into the heads of the editors at the *Physical Review?*

In fact, Fermi, along with Richard Feynman at Caltech, was pushing a different theory in which it was the high angular momentum of the strange particles that kept them from quickly decaying. Gell-Mann thought it was a bad idea, but Fermi sounded adamant. It was depressing that a scientist Murray so admired didn't seem to think much of his new idea. Then late one evening, Gell-Mann was in Fermi's office and saw a letter to an Italian colleague that his secretary had been typing. He couldn't resist reading it. The scientist had come up with some mathematical results that he thought supported the Feynman-Fermi model. Murray could hardly believe Fermi's reply. Casting doubts on his own theory, he said that Gell-Mann might have come up with a better answer. The cloud lifted, but he couldn't help being a little angry. Why hadn't Fermi told him he liked the idea? Why did he have to find out by reading someone else's mail?

When Pais read Gell-Mann's two preprints, he also believed he was on to something. This strangeness quality that Murray was still calling "y" was, Pais believed, one of the new quantum numbers he had been lobbying for. With the strangeness number, the ill-behaved cosmic-ray particles could be neatly united, just as isospin had allowed the proton and the neutron to become two faces of the same crystalline symmetry. Pais quickly found a way to express the idea in these terms and sent the manuscript to Murray in Chicago. The two started writing back and forth, Gell-Mann telling Pais that he was "quite enthusiastic about your new scheme." Around this time, the spring of 1954, Pais called from Princeton and invited him to come there and collaborate on a paper for a conference that July in Glasgow, Scotland. What Pais had in mind was an overview that would lay out the competing proposals for understanding the strange particles. Telegdi, who had encountered Pais in Europe, considered him manipulative, as sometimes jockeying for a little more credit than he deserved. He had warned Murray to watch out for him. But then Telegdi was famous for his dim views of people. Gell-Mann agreed to come to the Institute in May—anything to avoid the approaching Chicago summer—to put together a draft of the paper.

It was a grim spring in Princeton. The hearings "On the Matter of J. Robert Oppenheimer" were dragging on in Washington. Accused by the Atomic Energy Commission of being a Communist sympathizer, Oppenheimer would by June be stripped of his security clearance. There would be no more need for armed guards standing watch at his office; the safe of government secrets would be taken away. Never mind that the investigating board found no reason to conclude that Oppenheimer had been disloyal to his country. His arrogance and carelessness about following the letter of the law, and his indiscretion in having friends who were Communists, were enough to discredit him. Though most of Oppenheimer's colleagues testified resoundingly in his favor, Edward Teller, the anti-Communist hawk, made it clear to the board that he wouldn't trust the man with national secrets. Murray, who knew from his troubles with the AEC scholarship what it was like to be entangled in the mindless machinery of cold war security, resolved never to speak to Teller again.

For reasons that had nothing to do with either strange particles or the Oppenheimer affair, coming to Princeton that spring turned out to be one of those junctures Murray would later marvel at, when a single decision—or a complete coincidence—can cause your destiny to fork. One evening Pais invited some people to his place for a party. Goldberger was there, visiting from Chicago. And there was Gwen Groves, the daughter of General Leslie Groves of the Manhattan Project. Gell-Mann had met her briefly at Yale. It was not exactly a double date, but both had been part of a foursome, amused to be driving around in the general's car with its official-looking West Point emblem. Murray liked Gwen, but he was especially pleased to see that the guests also included a young Englishwoman named J. Margaret Dow. Murray had noticed Margaret on a previous visit to Princeton and found her absolutely stunning. That dirty-blond hair, those gray-blue eyes that didn't quite match—a broken symmetry, a touch of strangeness in her beauty. It turned out she was Gwen's best friend.

Murray had been too shy to talk to Margaret when he had seen her at the cafeteria, where she worked the cash register. Now they fell easily into conversation. Margaret had studied Greek archaeology at Cambridge University and even worked on an important excavation in Mycenae—just the kind of thing Murray might have done if Arthur hadn't been so discouraging. Now she was an assistant to an archaeologist at the Institute. (The family she boarded with ran the Institute's cafeteria, and she worked there mostly as a favor.) Margaret's parents had been teachers, and her father, like Arthur, had been unsuccessful in business; the family, like the Gell-Manns, had had to struggle at times to get by. Murray told her he was planning to visit Scotland to attend a physics conference, but that he was really looking forward to searching for puffins in the western isles.

"Puffins? Can I come too?" Murray could hardly believe it. She probably wasn't serious about running off to Scotland with a man she had just met at a party. But unlike the other women Murray had tried to court, she actually knew what a puffin was. That evening she went home and drew a perfect picture of one of the birds standing on a rock. She gave it to Murray the next day.

When the paper with Pais was mostly done, Gell-Mann left Princeton for Glasgow. Waiting around at LaGuardia Field, he worked up the nerve to call Margaret from a pay phone, just to say hello. A young woman on the other end of the line answered and said Margaret wasn't there. Well, could she find her and call her to the phone? No, that was impossible. Was she at work? Silence. He heard muffled voices in the background. Murray was wondering what could be going on, when Gwen Groves got on the line and told him Margaret had just learned that her mother was dying. She was on her way to the airport to fly home to England. "Well, *I'm* at the airport," Murray said. Maybe they were on the same flight to London. He hung up the phone and went to the various airlines to check their manifests (you could do that in those days). But he was on Pan American, she was on TWA. He left her a note, saying he would meet her on the other end. When she got off the plane, Murray was waiting. She phoned home and found that her mother had died sometime while Margaret was over the Atlantic. Murray would never forgive the family for not telling her earlier that her mother had cancer. They apparently hadn't even told her mother. They didn't want to alarm her.

He fed Margaret some ravioli at the airport, then rode with her to Paddington Station, where she boarded a train for Birmingham and her family home in Leamington Spa. Murray went on to Glasgow, where he presented a paper he had written with Goldberger on dispersion theory. Pais gave the paper on the strange particles. Among the biggest hits of the meeting was a series of spectacular pictures that Murray's friend Roger Hildebrand had taken of particles created in an accelerator blast and captured by a new kind of detector called a bubble chamber. In one of the shots, a pion was sucked into a nucleus, causing it to explode like a bomb. Weisskopf, Wheeler, Bethe, Dyson, Heisenberg, and Leprince-Ringuet were all on the Glasgow schedule. Fermi had to stay behind in

Chicago. The cancer that would soon kill him was starting to slow him down.

After the conference, Gell-Mann and Pais met for two days in London at the railroad hotel at Paddington to polish up their paper for publication. With strange particles falling from the sky like rain, they had to coin new names for some of them. One they dubbed the sigma. As they were trying to think of a good name for the cascade particle, the one with a strangeness of 2, the title of a silly fraternity drinking song popped into Pais's head: "The Sweetheart of Sigma Xi." Henceforth the "doubly strange" particle was called the xi. During the discussions, he tried to convince Gell-Mann that the shift in isotopic spin that accounted for strangeness should be described as nothing less than a new quantum number. But Murray resisted the idea, and it was left out of the paper.

From London, Gell-Mann headed north again, where he and Roger Hildebrand went hiking on the Isle of Skye. They stayed in a beautiful old hotel in the village of Sligachan, slogging through the moors and peat bogs when the rain wasn't too heavy. They climbed a peak, passing along the way a Scotsman in full highland regalia piling rocks into a ceremonial cairn. When it was too wet to hike, Murray sat at the bar trying to get the barmaid to teach him Gaelic.

One day when the weather cleared a bit, he and Roger headed for a small island in search of puffins. They hitched a ride as far as they could go. They ran through a pasture past some Aberdeen Angus bulls and climbed over a stone wall, and there they could see the island. They waved to an old fisherman smoking a pipe. Could he come across and get them? After ignoring them just long enough to establish that he wasn't taking orders from tourists, he sent his two sons over in a boat. The shore was so rocky the boat couldn't land. But the burlier son waded across in hip boots, carried Murray and then Roger to the boat, and rowed them across. There were no puffins, but Murray saw some strange birds with orange legs and long orange beaks—"gillipurnich," one of the fishermen called them, a kind of oystercatcher. Roger would never forget the smile on Murray's face when, back at the hotel, he emerged from the lounge. He had traced the etymology of the name back to its roots in ancient Gaelic.

One day at breakfast a servant brought Murray a letter on a silver tray. It was from Margaret. He phoned her in Leamington Spa and asked if she would meet him in Scotland after the conference to find those puffins. "Oh, I couldn't possibly," she said. "My brother needs me, my father needs me." Then she said she would come anyway.

She took the night train up to Glasgow. Murray met her at the station and bought her coffee, and off they went to the Scottish isles, taking a boat out to Ailsa Craig, a remote spit of land halfway between Ireland and Scotland, where the only other people were two men cutting curling stones. And there they saw a single puffin. It was their symbol from then on.

Margaret stayed behind in England with her family, and Murray returned alone to New York to begin a visiting professorship at Columbia that fall. He arrived on campus bursting with ideas. Leon Lederman, then a young Columbia experimenter, loved the way Gell-Mann was willing to entertain even the most bizarre possibilities. Nothing seemed too weird when it came to the enigmatic behavior of the subatomic world. One puzzle Gell-Mann returned to again and again was the mysterious neutral kaon. If, as Gell-Mann's scheme required, the K-zero and the anti–K-zero were really not identical but had different strangeness numbers, then what were some of the deeper implications? In trying to make better sense of the picture, Gell-Mann and Pais had come up with a novel idea while in Princeton, one they found so astonishing that they had decided against presenting it at Glasgow: Maybe what theorists had been calling the K-zero was not actually a single particle after all. Maybe it was a mixture—a quantum superposition—of two particles, which they dubbed the K_1 and the K_2. The anti–K-zero would then consist of a different mix of the K_1 and K_2.

While Lederman tried to think of an experiment that would confirm this odd prediction, Gell-Mann and Pais refined their equations. Living in a hotel near the Columbia campus, Murray would take the subway down to Greenwich Village to work in the fifth-floor walk-up on Perry Street that Pais rented on weekends to escape the dullness of Princeton. In late October they sent the paper off to *Physical Review*.

But this was to be the last of their collaborations. Egged on by Telegdi, Gell-Mann convinced himself that he was being used. What turned out to be the important ideas in the Glasgow paper had been mostly his own, but it was Pais who had given the presentation. As for particle mixtures, Murray contended that the idea had already been clear to him back when he was trying to sell Fermi on the whole idea of strangeness. True, Pais had done some elegant work on the notion of associated production (though Murray insisted he hadn't been influenced by it). But the Glasgow collaboration started to seem to the increasingly suspicious Gell-Mann like an attempt by Pais to coopt the strangeness scheme. And it didn't help matters that Pais had dated Margaret before Murray did.

Gell-Mann began bitterly complaining to other physicists that Pais had stolen his ideas. Goldberger took his friend aside and warned him that he was only making himself look stupid. It was useless. Murray had truly come to hate this man. Pais's friends affectionately called him "Bram," short for Abraham. Murray found that he could barely stand to utter Pais's name. "That Pais person," he would call him, through clenched teeth. Digging around for something nastier, he started using a nickname: the Evil Dwarf. (Years later in Geneva, Murray visited Telegdi, who was working at the huge accelerator at CERN. One day his wife said, "It's terrible. At any given time Val has to have somebody he hates." She seemed to expect Murray to sympathize a little. He looked at her very sweetly and said, "You know, somehow he always hates the right people." And Val, probably with a cigarette hanging out of his mouth, just sat there with his impish smile.)

When Pais heard about the insults and accusations, he was dumbfounded. What had he done to deserve this kind of attack? Gell-Mann's precipitous change in behavior was a puzzle he would never be able to solve. From then on at conferences, the two men passed each other by, staring straight ahead. That winter at the Rochester meeting of 1955, Pais pleaded with his colleagues not to use the term "strange particles." It wasn't dignified, he complained. It implied some kind of value judgment.

It can't have been easy for Pais. At an age when he had been hiding from Nazis, this brash young kid was becoming known—almost effortlessly, it seemed—as one of the best of the young theorists. It had taken two years of psychoanalysis for Pais to stem his postwar nightmares and finally dissolve the creative block that had delayed his career even beyond the war's end. Then, in his presentation in Holland, he had suggested the course physics would take—searching for new quantum numbers, higher symmetries. But before his talk had been much noticed, Gell-Mann, who was eleven years younger, had gone out and found strangeness—seemingly not realizing, as Pais had tried to convince him, that it was one of these new quantum numbers. Gell-Mann was the father of strangeness, but Pais considered himself the godfather.

When Margaret returned to the States in October, Murray moved back to Princeton, commuting to New York City for the job at Columbia. A month later, on November 11, 1954, he proposed and she accepted. He persuaded the Institute to give him another appointment that spring so he could be with his fiancée, and he extended his leave from Chicago.

It should have been a happier time. But Margaret's mother's death was soon followed by her father's. It would be a lonely wedding. And back in Chicago, Fermi was dying. He was just fifty-three years old. Gell-Mann and Frank Yang went to see him on his deathbed. When they entered the room, he was reading a book of stories about men who had overcome the seemingly impossible. He told them sadly that the doctors had given him only months to live. At his bedside was a notebook on nuclear physics that he hoped to revise for publication. "Don't worry, everything's O.K.," he assured them. He complained that he was spending all his time cheering up his visitors, that he had even had to cheer up the priest. "You two are doing good work," he said. Then he succeeded in cheering up Murray by criticizing Pais—or so Murray would remember. "It's up to you now," he told the two young scientists. Within weeks he was gone. Murray would always wonder if Fermi had been killed by his carelessness with radiation. He seemed to consider it macho not to worry about such things.

Fermi, Murray thought, had been the meson holding the Institute together. Now there seemed even less reason to return to gloomy Chicago. He and Margaret thought about where they would like to begin their new life together. He had been in touch with Feynman on and off since meeting him at a conference somewhere. Once, Feynman had come to Chicago from Caltech to lecture on his ideas about the strange phenomenon of superfluidity, the frictionless flow of liquid helium as its temperature approaches absolute zero. Feynman, as usual, had left everybody impressed with his madcap energy, barhopping all night with Murray and Roger Hildebrand, finally landing in a downtown burlesque hall. Now

Murray wrote to Feynman to ask about job prospects at Caltech. Feynman invited him to come and lecture about his work, the usual prelude to offering someone an appointment, and just before Christmas he and Margaret went to Pasadena.

Murray knew almost immediately that this was where he wanted to be. Feynman's intensity was amazing. Drumming his fingers against his desk or along the walls of the corridors of the physics building, he sometimes seemed impatient to get back to his bongo drums. He talked physics in the same rapid, staccato manner, firing off ideas in his exaggerated Brooklyn accent. He told Murray how much he had liked his paper with Low on QED at short distances. Feynman usually made a point of not reading papers—he preferred to do the derivations himself—so this was a high compliment. Then Feynman impressed Murray by taking the idea and generalizing it further, making it more powerful. They talked about everything. What if Einstein hadn't discovered general relativity, describing gravity as curved space-time? Would it be possible, Feynman wondered, to start from scratch now and reconstruct the theory from first principles? After all the pomposity Murray had encountered in Princeton—all that mathematical preciousness hiding mediocre ideas behind a smokescreen of fancy formalisms—Feynman seemed like the real thing.

Margaret returned to Princeton after Christmas, while Murray stayed into the New Year. Convincing her was going to be harder. She hadn't really liked Southern California. On stationery from the Athenaeum, the university's staid faculty club, Murray wrote her an airmail letter in green pencil, addressing it in red. "Dear Princess," it began. He tried to reassure her. He was struck by the "congenial character of the faculty and the students." And there was much they had missed in Pasadena—the Huntington Library, the nice shops, the Pasadena Playhouse. "Don't get scared, though," he wrote. "They haven't offered me anything yet. Also, I completely agree with you about the Los Angeles area. It's very depressing though not as depressing as Chicago."

And Caltech was not the only possibility. Across the country in Cambridge, Victor Weisskopf was pushing Harvard to hire Murray as a full professor, though he doubted the school would be quite so forthcoming. After all, Julian Schwinger, the other boy wonder, had to wait until he was twenty-nine for a full professorship. Murray was only twenty-five. Schwinger himself was enthusiastic about the prospect of having Gell-Mann as a colleague, telling Goldberger that he was glad to hear that Murray was less than ecstatic about Pasadena.

Chicago was also fighting to keep him. Telegdi thought it was just a matter of offering Murray more money. Goldberger, who had also been looking at other schools, promised his friend he would stay in Chicago if Murray would too. The school quickly sent Gell-Mann a letter offering to double his salary to $8,000. Then Caltech came through with an offer of $9,500 a year and full academic tenure. Murray accepted. It wasn't just the money. He wanted to work with Feynman.

Everything was happening so quickly. He had planned to spend a year first in Copenhagen at the Bohr Institute, but the draft board wouldn't allow it, even after Oppenheimer himself sent them a letter describing Gell-Mann as "one of the most brilliant, if not the most brilliant, of the small handful of young experts in theoretical particle physics." The draft board could understand being a student, and they could understand being a professor. But another postdoctoral visit wouldn't qualify for a deferment. They insisted that he have a real job. A tenured position at Caltech was enough to satisfy them.

Life, it seemed, was finally falling into place. Murray and Margaret were married in Princeton on her birthday, April 19, 1955. Gwen Groves was maid of honor, David Pines was best man. After the reception, Murray and Margaret went into Manhattan for a night at the Waldorf-Astoria. Then they drove across the country to Pasadena. Margaret was horrified by the desert, so different from the rolling green hills of England. She wasn't even particularly impressed with the Grand Canyon. But she was excited to be entering this new life.

Once they reached Pasadena, they stayed at a motel for a few days, then rented an apartment in a building called, outlandishly, the Fireside Manor Lanai. Margaret thought the name typified everything she hated about Southern California fakery. Even worse, it turned out, the building catered to young singles; some of the tenants would get drunk and jump off their balconies into the pool. But the Gell-Manns were only there for a month before leaving on a trip to Europe. Murray dropped Margaret off in England, then went to a conference in Pisa, where he finally gave a full presentation of his particle classification scheme, basing it on a new quantum number he was now unabashedly calling strangeness. And so began a progression that would ultimately lead to new quantum numbers called charm, color, top, and bottom. It turned out that—as often happens when a good idea is in the air—another physicist, Kazuhiko Nishijima in Japan, independently arrived at his own version of strangeness; he called it η-charge. But it was Gell-Mann's name that stuck.

Back in England, Murray bought a Hillman Minx convertible to ship to the States. But first he and Margaret took it on the ferry to France, driving on to Italy, Spain, Belgium, the Netherlands, Luxembourg, West Germany, Denmark (stopping in Copenhagen for a brief visit to the Bohr Institute), and Sweden. With so many professional and social obligations, it wasn't an idyllic honeymoon. Margaret's brother, John, stayed with them a few days in Copenhagen. Telegdi, who was also there, was amazed at how well Murray had learned to converse in Danish. And he was struck by how proud Murray was of the outstanding woman who had agreed to be his wife.

There was still some lingering resistance to the strangeness scheme. At the next Rochester meeting in 1956, Oppenheimer dismissed it as a "temporary kind of solution." But over the next few years, Murray's new quantum number became deeply rooted, as experimenters found the new particles his theory had predicted. A tradition was begun: They would send him the pictures, and he would hang them on the wall.

CHAPTER 7

A LOPSIDED UNIVERSE

After returning from Europe, Murray and Margaret moved out of the swinging-singles apartment complex and rented a house on Hidden Valley Road in Monrovia, just west of Pasadena. The suburb was on the edge of the wilderness, with the San Gabriel Mountains rising to the north and Mount Wilson Observatory perched overhead. The Gell-Manns were delighted to learn that there was a mountain lion living in the neighborhood. They would listen to it scream at night.

The genteel town of Pasadena was much better tamed, with its flowering trees and palm-lined avenues of mission-style houses— white stucco, red-tile roofs, yards sprawling with ice plant and roses. Bordering the town to the south were the mansions of San Marino, old ranchland converted into an enclave for rich Easterners who hated to be cold. The former estate of the railroad and real estate magnate Henry Huntington had been turned into a museum and botanical gardens. It was a pleasant place to pass a Sunday afternoon. One could look at rare manuscripts, including a Gutenberg Bible, admire Gainsborough's *Blue Boy* and other paintings, then stroll along the manicured paths winding their way through the palm garden, the private jungle, the cactus-studded desert, the rose garden, the Australian garden, the Japanese Zen garden—around the world in 200 acres. The encroaching sprawl of Los Angeles seemed far away.

The campus where Murray was now teaching was a garden in itself. From the ivy-covered walls of the East, he had arrived among the lilacs of Caltech. MIT had its grubby, fluorescent-lit Infinite Corridor tunneling from building to building. Caltech had its Olive Walk rambling from the center of campus, where the physics building stood, to the Athenaeum. Like Mount Wilson Observatory, this imposing piece of architecture was the brainchild of the astronomer George Ellery Hale. Modeled after the Athenaeum Club in London, it was a dark, woody showpiece of Italian Renaissance gaudiness. "The Mausoleum," Gell-Mann and other faculty members called it.

In those days, professors wishing to lunch there were required to wear coats and ties. Feynman liked to show up in shirtsleeves, making a great production of donning the tacky formal garb provided for ill-prepared diners. Gell-Mann, who took pride in dressing well, found the act a little irritating; Dick went to such lengths to play the clown. But it was hard not to be charmed by Feynman's childlike wonder. When Dick, then a bachelor, came over for Christmas dinner at Murray's house, he and Margaret presented him with a new set of bongos. He sat on the floor slapping out mesmerizing rhythms.

Caltech was paying Murray more money than he ever dreamed possible, and the figure increased rapidly after two other schools set off a bidding war. Harvard made another try but was quickly trumped by the University of Chicago's offer of a full professorship and a 26 percent salary increase, to $12,000—a deal that Caltech immediately matched, making him the youngest person ever given such an appointment at the school. Much as he hated Chicago, Murray knew he would miss working with Murph Goldberger, but he was settling into his new life. Before long he was supplementing his hefty university salary with consulting deals at Los Alamos Scientific Laboratory, in New Mexico, and at the RAND Corporation, the well-heeled Defense Department contractor located on the beach in nearby Santa Monica. Where else could you make $100 for a single day of thinking and giving advice?

First, though, he needed a security clearance, and he was haunted again by the old problems, and some new ones as well. What, he was asked during the investigation, had his car been doing parked at a Paul Robeson concert in Chicago back in 1952? Murray was incredulous. It seemed that the FBI had nothing better to do than send agents to record license plate numbers at performances given by radical black singers. Murray told them he hadn't even been in the country then. He had left for his first trip to Europe and loaned the old blue Chevy to his friend Art Rosenfeld. Rosenfeld had loaned the car to another young physicist, named Nina Byers, who apparently drove it to the performance. But when the investigators checked the passport records, they found that Murray hadn't actually departed for Europe until several days after the concert. They thought they had caught him in a lie. Murray apologetically admitted that he had forgotten about stopping off in Princeton first to give his strangeness talk, the one he felt Pais had belittled. Anything having to do with Pais seemed to give him a mental block as insoluble as concrete. He finally straightened out the misunderstanding. Contracts with California's burgeoning aerospace industry soon followed, and Gell-Mann would eventually find that a full third of his earnings came from consulting.

He was relieved to see that Margaret was starting to like the West. On weekends they would take off in the Hillman Minx, happily exploring the deserts and mountains. She had changed his life. He began to realize how rarely he used to think about anyone but himself. Before Margaret, he would say, he was like "a malfunctioning computing machine" or "an atom that is bounced around" by insentient forces. Now he felt like a person. His undernourished soul was starting to grow a little. He and Margaret would read poetry together in Latin. Murray especially liked Lucretius, with his poetic musings about the nature of reality—particle physics as it was in Roman times. Margaret considered her Latin better than his; after all, she was a trained classicist. Unlike her husband, she also read Greek. Of course, Murray could outdo her in half a dozen other languages, including mathematics. But in every sense, she was his intellectual equal, as well as his best friend. Inspired to try to be a better person, he started seeing an analyst. Maybe he could unearth the source of the melancholy that possessed him even in the best of times. Maybe he could find a way to dissolve his constant writer's block.

When he didn't have to struggle to make himself write a paper, he loved being at Caltech. Working with Feynman was, at first, everything he had hoped it would be. They argued nonstop—"twisting the tail of the cosmos," they called it—working themselves into a frenzy, gesticulating and shouting in the opaque jargon of their craft.

"No! You can't do that—you'll get the magnetic moment of the sigma wrong! . . . The decay mode won't be the right one. The branching ratio will come out wrong!"

"Yes! It will be perfectly all right! You don't understand what I'm doing!"

On the surface, creation appeared so unruly. But deep down scientists were discovering seemingly immutable laws. Gell-Mann's own conservation of strangeness had succeeded in explaining why certain cosmic ray particles lived so much longer than they had a right to: The strong force respected the dictate, the weak force did not. But these curious particles were still not behaving very well. In their marathon conversations, Gell-Mann and Feynman puzzled over recent experiments suggesting that some strange particles were defying another symmetry, one that most physicists considered sacrosanct. The law, called reflection symmetry, can be put metaphorically: A particle and its mirror image should act precisely the same.

Thinking in terms of symmetries was becoming second nature. Feynman would get his students into the swing of things by asking them to consider the symmetries of an ordinary alarm clock: If you pick up the clock and move it a few inches in any direction, it will keep right on ticking. The clock doesn't care where you put it. Or keep the clock in the same spot and let it run. All things being equal, it will operate the same now as it did in the past and as it will in the future. The behavior of the clock is said to be symmetrical, or invariant, through translation in both space and time. Feynman would quote the German mathematician Hermann Weyl: Symmetry is a change that leaves everything else the same. A sphere is symmetric under rotation because the operation does not alter its appearance.

The reason this example seems so trivial is that these symmetries are deeply embedded in the very nature of time and space. They are part of the cosmic rulebook, the way the world works. Some symmetries are not so obvious. No matter how fast or slow an object is moving, a light beam will always speed by it at the same velocity, 186,000 miles per second. This law, the so-called Lorentz invariance, is the foundation of Einstein's special theory of relativity.

To many physicists, reflection symmetry seemed just as basic. Hold the clock with its face to a mirror and watch its reflection. The 1 is swapped with 11, the 3 with 9, the 5 with 7. But since the hand is now moving counterclockwise, it still gives the correct time. The numbers are backward: Things may appear different in the looking-glass world, but the laws of physics seem to hold as surely as they do if we move the clock an inch in any direction, or listen to its ticking today or tomorrow. People could have easily decided many moons ago to make clocks so that they ran in the opposite direction. Flip over each gear, reverse the winding of the spring. In this might-have-been world, rotation to the left would be called clockwise. Nothing in physics forbids this anymore than it forbids people in

England and Japan from driving on the left side of the road. The screws we use to assemble our world are almost all threaded right-handed: You must twist them clockwise to drive them into a board. Nothing requires this. You can just as easily make left-handed screws, but society has adopted a different convention.

Reflection symmetry obviously applied in classical Newtonian physics, as witnessed by the clock and its backwards twin. Whether something is right-handed or left-handed, reflected or not, is arbitrary, the result of accident or decree. It was widely assumed that this symmetry—like the translation symmetries—was just as firmly established in the quantum world, governing the behavior of sub-atomic particles. It seemed that any physical system, no matter how small, should be indistinguishable from its mirror image, that there should be no inherent leftness or rightness.

Feynman liked to drive home the point with a story. Suppose we are trying to communicate with a Martian entirely through the use of radio signals. We want to tell him that our hearts are on the left side of the body, or that we drive our screws into wood by twisting them to the right. How would we convey this information?

On earth, we might explain, if we stand facing north and stick out our arms, the hand pointing in the direction where the sun sets is left. The Martian says, "O.K. We are also bilaterally symmetrical and have two arms and two legs. And our planet orbits the sun in the same direction as does earth. But we're confused by this idea of north. Which way should we face so that the hand pointing to the sunset is what you call left?" If Mars had a magnetic field like earth's, we could tell the Martian to take a magnetized needle and suspend it on a pivot. It would line up along the field in a north–south direction. "Fine," he says. "But which end of the needle is north and which is south?" We look at our own compasses and say, "The end of the needle that is painted red points north." "Great," the Martian says. "So which end do we paint red?" "Well, the one that's pointing north." As the story shows, when stripped from their local context, the words "left" and "right" or "north" and "south" seem to be meaningless conventions. There appears to be no innate phenomenon called handedness.

Remember that all communication in this story must be over the radio, and for good reason. We could conceivably send the Martian one of our screws with instructions to drive it into a solid object and note the direction it turns. That is what we call right. Or we might somehow send a light beam polarized so that all the photons are spinning either left or right. But that would be cheating, merely establishing our arbitrary convention on another planet like some kind of infection. We could have the Martian look at the Big Dipper: Viewed right side up, the four stars forming the bowl are to the right of the handle. But that also violates the spirit of the game. The point is that there is no way to derive leftness or rightness purely through natural law. Or so it was widely believed.

If there was something about the laws of physics, Feynman would say, that caused little hairs to sprout on the north pole of a magnet wherever it was in the universe—if there was some way to always tell one side from the other—then reflection symmetry would be violated. The universe would care which was left and which was right. To the surprise of the physicists, this turned out to be true. Electromagnetism respects reflection symmetry, as does the strong nuclear force. But in 1956, shortly after Gell-Mann arrived in Pasadena, physicists were starting to find hints that the weak force was an exception to the law.

One of the great truths of modern physics, established by the mathematician Emmy Noether, is that every symmetry is intimately associated with a conservation law. Saying that the laws of physics are the same when translated to a different point in space is equivalent to saying that nature respects the law of conservation of momentum. Saying that the laws are symmetrical when translated in time is equivalent to saying that nature respects the conservation of energy. Though far from intuitively obvious, the mathematics demands that these statements be true. The laws must be symmetrical under rotation if angular momentum, or spin, is to be conserved. The symmetry of reflection, it turns out, implies a law called conservation of parity. If nature really doesn't care which is left or right, the equations assure us, then something called parity must be conserved.

Parity is a quantum number that tells what happens to a particle when you reverse its coordinate system—the grid of x-, y-, and z-axes that is being used to describe it. Simplifying slightly, parity describes what happens when you reflect something in a mirror. (To be absolutely precise, this would be a mirror that reversed not only left and right and inward and outward, but also up and down.) There are two possibilities. A particle might look the same as its reflection, as would a sphere or a cylinder. Or it might be flipped around backwards, like an arrow pointing left or right. Particles for which the reflection and the thing being reflected are identical are said to have even parity and are assigned a parity number of +1. Particles that are reversed in a mirror have odd parity and are assigned the value −1. The round gears inside a clock have even parity: They are indistinguishable from their reflections. The numbers on the dial face have odd parity because they are flipped around. Still, whether something was even or odd, it was assumed that it and its reflection would obey the same physical laws.

Since parity is a quantum number, we expect it to be governed by a conservation law: In any particle interaction, the total parity must remain the same, as with electric charge or spin. As it happens, parity is a multiplicative rather than an additive quantum number. Multiply the parity of the particles going into the reaction. Then multiply the parities of the particles coming out. The numbers are supposed to match. And according to Noether's theorem, a conservation law must imply a symmetry. If parity conservation is true, then so is reflection symmetry—the requirement that left- and right-handed systems cannot be told apart.

In the early 1950s, an Australian-born physicist named Richard Dalitz was examining K-mesons, the so-called kaons, when he noticed something that should not be. Like all strange particles, kaons decay slowly, through the weak force. Sometimes the positively charged kaons crumbled into three pions, sometimes into two pions. That in itself was not so bad. A particle can have various "decay modes," breaking down into different sets of particles. But with the kaons, the arithmetic of parity conservation didn't work right. A pion has a parity of −1. So three pions ($-1 \times -1 \times -1$) have a total parity of −1. By parity conservation, the original particle would also have to be of parity −1. But two pions (-1×-1) have a parity of +1. The initial particle would have to be +1. So which was it? A particle might conceivably have two different ways of decaying but not two parities—assuming, that is, that parity conservation and thus reflection symmetry are truly obeyed by the weak force.

Confronted with this puzzle, physicists decided there must be two different kinds of kaons, the tau and the theta, each with different parities. One decayed into three pions and the other into two pions. But the two particles otherwise seemed indistinguishable. They had the same mass; they seemed to live the same amount of time before breaking down. The idea that the tau and theta were two different particles was beginning to seem like a stretch, an ad hoc assumption to preserve the idea of reflection symmetry.

For several months Murray had been trying to make sense of this. Could there be some elegant explanation for these almost-identical twins—a way to show that the seemingly identical masses and so forth were not meaningless coincidences but the result of some deep law? Maybe the tau and the theta were two halves of a doublet. Like the nucleon, which can have either up or down isospin, manifesting itself as a neutron or proton, maybe the theta and tau were the opposite faces of a single particle. The nucleon was an isospin doublet; this hypothetical particle would be a parity doublet. In some artificial space, you could turn it one way and it would be a particle with even parity, another way and it would be a particle with odd parity. There was still a problem, though. Gell-Mann's scheme explained how the theta and tau could have the same mass and even the same spin. But the identical lifetimes remained, as he ruefully put it, a "miracle."

He had discussed the idea with Pais before their falling out and had given a talk on parity doublets at Berkeley. But as usual, he hadn't gotten around to writing the idea up. What if it turned out to be wrong? Once an idea was in print, there was no way to take it back. It was out there to embarrass you forever.

Others were not so timid about their speculations. In the days before the Internet, scientists routinely mailed out preliminary preprints of forthcoming articles. Since they had already been accepted for publication, priority was established; the ideas could be safely shared. Early in 1956, Murray was thumbing through some preprints of articles scheduled to appear in the *Physical Review* when he saw that he had been scooped. T. D. Lee and Frank Yang, his old colleagues from the Institute for Advanced Study, were proposing their own version of parity doublets. Murray was furious, grumbling to anyone who would listen that they had stolen his idea. Ever since the blowup with Pais, Murray's friends had noticed how much more competitive and aggressive he had become. His verbal attacks on some of his colleagues were starting to border on viciousness. Goldberger again tried to play Dutch uncle, warning Murray that he was making a fool of himself. You don't get credit for what is in your notebooks, Murph would tell him. You have to write up the ideas.

After Murray insinuated during a visit to Berkeley that his theory had been pirated, word got back to Lee and Yang. They sent him a stern letter warning him to back off. He realized he had gone too far and quickly mailed an apology. "I do not believe either of you capable of stealing ideas," he wrote, "and I do not think that you learned the contents of your paper from me." But he couldn't resist adding that he *had* told Lee about the idea, though "clearly he was not listening or forgot." A month later, Lee and Yang sent back a terse note accepting his apology and saying they considered the matter closed. It was a big fuss over an idea that turned out to be wrong. (Coincidentally, the Lee and Yang article was published on April Fool's Day.)

That spring Gell-Mann, Feynman, Yang, and Lee were among 200 theorists and experimenters who converged on upstate New York for the 1956 Rochester conference, finally held in a more hospitable season, when Rochester was not buried in snow. The antiproton recently had been discovered at the Berkeley accelerator. In the coming months the experimenters Clyde Cowan and Frederick Reines would announce that two and a half decades after it was invented by Pauli and Fermi, the neutrino had been found. So many pieces of the subatomic puzzle were falling into place. But what was one to make of the theta and the tau? As physicists began to build more powerful accelerators, they didn't have to wait around for heaven to send them occasional strange particles in the form of cosmic rays. They were able to create the specimens they needed.

By now Dalitz, who had first drawn attention to the puzzle, had graphed the results of some 600 weak decays on a diagram that became known as a Dalitz plot. The theta and tau continued to differ in parity but were otherwise indistinguishable. Declaring them to be two different particles was seeming more and more suspect.

During the conference, Feynman and his roommate, an experimenter named Martin Block, mulled over the problem. Maybe there *is* no difference, Block ventured. Would it really be so bad if the weak force violated parity when it caused the K particles to decay? Feynman started to blurt out an objection, then stopped himself. He thought for a minute and said he wasn't sure. If you could violate parity, then, by implication, reflection symmetry would be violated as well. As far as the weak force was concerned, there would be a difference between left and right. That might lead to all sorts of bad consequences. Or maybe not. He would have to think about it. Saturday morning, the last day of the conference, Oppenheimer was chairing a session called "Theoretical Interpretation of New Particles." That would be the appropriate time to bring up this immodest suggestion. "Why don't you ask the experts?" Feynman said. Block replied, "No, they won't listen to me." This was, after all, a theoretical session, and Block was an experimenter. "You ask."

First Feynman ran the idea by Gell-Mann. Was Block's proposal as crazy as it sounded? Murray was still putting his bets on parity doublets—the notion that theta and tau were two sides of the same coin. But he didn't think parity conservation was necessarily sacrosanct. When Gell-Mann was still in graduate school at MIT, a professor named Herman Feshbach had assigned his students a problem: Prove that when you transform the coordinates of a system—when you take a physical process and reflect it in the symmetry mirror—parity is conserved. After struggling in vain with the calculations over a weekend, Murray concluded that the proof was impossible. He came in on Monday and turned in his answer: Parity conservation was not a fundamental truth; it was an empirical phenomenon. Sometimes parity might be conserved, sometimes it might not—it depended on the nature of the force involved, on its structure. Later, in Chicago, Fermi had told him that he didn't believe that parity necessarily had to be respected in beta decay. And Dirac didn't seem to believe in it.

So, Gell-Mann told Feynman, it would be difficult to know whether the weak force violated parity until the force was better understood. Fermi had come up with a theory of one kind of weak interaction, beta decay. Physicists hoped this and other weak interactions—the decay of strange particles like the theta and tau into pions, the decay of pions into muons, and muons into electrons, and so forth—were all versions of the same phenomenon. They all seemed to have the same tiny strength. If they turned out to be completely different forces, nature would be horribly ugly. But still, no one had come up with a theory uniting all these interactions into one all-encompassing weak force—a Holy Grail reverently called the Universal Fermi Interaction. It was clear that the electromagnetic force conserved parity. That followed from its structure. And the strong force, for all its mysteries, had better conserve parity, or nuclear physics would be a total mess. But the weak force, Murray argued, was still an open question. Feynman said something like, "O.K., that's my opinion too. Since we both agree, it must be true, so I'll get up and tell the meeting."

On Saturday Oppenheimer started off the session, and before

long Yang launched into a lengthy talk on the new particles, including his and Lee's proposed parity-doublet solution to the tau-theta puzzle. Unconvinced, Robert Marshak, the organizer of the conference, described what he considered a less radical alternative: If you assumed the existence of a hypothetical particle with a spin of 2, you could cook up a scheme in which it decayed into either two or three pions without violating parity. But no one had yet seen such a particle. Dalitz threw cold water on the idea. But he conceded that, in a stretch, Marshak's scheme might work. Oppenheimer, in his best self-parody, intoned: "The tau meson will have either domestic or foreign complications. It will not be simple on both fronts." It wasn't clear what he meant, other than that this was a very confusing situation.

When it came Murray's turn to speak, he launched into an explication of his rendition of parity doublets, taking pains to show that the theory was not quite identical to Lee and Yang's. (In his version, neither the strong nor the electromagnetic force could tell the difference between the theta and tau. In the Lee and Yang theory, electromagnetism might recognize slight differences in the two particles' masses.) But there were grander things on his mind. He told the audience about an idea he and another physicist had come up with that might help unify the various weak forces. Physicists customarily described the weak interactions—beta decay and interactions involving pions and muons—with what was known as the Puppi triangle. To include the newfound strange particles in the framework, Gell-Mann had generalized the triangle into a tetrahedron. But he admitted that much more would have to be understood about the nature of the weak interactions before they could be melded into one.

Yang conceded that physicists needed to keep an open mind about how to explain the tau-theta mystery. Taking that as a cue, Feynman stood up and said he would like to present an idea suggested by Martin Block: that parity was not conserved. The statement didn't exactly bring down the house. Yang coyly said that he and Lee had looked into the idea without reaching any definite conclusion. And then the discussion quickly turned to other, wilder possibilities. Someone suggested that maybe the puzzle could be solved if one supposed that, during theta and tau decay, a particle was emitted that had no mass, charge, or momentum but simply served to carry away "some strange space-time transformation properties." The speaker admitted that the idea seemed "nonsensical." But who knew? Murray thought the discussion was getting out of hand. Perhaps, he said, they should keep their minds open to less radical alternatives. After some more debate, Oppenheimer said he "felt that the moment had come to close our minds." Precious as usual, he added: "Perhaps some oscillation between learning from the past and being surprised by the future . . . is the only way to mediate the battle."

Gell-Mann and Feynman went back to Pasadena excited that parity might indeed be violated. On a train heading in the other direction, back to New York City, Abraham Pais and Frank Yang bet John Wheeler a dollar that the theta and tau would turn out to be different particles.

Murray was still weighing the possibilities a few weeks later when he joined the first large group of physicists to visit the Soviet Union since World War II. After passing the scrutiny of the State Department and the AEC, about a dozen theorists and experimenters, including Gell-Mann's mentor Victor Weisskopf, his nemesis Abraham Pais, Dyson, Marshak, Luis Alvarez, Wolfgang Panofsky, and Robert Wilson, set off for a journey behind the Iron Curtain in May 1956. After so many years of cold war isolation, Murray felt as though he were embarking on a visit to another planet. A television crew came to the Los Angeles airport and filmed him and Margaret kissing good-bye. Margaret was pregnant now with their first child and went to stay with the Pines family in Princeton.

The final leg of the flight was from Copenhagen. When the scientists landed in Moscow, after midnight, the eminent Russian physicist Lev Landau and a group from the Soviet Academy of Sciences met them on the runway. The Americans were housed in the lavish but stodgy Hotel Moskva, near Red Square. (A statue of Stalin standing in the lobby would be removed just days later—the purge had begun.) The visitors were treated as honored guests—a visit to the Bolshoi Ballet (with champagne and caviar at intermission), a tour of the Kremlin. Many of them came away impressed that, under Khrushchev, the Soviet nuclear program seemed to be shifting to peacetime uses. But that was probably the impression they were supposed to get from what was surely a carefully orchestrated visit.

Inside the lecture hall, cold war politics quickly became irrelevant. The physicists listened through headphones as the Russians' impassioned arguments were translated on the fly. Raising their fists to emphasize a point, they sometimes seemed on the verge of violence. But it was just the Russian style. Gell-Mann had found the Soviet physics journals deeply mysterious, but now as he watched the fights unfold, he began to understand who was who in the warring intellectual factions. Landau railed against field theory, declaring that it was fatally flawed and must be replaced with a whole new kind of framework—something based perhaps on dispersion relations and S-matrices. Murray wasn't convinced, but he resolved to give the subject more thought.

The tau-theta puzzle was such a hot topic that when Gell-Mann lectured on it, dozens of scientists packed the room and even more tried to crowd inside. He had to give the talk twice. He laid out all the known possibilities. There was the scheme by Marshak that required spin-2 particles. That was too ugly, Murray said. And there was his own favorite: the idea of parity doublets. But he also allowed that there was no reason to believe that the weak force must respect parity conservation. If it was granted a waiver, then the puzzle would be solved. Some of the Soviet scientists found the idea outrageous. Why not just throw out rotational invariance and Lorentz invariance, one of them declared. The symmetries were there to keep physicists honest. Otherwise theorizing would be a free-for-all.

While Murray was in the Soviet Union, the question of whether the weak force might indeed violate parity kept eating away at Lee and Yang. Each week Yang would drive in from Brookhaven to meet with his colleague, who was now at Columbia. One day he had to move his car by 11 a.m. to avoid getting ticketed or towed, so they parked near 125th Street and Broadway and went into the White Rose café to talk. Then they moved to a nearby Chinese restaurant for lunch. Arguing intensely, sometimes shouting at each other, they took turns playing devil's advocate, switching positions to see if their arguments held up. Lee found the give-and-take exhilarating—harmony and opposition complementing each other. The whole world seemed to lie in front of them, Lee felt, with new realms to explore. They emerged from the restaurant with the strong suspicion that there was no evidence one way or the other for whether beta decay respected parity. They doubted that something as seemingly fundamental as reflection symmetry was violated by weak interactions, but someone, they thought, had better find out.

One of the world's foremost experts on beta decay worked just a few floors above Lee's office at Columbia: Chien-Shiung Wu, an ambitious experimenter whose no-nonsense style had earned her the nickname the Dragon Lady. One day after his lunch with Yang, Lee asked Wu whether anyone had actually done an experiment to see if weak interactions respected parity. Wu didn't know. She said most people were so sure that parity was conserved that they considered it a waste of time and money to test for the obvious. But maybe if he and Yang scrutinized past experiments, they would find a clue. She handed him an imposing volume of experimental data called *Beta- and Gamma-Ray Spectroscopy*, which included just about everything that had been done on beta decay.

For two weeks Yang and Lee pored over the book, tediously analyzing the data and rederiving the formulas, then sharing their results by telephone. They reached the end of the tome without prying loose a shred of evidence for parity conservation. In mid-June they submitted a paper to *Physical Review* entitled "Is Parity Conserved in Weak Interactions?" The matter was still very much up in the air, they concluded, not just for beta decay but for all the weak interactions. The editors of the review, who didn't like titles with question marks (any more than they had liked Murray's choice of terms for "curious" or "strange" particles), blandly renamed the paper "Question of Parity Conservation in Weak Interactions." Lee and Yang suggested a way to resolve the matter: Experimenters would observe a weak decay and its mirror image and see if there was any difference.

Once the paper was written, they immediately went on to other things. They expected experimenters to take up the challenge and show that parity is indeed conserved. With that avenue closed, they could look for other solutions to the tau-theta puzzle, like Lee and Yang's (and Gell-Mann's) parity doublets. One of the first to bite was Wu. She and her husband, a physicist named Chia-Liu Yuan, had booked passage on the *Queen Elizabeth* to celebrate the twentieth anniversary of their departure from China. The plan was to take in a physics conference in Geneva and then lecture in the Far East. Now she found herself in a dilemma. Like Lee and Yang, she thought parity would probably turn out to be conserved. But what if it wasn't? She didn't want to blow her head start and be upstaged on this crucial experiment. Her husband would have to take the trip alone.

Fermi had shown that in beta decay, a neutron changes into a proton, with an electron and a neutrino (strictly speaking, an anti-neutrino) flying in the debris. When this happens to one of the neutrons in the isotope cobalt 60 (with twenty-seven protons and thirty-three neutrons), it changes into the element nickel 60 (with twenty-eight protons and thirty-two neutrons). So suppose you take a bunch of these cobalt nuclei, cool them to near absolute zero to dampen thermal vibrations, and use a magnetic field to ensure that they all line up the same way, that they spin in the same direction. Then watch them decay. If parity is conserved, electrons will come shooting out equally in two directions. If the process is reflected in a mirror, both images will look the same. But if more electrons come out one way than the other, the symmetry will be broken. The same experiment viewed in a mirror will produce the opposite result. Since Columbia didn't have the proper equipment for this delicate experiment, Wu arranged to collaborate with scientists at the Bureau of Standards in Washington, D.C., and began commuting back and forth.

Meanwhile in Chicago, Gell-Mann's old friend Val Telegdi saw a preprint of Yang and Lee's paper and proposed a test to see if parity was violated in another kind of weak interaction: pions decaying into muons, which in turn decay into electrons. His colleagues were unenthusiastic. Of course parity was respected. Why waste the time? The stubborn Telegdi decided to do the experiment anyway with a postdoc named Jerry Friedman. They were working on the project when, in September, Telegdi's father died. He went to Milan over Christmas vacation to help his mother. He didn't know he was competing with Wu. Soon others would join the game.

In early January, Wu gave Lee some exciting news: Though they still needed to double- and triple-check the results, she and her colleagues were pretty sure parity was being violated in beta decay. Lee immediately called Leon Lederman at Columbia. For months Lee and Yang had been trying to convince Lederman to run a parity test on pion decay. But he figured that any effect would be so tiny and open to interpretation that trying to measure it wouldn't be worth the grief. Lee told him that if Wu was right, they were not talking about something subtle. It could be close to a 100 percent effect. This was starting to sound interesting. The next day, at a weekly physics department lunch at the Shanghai Cafe, Lee, Lederman, and a group of scientists mapped out the plan. Lederman and his colleague Richard Garwin were already experimenting with pions and muons at the Columbia accelerator in Irvington-on-Hudson,

1. Murray's parents, Arthur Gell-Mann and Pauline Reichstein Gell-Mann.

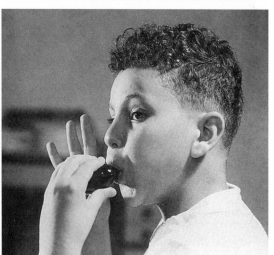

2, 3. Growing up on 15th Street in Manhattan.

*4. Finding order in
the world of music.*

*5. With his mother,
Pauline, in
Central Park.*

*6. Brother Ben
(seated) taught him
about science.*

*7. Arthur, it seemed
to Murray, was a
disappointed man.*

MURRAY GELL-MANN   1937

President of Debating Club (1); Shakespeare Club (2); Soccer Team (2); General Organization (3, 4), Secretary (4); Chairman of Assembly Committee (4); Manager of Tennis Team (4), Asst. Manager (3); Manager of Chess Team (4); News Editor of Columbia News (3), Associate Editor of Magazine (3), Editor (4); Secretary Laws Committee (4); Math. Club (4); History Club (4); Virgil Society (2).

8, 9. Graduating from high school at fourteen, Murray, as class valedictorian, delivered a commencement speech that caught the eye of The New Yorker.

"And now, if I may be allowed to digress, a few words about the fourth term ..."

10. Becoming a Yalie.

11. From left: *Niels Bohr, Werner Heisenberg, and Victor Weisskopf, in 1937. (The woman with them is unidentified.)*

12. Some of the luminaries at the Shelter Island conference of 1947. Standing, left to right: *Willis Lamb, John Wheeler;* seated, left to right: *Abraham Pais, Richard Feynman, Herman Feshbach, Julian Schwinger.*

*13. Don't fall in love
with fancy mathematics,
Weisskopf admonished
his students.*

*14. Enrico Fermi.
If he didn't have
the equipment he
needed, he would
turn it on a lathe.*

15. Geoffrey Chew. According to his bootstrap theory, all particles were interconnected like the nodes of a fishing net.

16. Freeman Dyson. A revelation on a cross-country bus trip helped save quantum field theory.

17. T. D. Lee (left), and Frank Yang. Finding a broken symmetry brought them a Nobel Prize.

18. Julian Schwinger. His brilliant theories were presented in a language all his own.

19. Marvin Goldberger in Trieste, 1969. Murray thought "Murph" was absolutely amazing.

20. Valentine Telegdi. "At any given time," his wife once said, "Val has to have somebody he hates."

21. Abraham Pais (right) with Robert Oppenheimer.

22. Who was smarter, Murray Gell-Mann or Richard Feynman?

23, 24. *Feynman loved to get under Murray's skin. Gell-Mann thought Dick was too preoccupied with generating anecdotes.*

25. *With Yuval Ne'eman (left). The same idea, the Eightfold Way, crystallized simultaneously in two minds.*

26. *Arguing with Lev Landau. Was quantum field theory really dead?*

27. *The omega-minus. Vindication of the Eightfold Way lay hidden in the lines and squiggles of a bubble-chamber photograph.*

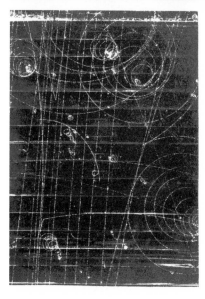

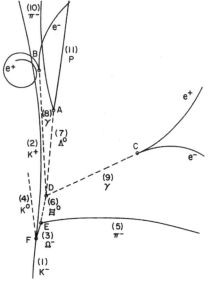

28. With his first wife, Margaret, and their son, Nicky. Finally, Gell-Mann had won "the Swedish Prize."

29. On a family picnic near Aspen: Murray, Lisa, Nicky, Tengi, and Margaret.

30. With Margaret in the San Gabriel Mountains, months before she died.

31. The poet and the physicist. Murray with his second wife, Marcia Southwick.

32. Reminiscing with his brother, Ben, in Illinois.

33. "How does complexity arise from simplicity?" Teaching at the University of New Mexico.

New York. All they had to do was rejigger the apparatus. The lunch was on Friday. By the end of the weekend, they had shown that parity in pion decay was violated to an outrageous extent.

The next day, January 9, around 2 a.m., Wu and the Washington team finished their cross-checks and confirmed their own results. They opened a bottle of Lafite-Rothschild for a toast. When some of the other researchers at the Bureau of Standards came into the lab the next day and saw the cups and bottle in the trash can, they knew something important had happened.

Looking back, physicists realized they had seen parity violation before but dismissed it as an anomaly or explained it away. What they had thought was noise—a dirt effect—was the true signal. It was soon discovered that because of parity violation, neutrinos, unlike wood screws and lightbulbs, cannot be made arbitrarily with either left-hand or right-hand threads. A photon, the carrier of electromagnetism, can spin either clockwise or counterclockwise. Moving through space, it can be thought of as tracing out a right-hand or left-hand spiral. But a neutrino, intimately related to the weak force, will always trace a left-hand spiral. Nature, at this level, is lopsided.

As Feynman had put it, the magnet had indeed grown hairs. Now we could tell the Martian which was north and south or left and right, which side our hearts were on, how we turned our screws. Just ask him to look at a neutrino. It is always left-handed.

A clock operating according to classical mechanics would still respect reflection symmetry. But imagine a clock that was driven by beta decay: Whenever an exuded electron hits a detector, the second hand ticks. Place the apparatus next to a polarized cobalt source, on the side where the electrons are streaming out, and it will keep right on ticking. But put the clock on the other side and it won't run.

Telegdi, returning to Chicago from Europe in mid-January, heard the rumors about the Wu and Lederman experiments. He rushed off a paper to *Physical Review,* hoping that it would run with the other reports. But it arrived a couple of days too late. Anyway, the editors said, it was unclear, confusing, and inconclusive—not nearly as strong as the Lederman experiment. The diplomatic Gregor Wentzel tried to intervene. Telegdi also called Eugene Wigner. "What do you want me to do?" Wigner said in Hungarian. "I am only the president of the American Physical Society" (the publisher of the journal). *Physical Review* ultimately agreed to run the paper in the next issue, with a footnote saying it had been delayed for "technical reasons." The way it was worded, it was unclear whether the technical reasons were supposed to be Telegdi's or those of the journal's production staff. Gell-Mann sympathized with his friend's complaints. The *Physical Review* could be a pain. Telegdi was so mad he quit the American Physical Society.

The downfall of parity became a media sensation. (Sometimes the news was garbled. The physicist Anthony Zee, then just a boy, remembers his father telling him that he had heard from a friend that two Chinese-American physicists had overthrown Einstein's theory of relativity.) With history looking over their shoulders, everyone, it seems, had something profound to say. Anticipating the news, Pauli wrote his famous line to Weisskopf: "I do not believe that the Lord is a weak left-hander." In answer to a cable from Yang reporting the news, Oppenheimer replied with just three words: "Walked through door." The debonair Julian Schwinger, who hated the idea of parity violation, conceded: "Gentlemen, we must bow to nature." And Gell-Mann annoyed his colleagues by insisting that he had never thought there was anything special about parity conservation anyway.

It soon became clear from analyzing the experiments that the weak force also violated another symmetry called charge conjugation. The notion here is that one can take something and replace each particle by its antimatter equivalent. The thing and the anti-thing will behave the same. A classical clock made from antimatter will work just fine. But the cobalt clock, driven by weak decay, would be another story. Since the positrons would fly out in the opposite direction from the electrons, an antimatter clock wouldn't run. However, take the antimatter clock and reflect it in a mirror, and the combined symmetry violations will cancel each other out. Though charge conjugation (C) and parity conservation (P) are individually violated, it seemed that CP conservation would be obeyed.

While all this was going on, Murray had other things to think about. On October 21, 1956, Elizabeth Sarah Gell-Mann, who quickly came to be called Lisa, was born three weeks premature, weighing barely six pounds but healthy and curious as could be. Margaret thought she looked like Murray, with her hair converging at the top of her forehead in a widow's peak. He would rock her in the chair provided by the hospital, singing lullabies in different languages—they had been collecting them for months. Grandfather Arthur sent the baby a magic square to play with, and Margaret promised to keep it until her daughter was old enough to appreciate the arithmetical subtleties. Murray and Margaret didn't have an address for brother Ben, who had recently moved to Carbondale, Illinois, to work as a newspaper photographer and reporter, but they sent a birth announcement for Arthur to forward.

Though Gell-Mann had been on the sidelines during the downfall of parity, he was now ready to join the effort to develop a full-scale theory of the weak force. He had hinted in his talk in Rochester that he was thinking about how to unify the various weak interactions and show that they are identical, just as the force that causes apples to fall to earth is one with that which holds the planets in orbit around the sun. But like other theorists, he had been assuming, for lack of contrary evidence, that parity probably was conserved. Now that this mistaken belief had been discarded, he and the rest of the pack could get on with the job.

In quantum theory, every particle can be represented by a differently shaped wave. When a particle decays by the weak force—a neutron disintegrating into a proton, or a pion into a muon—one kind of wave is converted into another. These transformations can be catalogued as five different types—scalar, vector, axial vector, pseudoscalar, and tensor, or S, V, A, P, and T. Some of these terms are familiar from high school physics. A scalar is a simple quantity, like temperature or the number of fingers on a hand. A vector quantity, like velocity, has not only a magnitude but a direction: An object is moving five miles per hour on a north-northwesterly trajectory. Scalars and vectors act differently under reflection. Hold up your hand to a mirror and it still has five fingers. But move your hand into the mirror and your double will seem to be pushing its hand outward; the vector is reversed.

An axial vector is an arrow like the one describing whether a rotating object is spinning "down" (clockwise) or "up" (counterclockwise). It is not reversed in a mirror. (To see why this is true, remember the right-hand rule. Hold your right hand with the fingers curling counterclockwise. The thumb points up. Now look in the mirror. Though the fingers are curling clockwise, the thumb—the axial vector—still points up.) Pseudoscalar refers to directionless, seemingly scalar quantities that are changed by reflection. Stand in front of a mirror and hold up your left hand. There are still five fingers, but the sign of the quantity has changed from left hand to right hand. Tensor refers to a more abstract quantity that resists metaphor. Details aside, the important point is that all of these terms can be used to describe different kinds of interactions—the various ways one kind of particle can be changed into another.

Finally, each kind of transformation can also be associated with a different force-carrying boson. Electromagnetism, for various reasons, is a vector interaction and its carrier, the photon, is called a vector particle. Or, as Murray preferred to say, electromagnetism involved a "vector current," a flow of photons. Other types of particle interactions could also be thought of as involving currents. During beta decay, for example, a neutron turned into a proton, acquiring positive charge and reversing its isospin. This flow of quantum numbers from one particle to another was a current. In the interaction, an electron and antineutrino also were created. That was another current.

The trick to unification was to find the right pair of currents, S, V, A, P, and T, that would embrace not only beta decay, but also pion decay, muon decay, and any other weak interactions. To further complicate matters, there were two different kinds of beta decay. (One preserved the total spin of the protons and neutrons in a decaying nucleus; the other changed the spin.) The first seemed to involve either a V interaction (like the one in electromagnetism) or else an S interaction; the second either an A or a T. So a theory encompassing both kinds of beta decay might be, for example, VA—some combination of vector and axial vector. Or VT, SA, or ST. If the other weak interactions were indeed all manifestations of the same force, then they too would have to fit into one of these four patterns.

And there was the rub. Recent experiments seemed to say that beta decay involved S and T. But the Lederman and Telegdi experiments suggested that muon decay was some kind of VA interaction. If beta decay involved S and T, and muon decay involved V and A, then they were as immiscible as oil and water. But the experiments were extremely hard to interpret. Wu was pretty sure she had demonstrated that beta decay was V and T. But according to the rules of the game, V and T were incompatible; if she was right, then not even the two kinds of beta decay were caused by the same force. At the Rochester 7 conference, held in 1957 just after the parity bombshell, T. D. Lee lamented that maybe there was no Universal Fermi Interaction after all.

Sitting in the audience was a very frustrated young graduate student named E. C. George Sudarshan who had just come from India to the University of Rochester to work with Marshak. After poring over the data, Sudarshan and his adviser had become convinced that the only framework into which all the weak interactions would fit was V and A—more specifically a combination described as V minus A. Indeed they had convinced themselves that either their theory was right and the experiments were wrong, or there could be no Universal Fermi Interaction, just all these feeble forces with coincidentally similar strengths. Sudarshan was dying to present the results at the Rochester conference, but Marshak wouldn't let him. It would be inappropriate for a graduate student to address the conference, Marshak said, and he himself intended to deliver a talk on another subject. They discussed the possibility of having a visiting professor at the university unveil the theory, but for some reason that didn't materialize. Considering that Marshak was the chairman of the conference, he surely could have found some way to get the idea onto the schedule. But he was nervous that V-A contradicted the experiments. It is hard to escape the conclusion that he just didn't think the theory was very important.

That summer the RAND Corporation invited Marshak to Santa Monica, all expenses paid, to do some consulting. He brought along Sudarshan and another graduate student as assistants. At a desk at UCLA, Sudarshan continued to carve away at the theory, describing it in a paper that Marshak had finally agreed to present for them that September in Italy. After a systematic examination of all the experiments, Sudarshan was able to show that their results were incompatible. At least some of them had to be wrong. He and Marshak were on the verge of being the first to unify the weak interactions. Then, through a quirk of fate, they crossed paths with Murray Gell-Mann.

As early as the Pisa conference in 1955, Murray had been toying with the possibility that the decay of strange particles involved an axial current and that maybe beta decay did too. But the French physicist Louis Michel told him that the experiments all pointed toward S and T. Murray felt stupid and dropped the idea. The fire was rekindled when he ran into Marshak at RAND. Marshak had heard that Felix Boehm, one of Gell-Mann's Caltech colleagues, was doing experiments on weak interactions. Did his work support the possibility of V-A? Gell-Mann agreed to set up a lunch date to exchange information.

Murray was fascinated by Sudarshan's name, which means "good to look upon" in Sanskrit. A Syrian Christian, he had been born with the name E. C. George, but adopted "Sudarshan" after marrying a Hindu. Murray was delighted to learn from him that some of the experiments contradicting V-A might well be wrong. For his part, Sudarshan was pleased that Murray seemed so cordial and appreciative of his work. It was hard not to be a little in awe of Gell-Mann, who, just a couple of years older than Sudarshan, had recently been promoted to full professor at Caltech. Sudarshan was even happier to hear that Boehm's experiments could indeed be taken to support V-A. He and Marshak anxiously asked Gell-Mann if he was planning to write a paper. Probably not, he said. He and Art Rosenfeld, his old colleague from Chicago, were in the midst of producing a long, laborious review of weak interaction physics. Murray thought he would content himself with a paragraph suggesting that V-A was the last stand for the Universal Fermi Interaction—the UFI, or "oofi," as they liked to call it. Either all the weak forces fit into this mold or they were just a bunch of different interactions. Of course, he would give proper credit to Sudarshan and Marshak.

Afterward Murray and Margaret headed off for a vacation in the Siskiyou Mountains in the far northern reaches of California. Their daughter Lisa stayed home with a baby-sitter. It was a disappointing trip. They had paid a guide to take them by horseback into the wilderness and drop them off in a remote locale. But when they got to their destination, there were other campers. It was too crowded— not what they had wanted at all. But if the solitude wasn't total, at least the scenery was nice. They went on walks through the woods until the guide came to retrieve them. When they returned to Caltech, Murray found that Feynman, just back from his own vacation in Brazil, was running around all hyped up about V-A.

Feynman too had been thinking about weak unification for months. Like Murray, he felt at an impasse because of the confusing results of the experiments. That summer he had passed through New York City and asked Lee which ones he should believe. Lee suggested that the best way to decide was to flip a coin. Home again in Pasadena, he heard from Boehm that everything was up in the air again, that Murray had been thinking that V and A could work after all. A light went on in Feynman's head. If V-A *was* correct, then he could see how to fit all the weak forces into the same pattern. He was so excited that he convinced himself that he was the first person on earth to perceive this new law of nature. He had always felt that his most famous work, on renormalizing QED, was merely a matter of repackaging other people's discoveries, using his ingenious Feynman diagrams. He was obsessed with the desire to discover something really new.

Gell-Mann couldn't believe the bombast. Feynman seemed to think Murray had just come up with some dumb idea about V-A without realizing the obvious: This was the key to unifying the weak forces. What could Dick have been thinking? Even worse, Feynman was planning to write up the theory and submit it to *Physical Review*. Well then, Murray thought, I'm going to write it up too. He felt bad since he had told Marshak and Sudarshan that he wasn't going to compete with them and write a full-scale paper. But it wasn't exactly what he considered a solemn oath. If Dick was going to rush into print, he had no choice. It was simple self-preservation. Murray had irritated enough people by saying he had thought of this idea or that idea but was too busy to publish. This time he would make himself follow through.

Feynman and Gell-Mann were sitting in their offices playing dueling physics when the department chairman, Robert Bacher, intervened. He wasn't about to allow his two top theorists to submit papers on the same thing. It would look bad for the school. So Dick and Murray were forced to collaborate. Feynman was almost finished with his paper. Gell-Mann had barely started. And so the final product bore Feynman's stamp. It used a kind of formalism that Murray considered garbage. Feynman agreed to take full responsibility for it: "One of the authors has always had a predilection for this equation." But Murray managed to get in some of his own ideas as well. And he made sure the paper ended with an acknowledgment: "One of us (M.G.M.) would like to thank R. E. Marshak and E. C. G. Sudarshan for valuable discussions." Sudarshan and Marshak's paper had not yet been published, so there was nothing specific to cite. But he didn't want to be accused of doing what he felt others had done to him: neglecting to acknowledge informal contributions. And he also felt a little guilty.

As soon as the preprint went out, the accolades came rolling in. Goldberger found the theory "exceedingly nice. If it is not correct then nature really is untrustworthy." Feynman sent a copy to Lee. Scribbled in the margin was a single sentence: "I have flipped my coin, and this is the answer." Lee would later ruefully tell people that he should have borrowed Feynman's coin. But after the Nobel Prize—one of the fastest ever given—was awarded that fall to Lee and Yang for their paper on parity violation, Lee had little reason to feel regretful.

Around that time, Sudarshan, by then a fellow at Harvard, was sitting in on a lecture by Yang when he was startled to hear him mention V-A, attributing it to Richard Feynman. Then another young Harvard physicist, Sheldon Glashow, told Sudarshan that Feynman and Gell-Mann had submitted a paper to *Physical Review*. He called Marshak, who told him not to worry; their paper had been presented at the Italian conference and would appear in the proceedings. They had established their priority. But the mill at the *Physical Review* ran faster, and the Feynman and Gell-Mann paper was the first in print. When talking about the theory, Murray went out of his way to mention Sudarshan as the first person to suggest V-A (also giving credit to a Cornell graduate student, J. J. Sakurai, who, it later became known, had come up with the theory independently). And he was always ready to write glowing recommendations for Sudarshan, praising him as a theorist full of "originality, imagination, and physical insight."

But for years everyone cited Gell-Mann and Feynman (whose paper many felt was more deeply insightful) and ignored Sudarshan and Marshak. As late as 1959, Marshak had to publicly remind Feynman at an American Physical Society conference that V-A was not the Feynman–Gell-Mann theory. And he was hurt that it was not until 1970 that Murray approached him to apologize for the widespread misunderstanding. Sudarshan never got over his feelings of bitterness and disappointment.

Telegdi was skeptical at first. If Feynman and Gell-Mann were right, one of his own experiments was flawed. He dashed off a letter to them. "This F-G theory of beta-decay is no F-G," he wrote. "If I were you, I would amend it before it gets printed." Telegdi loved to get on people's nerves. He had written to Murray a year before asking if there were any janitorships available for him at Caltech. Murray liked Telegdi. He didn't consider him an ace experimenter, but he was bursting with ideas, like his fellow Hungarian Leo Szilard. Val had an intuitive feel and an unorthodox approach that gave him an unusual grasp of complex problems, Murray thought. He could be warm, friendly, even generous. But he seemed to have a need for enemies, creating them when they weren't immediately forthcoming. You either liked him or hated him.

When Telegdi checked his results, he realized that he had made a mistake. Two days later he sent an apology to Gell-Mann and Feynman. The experimenters liked to think that the theorists were just there to explain what the laboratories had found. But sometimes an experiment could be so ambiguous and open to interpretation that the experimenters needed a theory to focus their attention.

With strange particles and now the weak theory coming under control, the theorists, particularly Feynman and Gell-Mann, were triumphant. Under the headline "Genius at Cal Tech," *Newsweek* had recently declared that the "dapper, cocky" Gell-Mann was many physicists' choice for "brightest light in his esoteric field." With a "Madison Avenue fastidiousness about his clothes" and his "boyish face and quick tongue," the writer observed, Gell-Mann stood out at physics conferences. "He has, as well, a kind of sharp intellectual fussiness that has more than once wounded his colleagues." One of them summed up Murray like this: "He's very patient with ignorance, totally impatient with stupidity." Around this time he was quoted in the *New York Times* when he testified before a joint congressional subcommittee on basic research at the Atomic Energy Commission. Speaking about antimatter, he suggested that perhaps a bit of the stuff could be brought back from outer space. "Of course," he said with a straight face, "it probably would have to be contained in an antibottle." When a congressman asked what would happen if the two kinds of matter were mixed, Murray dramatically replied: "Catastrophe. A complete release of energy."

Job offers kept rolling in from places as varied as the University of Pennsylvania, which offered him its Donner professorship, and the University of New Mexico, which wanted him as chairman of its modest physics department. Gregory Breit sounded him out about a professorship at Yale. Goldberger, after relocating to Princeton University, tried to persuade Murray to join him either there or next door at the Institute for Advanced Study. Fan mail started arriving in his box at Caltech. A woman in Oregon, after reading about this young genius who had shaken the physics world, wrote for advice on what to do with her son, whom she considered a prodigy. Gell-Mann composed a long, thoughtful reply: "I would guess that it is pretty much hopeless to expect real social adjustment. . . . But child prodigies often seem to straighten themselves out when they reach the age of 21 or so, even if they have been queer and unhappy for many years." He strongly advised her not to let her son skip grades lest he "suffer from constant association with children who are bigger and stronger than he is and resentful of his brilliance."

A minister in Pennsylvania, inspired by Murray's congressional testimony, wrote to ask his opinion of a passage in the New Testament book of Hebrews, chapter 11: "Through faith we understand that the worlds were framed by the word of God, so that things which are seen were not made of things which do appear." Was this a veiled reference to antimatter? With Margaret's help, Murray looked up the passage in the original Greek and decided that it should read "the eons were framed" not "the worlds," thereby undermining the reverend's hopeful interpretation.

Inevitably, he started getting a steady flow of dense correspondence from self-educated freelance physicists asking him to evaluate their theories of everything. One wrote to complain that Gell-Mann had declined to meet with him about his new theory unifying "all quantum, electromagnetic, and classical physics." He bitterly denounced a recent Gell-Mann paper as "a feeble and puerile piece of nothing." Such were the annoyances of fame.

Time magazine soon followed *Newsweek* with its own accolade. With the country agonizing over two Russian sputniks circling overhead, one with a dog named Laika trapped inside, the magazine decided it was a propitious time to put Edward Teller on the cover, standing in front of an equation-filled blackboard. The headline: "U.S. Science: Where It Stands Today." Were the Russians irretrievably ahead? Gell-Mann still hated Teller for undermining Oppenheimer and for his hawkish views on the cold war. Murray had been to Russia; he knew the people there weren't evil. But it was nice to be mentioned in the magazine's list of "nine leading lights" in American science. The roster included some older physicists: Rabi, Ernest Lawrence, Oppenheimer. In the middle generation were Glenn Seaborg, Luis Alvarez, and the mathematician Claude Shannon. Toward the younger end of the spectrum were two thirty-nine-year-olds, Julian Schwinger ("heir apparent to the mantle of Einstein") and Feynman, depicted as a bongo-playing beatnik of uncanny mathematical powers who liked to pick locks in his spare time. Youngest of all was twenty-eight-year-old Murray Gell-Mann, shown in a photograph with his head cocked to one side, eyebrows raised quizzically, looking down, birdlike, through black-framed glasses. *Time* described the Caltech physicists at work: "At the blackboard the two explode with ideas like sparks flying from a grindstone, alternately slap their foreheads at each other's simplifications, quibble over the niceties of wall-length equations, charge their creative batteries by flipping paper clips at distant targets." As never before, scientists were becoming celebrities.

In the *Time* article, Murray appeared to project an air of eloquent stoicism. "You might say there is a sort of truce between Nature and our understanding of her," he was quoted as saying. "But Nature is not obligated; she has made us no promises." He quickly fired off a bemused letter to the editor: He hadn't made the remark, nor did he have any idea what it meant. Feynman used some of his five and a half column-inches of fame to explain why he had recently decided to become a teetotaler: "I got potted in a Buffalo bar one night and wound up with a lulu of a black eye."

It seemed that these two new stars, with their clashing styles, were now joining forces to take on nature's mysteries. It wasn't apparent to the outside world how strained and cobbled together their collaboration had really been.

CHAPTER 8

FIELD OF DREAMS

After a theory is done, with most of the loose ends tied up into a neat bow, it seems in retrospect like a self-evident truth. All the false starts and years of muddling are soon forgotten, and the past appears as a steady march toward what was obviously true all along. How could anyone have thought differently? With each passing year it becomes all the more difficult to appreciate the feeling of confusion, the sense of staring into a void, that gripped scientists in the dark ages before understanding dawned. So it would eventually be with the Standard Model, the reigning theory of particle physics (perfected in the late 1970s) in which three very different phenomena—electromagnetism, and the weak and strong nuclear forces—were joined into a single frame.

But in the late 1950s, physicists still had only one piece: quantum electrodynamics, the field theory that explained how electromagnetism worked. Spooky action-at-a-distance had been eliminated by showing that the massless photons were the carriers of the electromagnetic fields that reach endlessly across space. Work by Feynman, Dyson, Schwinger, and Tomonaga had eliminated the infinities that originally popped up in the equations. The theory had been "renormalized," in the cumbersome jargon, and was appropriately called QED. Like the seal of approval *quod erat demonstrandum* ("that which was to be shown"), put at the end of a mathematical proof to announce that it is done, the name "quantum electrodynamics" bore the hallmark of perfection. For Murray Gell-Mann and his colleagues, the greatest hope was that the other forces could be described in this elegant manner—as quantum fields carried by particles something like photons. But this vision was still far from being realized.

First of all, the strong force remained maddeningly resistant.

Physicists had the roughest sense of what its quantum field theory might look like, with pions as the carriers. But the problems with the infinities just wouldn't go away. When physicists tried to describe a strong interaction using one of those powerful equations called a perturbation expansion—the syntax that worked so beautifully with electromagnetism—each successive correction grew larger and larger until the calculation was overwhelmed by error. Throughout the 1950s and into the 1960s, Gell-Mann continued to play an important role in developing the technique he and Goldberger called dispersion theory—an alternate tool for describing particle interactions. Dispersion theory allowed physicists to make some predictions about what happens when the strong force does its work. And it helped tease out hints that some hoped would eventually be used for constructing a full-blown theory. But no one had succeeded in putting it all together. A few physicists were beginning to doubt that the task was even possible. Perhaps quantum field theory only worked for electromagnetism. Perhaps a whole new kind of theory was needed for the other forces.

The situation with the weak force was not much better. Feynman and Gell-Mann (along with Sudarshan and Marshak) had taken an important step, pulling the V-A theory from the alphabet soup of conflicting possibilities. Different kinds of weak interactions now fit into the same template, in which a so-called vector current (like the one for electromagnetism) was intertwined with an axial current. The theory still didn't quite accommodate the weak decays of strange particles. But Murray was working on that.

For all its beauty, though, V-A was far from being a fully developed field theory. It too suffered from the kind of infinities that plagued the equations for the strong force. In some ways, the situation was even worse. With the strong force, physicists at least thought they knew that it was transmitted by the pions (a case of ignorance-as-bliss—the mechanism would turn out to be more subtle). With the weak force, physicists didn't have a clue as to what kind of particles acted as carriers, what the bosons were.

Of course it was possible that all three theories would turn out to be wildly different in style—that physics (like biology) was a hodge-podge of Rube Goldberg devices, that the universe was not crafted by a great mathematician but thrown together by a backyard mechanic with spare parts from his garage. But hardly anyone believed that could be true. The arbitrary laws that governed how people should behave might conflict and even contradict one another. What was legal under one state's law might be illegal under another's. But the laws of the universe needed to work in close cooperation. When a neutron, to take a simple case, decayed into a proton through the weak force, the particle's charge was changed. So electromagnetism was also involved. Furthermore, the neutron and its decay product, the proton, were related, through the strong interaction, as two opposing faces of the nucleon doublet, with their opposite isospins. Weak force, strong force, and electromagnetism were clearly woven together. The question was exactly how.

In trying to bring these forces under the same constitution, physicists were not completely in the dark. There was a growing suspicion that if there was a common framework, it would be not only quantum field theory but quantum field theory of a special kind—one involving a powerful idea called gauge symmetry. Emmy Noether had shown that every conservation law is intimately related to a symmetry—that invariance under translation in space (the fact that a clock works as well here as there) implies the conservation of momentum; that invariance under translation in time (that the clock works as well now as then) implies conservation of energy. Other conservation laws arise from more abstract symmetries. The conservation of electric charge is related to something called a local gauge symmetry.

To get a feel for what that means, remember the notion of wave-particle duality: Every particle can be represented by a quantum wave. Like all waves, whether rippling through water or mathematical space, the quantum wave representing an electron has a certain phase. Two waves perfectly superimposed, one on top of the other, are said to be in phase. If one wave lags behind, the two are said to be out of phase. But which wave is ahead, and which behind? Which should be the gold standard, the wave whose phase is set to zero? The choice is as arbitrary as the one by which zero longitude was assigned to the imaginary meridian running through Greenwich Observatory in England. We could pick any one electron in the universe and use it as the reference point, the Greenwich mean time, then express the phase of every other electron in its terms. Or we could pick another electron and use it as the reference point. Each time we changed the standard, the numbers assigned to every electron everywhere would change, as though each had been simultaneously shifted, in lockstep, a little to the left, a little to the right.

It is obvious that throughout all these transitions, the universe would remain the same. Changing the gauge is no more disruptive than going on daylight saving time. The number expressing the phase of a wave is not an inherent, intrinsic quality. Its value is arbitrary, depending on the gauge you choose. But what *is* an objective, enduring feature of the universe is the *relationship* between the waves. Whatever gauge is chosen, any two waves are either in phase or separated by the same number of degrees. If everybody moves his clock by the same amount, it is still three hours earlier in Pasadena than in New York. Again, it is the relationship and not the actual number that is fundamental.

A symmetry, recalling the mathematician Hermann Weyl's definition, is something that is conserved while something else changes (like the silhouette of a sphere, which remains constant as the ball is rotated). Shift all the phases of the electrons simultaneously by the same amount and in the same direction, and the relationship between the waves is preserved. This is called a global gauge symmetry. And it is a good thing that it holds true, for it turns out that the charge of the electron is intimately tied to the phase of its wave. If global gauge symmetry did not hold, then the charge of the electron would absurdly change depending on how you chose to measure phase. Either God would have to decree which was the true phase-zero electron, the standard to base the measurements on, or physics would be a mess.

That electric charge, like the time of day, is invariant under a global gauge transformation is not surprising. But what Weyl and the physicist Fritz London proposed in the 1920s and 1930s was that charge must also be invariant under a much stranger phenomenon called a *local* gauge transformation. Suppose that we do not just change the standard used to measure all waves, but that we use different standards in different parts of the universe, applying them however we choose. After all, who is to say that all scientists, whether at MIT, Caltech, or the University of Andromeda, should have to agree to use the same measuring stick? If the number expressing the phase of the wave has no intrinsic value, but only a value in relationship to all other waves, then we should be able to use any gauge we wish anywhere we want and to change our minds at anytime.

But getting a local gauge symmetry to hold is not as trivial a matter as it is with global gauge symmetries. Imposing a global gauge symmetry is equivalent to shifting all the electron waves simultaneously. Imposing a local gauge symmetry is equivalent to randomly shifting the phases of electrons individually, not in lockstep but willy-nilly—this one to the left, this one to the right. Since charge is related to phase, the result would be that different electrons would have different charges. The law of conservation of charge would be violated. And that just won't do. The very structure of physics is based on the assumption that the charge of the electron is always, everywhere, −1. How can we preserve local gauge symmetry—the freedom to use any convention in defining the phase—and keep conservation of charge?

London showed that one can do so with an ingenious mathematical trick, a special kind of field whose job is to smooth out the charge differences caused by haphazardly shifting the phases. The field essentially communicates with every electron in the universe, keeping track of the various phase shifts and counteracting them, ensuring that charge remains constant and universal. Since the field must extend infinitely in all directions, it must be carried by massless bosons traveling at the speed of light. Why not assign the job to the photons? In fact, this gauge field turns out to be none other than the electromagnetic field. They are one and the same. Defined this way, electromagnetism has an important new purpose. It arises to *enforce* the conservation of charge, no matter what conventions we use. It is as if all the people on earth were free to set their clocks anyway they wished. But an invisible field would arise, twisting the hands on the dials, ensuring that it is always three hours later in New York than in Pasadena.

Once you get the hang of it, this weird, utterly counterintuitive idea—the local gauge symmetry of electromagnetism—starts to make a lot of sense. When it comes to charge, the labels "positive" and "negative" are also arbitrary. We could redefine electrons as having positive charge and positrons as having negative charge, and physics wouldn't change. Positivity and negativity are just labels, not intrinsic qualities. What matters again is the relationship. If someone at Caltech wanted to define the electron as negative and someone in Moscow wanted to define the electron as positive, it wouldn't affect the way the world works. Local gauge symmetry, enforced by the electromagnetic field, will ensure that whatever the choice of convention, the relationship—positive is the opposite of negative—is maintained throughout the universe.

A commonsense analogy to this abstract concept is the notion of the "invisible hand" in economics. In the marketplace, the prices of goods fluctuate at random depending on how storekeepers decide to price their wares. But the prices never stray very far. You don't find a head of lettuce selling for one hundred dollars at one store and fifty-nine cents next door. The same items tend to go for approximately the same price. The invisible hand of supply and demand levels out the differences. The invisible hand of gauge symmetry reaches down and guarantees that the charge of the electron is constant.

Before London and Weyl, no one could say why there are forces in the universe. They were just there, and the best physicists could do was to describe them. Now it seemed that the forces—electromagnetism, at least—had a purpose: to enforce symmetries. In fact one could go further and propose that symmetries, not forces, are the deepest of all phenomena.

During the summer of 1954 at Brookhaven, Frank Yang and his officemate, a physicist named Robert Mills, tried to see whether they could expand the idea of a local gauge field to explain not just electromagnetism but the strong nuclear force as well. What is arbitrary in this case are the labels "proton" and "neutron." The strong force can't tell the difference between the two particles; it sees them both as nucleons and grips them just the same. Nothing stops us from relabeling every neutron in the universe as a proton, and vice versa. Nuclei won't go flying apart because of the change in convention. Whatever we call them, the relationship between the proton and neutron will endure. In other words, *global* gauge symmetry holds true. But suppose some demon goes through the universe arbitrarily twisting one nucleon this way in isospin space, another nucleon that way, randomly changing the labels. What Yang and Mills tried to show was that a local gauge field would arise to counteract the changes, to ensure that with any choice of labeling, the strong nuclear reactions would be unaltered. If so, then the strong force, like electromagnetism, could be mathematically expressed in this elegant style. But there was a big problem. Since gauge fields have to reach everywhere, propagating their effects at the speed of light, they must be carried by massless bosons. And the carriers of the strong force were thought to be pions, which were quite massive. In fact, it seemed that the pions had to be massive to explain why the strong force was so short in range. The two requirements clashed.

Yang and Mills put the theory aside, but their basic idea survived. The notion of local gauge symmetry was just too elegant to abandon. Fitting all the forces into this form—making them into what were soon being called Yang-Mills theories—would not just be aesthetically pleasing. The success in renormalizing QED had been related to the fact that it was gauge-invariant. Coming up with gauge theories of the other forces might help ensure that they too could be tamed.

The loftier goal behind all this was the grand tradition of unification. Newton had shown that the forces pulling an apple to the ground and holding the moon in orbit around the earth were the same. Both were examples of gravity. James Clerk Maxwell had shown that electricity and magnetism were different manifestations of a unified force called electromagnetism. With V-A, physicists had come close to unifying the various weak forces—all were versions of the "oofi," the Universal Fermi Interaction. Few physicists dared hope that all the forces affecting subatomic particles would mesh together into a single, seamless whole, that they would be melded as neatly as electricity and magnetism. But at least, they hoped, the three forces would have the same mathematical style. (Gravity was too weak to matter on the tiny scale inside atoms and could be politely ignored. Anyway, the shape of the reigning theory of gravity, general relativity, was so different from the quantum theories of the particle forces that the two seemed irreconcilable. When Einstein tried to mix them together, he ended up with stains on his sweatshirt.)

Finally, as the 1950s drew to a close, there was a deep feeling that an understanding of how the forces were related would point the way toward categorizing the dozens of different particles that continued to pour from the sky and the accelerators in a random spray. Oppenheimer talked about the subatomic zoo, but physicists had barely begun putting the beasts in labeled cages. Science writers of the time liked to say that particle physics was in need of a Mendeleev, the Russian scientist who had arranged the dozens of different types of atoms into the cells of his periodic table of the elements. Line up the elements in rows, read down the columns, and you would find groups of substances with similar properties.

When it came to the subatomic particles, the Mendeleevian task had just begun. The matter-making particles—the fermions— had been divided into three broad classes. There were the lightweight particles called leptons, like the electron and the neutrino. There were the middleweight mesons. And there were the heavier particles, the baryons, like the neutron and proton. Together the baryons and mesons were called the hadrons. The hallmark of hadrons was that they felt the strong force, and so Gell-Mann and Feynman called them the stronglies. Leptons ignored the strong force.

Some progress had been made in refining this crude taxonomy. The neutron and proton had been unified by Heisenberg into the nucleon. Twirl it one way and you got a neutron; twirl it the other way and you got a proton. While the nucleon was like a two-sided coin, the pion (with its positive, negative, and neutral versions) was like an equilateral triangle, with each of the three corners representing a different particle. With his strangeness scheme, Gell-Mann had arranged the exotic sigmas, kaons, lambdas, and xis into their own doublets and triplets. But why did the particles fall into these particular patterns?

In his *Scientific American* article in 1957, Gell-Mann had declared that the present understanding of the subatomic particles was like that of the elements after Mendeleev had crafted his periodic table. Mendeleev had charted the similarities, arranged the elements into groups. But an understanding of *why* the elements lined up just so hadn't come until the invention of atomic theory. Only then was it established that each successive element had one more proton in its nucleus, hence one more counterbalancing electron in its shell, giving atoms their different chemical characters. Now an ordering principle had to be found for the subatomic particles. It was clear that there were deep patterns, but no one had yet glimpsed the underlying mechanism.

"Are all the particles we have mentioned really elementary, or are some of them just compounds of other particles?" Gell-Mann wrote (with the help of his coauthor, Ted Rosenbaum). "If so, which are elementary and which are not? Why has nature chosen to use this particular set of particles to build the material world? . . . These and many other such puzzles seem to lie entirely beyond the power of our present theories. Shall we ever know the answers? Every physicist has an abiding faith that we shall. But it will probably require some wholly new ideas. . . . It is likely to be quite a while before the particle physicist finds himself out of a job."

In fact, while the article was in press, Gell-Mann was already working on an exciting new possibility—a taxonomy in which the various doublets and triplets of particles, both ordinary and strange, would turn out to be pieces of a larger jigsaw puzzle. Every particle that felt the strong force would be arranged into an all-embracing pattern. "Suppose that the eight baryons are really eight states of the same particle," he proposed at the 1957 Rochester conference. A supermultiplet! If the nucleon was like a coin and the pion like a triangle, then this supermultiplet would be like an octahedron. Flip it different ways in some abstract space and it would become eight different particles. Julian Schwinger, who was in the audience, said he agreed with Murray completely and in fact was working on a similar scheme.

But neither succeeded. A few months later, when Murray's paper "Model of the Strong Couplings" appeared in the *Physical Review,* it was dripping with pessimism. In its existing form, the language of field theory was clearly inadequate for such an ambitious scheme, Murray lamented. He said there were "grave doubts" about whether conventional field theory could be used with the strong force at all. But he figured it was worth attempting: "We shall try to see to what extent the new may be like the old." He admitted that the experimental evidence for his model was weak. But he hoped he could use it to make predictions that could later be verified. "The likelihood of success may not be great," he allowed. The conclusion of the paper was less than rousing. "Supposing that the model we have presented has elements of truth," he proposed, and then tapered off from there. Something was missing, and it would be a few more years before he figured it out.

Gell-Mann loved to turn over these questions in his mind. When a freelance writer from Cincinnati asked for his opinions about science and elementary education, he described the excitement of his craft: "If a child grows up to be a scientist, he finds that he is paid to play all day the most exciting game ever devised by mankind." Nor was it just particles that fascinated him. Gell-Mann and Luis Alvarez, an experimenter who had traveled with him to Russia, collaborated on an effort to explain the iridescence of hummingbird feathers. Was it, Alvarez wondered, the result of "metallic colors" or "interference colors from some sort of a lattice of spheres"?

Even with the success of V-A and strangeness, Murray was still afraid to push ideas that might flop. But for all his lingering insecurities, something was changing in the way he presented himself to the world. A certain amount of prickliness had always lurked behind the easy camaraderie that had made him so popular at school. That was considered part of his charm. But since his rift with Pais and the spat with Yang and Lee, Gell-Mann was becoming harsher and harsher in his judgments of others.

In the spring of 1957, while Murray was working on the baryon symmetry scheme, a newly graduated physicist named Richard Norton arrived at Caltech from the University of Pennsylvania. Norton's thesis on dispersion relations had caught Gell-Mann's eye, so he had invited him to work as a postdoc. His first day on campus, Norton was loitering around the department mailboxes when Gell-Mann walked up and spotted him for the first time. He seemed to stare right through him. "Who are you?" he asked. Gell-Mann was not exactly rude, but Norton immediately felt intimidated. Clearly this was someone who didn't have much time for social amenities. In the coming weeks, Norton found it impossible to converse with Gell-Mann on a one-to-one level. He was always trying to upstage you, correcting your pronunciation or your misperception of some event. What had seemed charming and unassuming coming from a child was annoying and even insulting in an adult. Murray got away with his rudeness, Norton thought, because he worked harder at theoretical physics than just about anyone. Gell-Mann would stride into Norton's office and start asking pointed questions. He would jot down notes, completely focused, to make sure he really understood the work. It was almost flattering. The intensity even extended to the strength of the martinis, as powerful as Oppenheimer's, that Murray served at the parties he and Margaret held at their house. He kept them ice-cold in the freezer. On one occasion, Norton nervously quaffed so many that he passed out on the Gell-Manns' front lawn. A neighbor knocked on the door in the early hours of the morning and told Murray he was afraid there had been a gangland killing and the body had been dumped at his curb.

The one person who could stand up to Gell-Mann was Feynman. At the weekly seminars on theoretical physics, he was always teasing his more serious colleague. Murray would strike back, wielding some fact Feynman didn't know. This wasn't hard to do. Dick, it sometimes seemed, knew everything about physics and almost nothing about anything else. One day Feynman asked idly where the word for "dollar" came from. Of course Murray knew the answer: from the German *thaler*, a sixteenth-century Bohemian coin. Didn't everyone know that? Feynman immediately realized he had made a mistake, giving Murray an opening to display another bit of linguistic trivia. "Murray," he said in mock exasperation, "in a hundred years nobody will know whether your name is hyphenated or not." It was no use trying to get Feynman's goat. The man was unflappable.

Dick could be annoying, Murray thought, but when it came to arguing physics no one was a better sparring partner. The intellectual heat from the fission and fusion of ideas kept them aroused for hours. Lately they had been preoccupied with the weak force. If they were going to succeed in turning V-A into a real quantum field theory, they would have to figure out what kind of particles might be acting as carriers. For a while they toyed with a model in which the weak force was transmitted by two hypothetical bosons, the X-plus and X-minus. Uxyls, they liked to call them. When weak decay caused one particle to change into another—a neutron into a proton, for example—the uxyls would ferry around the tiny scraps of charge. In accordance with the reigning aesthetics, the uxyls would arise, as did the photon in electromagnetism, to enforce some kind of Yang-Mills gauge symmetry.

They were also pondering another mystery about the weak force. According to V-A, the muon (the electron's fat cousin) should decay into an electron, with a neutrino and an antineutrino emerging as by-products. These antimatter twins should then come together and self-annihilate in a flash of gamma-ray photons. Physicists had been looking for the photon signature but couldn't find it. Maybe, Gell-Mann and Feynman speculated, there were two different kinds of neutrinos—they jokingly called them red and blue. One was associated with the electron and one with the muon. When the muon decayed, it would produce a regular "red" neutrino and a "blue" antineutrino. Since these were not direct antimatter opposites, there would be no self-annihilation and no flash of light.

Feynman and Gell-Mann presented some of this work-in-progress at a meeting of the American Physical Society at Stanford around Christmas 1957. Murray was particularly fond of red and blue neutrinos and was hurt when, a few months later, Feynman lost interest in the idea. They had even started to write up the theory, but Feynman refused to sign off on it. When he and Gell-Mann heard another physicist propose a similar idea at a conference in Gatlinburg, Tennessee, Feynman ridiculed it. Murray was baffled by his friend's behavior. But he lacked the self-assurance to send off the paper by himself. When Leon Lederman, Melvin Schwartz, and Jack Steinberger later established the second neutrino's existence in an accelerator experiment, Murray was quick to point out that he had thought of the idea earlier, and he blamed Feynman for not encouraging him to follow through. But without a published paper, he had little to back him up. Lederman had heard his presentation at the Stanford conference, Murray grumbled, but he didn't give him credit.

For the next year, Gell-Mann kept playing with the weak force, trying to find its bosons. It soon became clear that two uxyls were not enough. When a decaying particle changed its charge, positive or negative uxyls were involved. But sometimes during a weak decay, the charge of a particle remained the same. So in addition to positive and negative uxyls, there must be neutral ones. As he thought about the scheme, it grew even more baroque. The uxyls could be thought of as carriers of the hypothetical weak currents, just as the photon was the carrier of the electromagnetic vector current. But there were some kinks in the scheme. Sometimes the neutral weak currents, presumably carried by chargeless uxyls, changed a particle's strangeness, and sometimes they left it intact. So were there two kinds of neutral X-particles? Yang and Lee thought so and called them "schizons." Nobody could get the mechanism to work.

As usual, everyone was competing with Schwinger, who was also working on a Yang-Mills–type theory of the weak force. He too was assuming two different kinds of neutrinos. And he was telling his students at Harvard that he thought the weak force was carried by a triplet of particles: two charged bosons something like Gell-Mann and Feynman's uxyls and a neutral particle which Schwinger thought was the photon, performing double duty. Using the photon as a carrier for both forces raised an intriguing possibility. Maybe the two could be unified as Maxwell had done when he tied together electricity and magnetism. But there was a problem with the boson masses. The two X-particles, it seemed, would have to be very massive to account for the short range of the weak force. But the photon was massless. Somehow a symmetry between the X's and the photon had become broken. How could you get the three to work together in a mathematically pleasing theory? Schwinger wasn't sure. The problem would not be solved for years.

Murray too was thinking that understanding the weak force would ultimately mean unifying it with electromagnetism. After all, electromagnetism was a vector force, like the V in V-A. In a 1958 paper, he tried to draw an analogy between electromagnetism and the weak interactions—"weak magnetism," he called it. But the pieces didn't quite fit together.

As the physicists floundered around, more and more of them began to worry that there was something wrong with their most basic assumption. Maybe quantum field theory really wasn't the right framework after all. For all the success of QED, some physicists still found renormalization just so much hocus-pocus—arranging the infinities so they conveniently canceled out, plugging in experimental values to replace the infinities that remained. And with the weak and strong forces, even this mathematical legerdemain wasn't working. Like Heisenberg in the 1930s, some physicists were calling for the abandonment of the whole notion of invisible forces and virtual particles, suggesting that science should deal only with what could be observed: Two particles had a certain set of quantum numbers going into an interaction, and another set coming out. What happened in between was unknowable and possibly irrelevant. Maybe all you could do was describe the before and the after using the mathematical device called the S-matrix. The flamboyant Russian theorist Lev Landau was espousing more vigorously than ever the radical proposition that field theory itself was rotten to the core. All field theories, he claimed, inevitably included negative probabilities—what he called "ghosts." The S-matrix seemed to Landau like a powerful redeemer.

Gell-Mann wasn't so sure. At the Rochester conference in 1956, he and Goldberger had described the power of dispersion relations, their improved version of the S-matrix. They told how their technique let you do field theory "on the mass shell," avoiding negative probabilities and other chimeras that dwelled in mathematics' never-never land. Murray, paying homage to Heisenberg, had acknowledged that their work was reminiscent of the S-matrix. But could it really be developed into a *replacement* for field theory? Did field theory need replacing? He had tried to argue the issue with Landau on his visit to the Soviet Union. But it was like talking to a brick wall. Murray still liked to believe there was a proper field theory out there—it just hadn't been found yet. Dispersion relations might be a means of zeroing in on it. Then again, maybe Landau was right. Maybe what was needed was not a prosthetic device for field theory but a new kind of physics.

It was a wrenching dilemma. The idea of a Yang-Mills field theory, in which the forces arose from basic, deep-rooted symmetries, seemed so beautiful. If field theory was eventually vindicated, Murray didn't want to be seen as part of Landau's radical fringe. On the other hand, he was one of the main developers of dispersion relations. If field theory was overthrown, he wanted credit. And so he wavered. "We all have our doubts about field theory," he told a group of colleagues. "For most experimental physicists these doubts assume enormous proportions." Then he added, "I *some-what* share their feelings." At other times he ventured closer to the edge. In the fall of 1958, in an application for a National Science Foundation fellowship, he boldly declared that he and his students at Caltech were "trying to show that the methods of dispersion theory can replace quantum field theory entirely." But later in the application, he softened the tone a bit: It will be necessary *either* to modify field theory or to "replace it entirely by a theory with more physical content."

By the following summer, when the Rochester conference was held in Kiev, pessimism about field theory was spreading. On the way to the meeting, Murray met Roger Hildebrand, his friend from Chicago, in Belgium and then stopped off in Moscow. They met for a Chinese dinner—it was just like old times—and talked physics before heading off to Kiev.

The usual spontaneity of the Rochester sessions was muted by the need for simultaneous translation. But nothing could deter the flamboyant Landau from promoting his new vision. The Indian-born physicist Abdus Salam, who had been working on Yang-Mills theories, was startled when Landau, wearing a loud colored shirt, strode up to him and declared: "Aren't you *ashamed* of yourself?" Salam didn't know what he was talking about. "Aren't you a believer in field theory?" Landau said in mock astonishment. For Landau, opposing field theory seemed almost as much a matter of honor as challenging the ruling Soviet regime. Gell-Mann was put off by Landau's bluster. It was harder than ever to have a rational discussion with the man. His ideas seemed to be hardening into dogma.

With the 1960s approaching, everything in physics was up in the air. It seemed likely that Murray, not quite thirty, would be one of those who finally helped figure it all out. In February 1959 a committee that included Eugene Wigner and Victor Weisskopf selected him as the first recipient of the Dannie Heineman Prize for mathematical physics from the American Physical Society and the American Institute of Physics. Since Heineman, the benefactor of the award, was eighty-seven years old, the banquet was quickly planned for May. When, after ten days, Murray had not acknowledged the prize, an exasperated Karl Darrow of the American Physical Society fired off a second letter: "Perhaps you are still paralyzed with joy by the news . . . but it seems more likely that you have been out of town." Gell-Mann quickly and apologetically thanked the society for the honor, which soon became as prestigious in physics as a Fields

Medal is in mathematics. He was photographed accepting it personally from Heineman himself. "The photo is certainly a splendid one of Mr. Heineman," Gell-Mann later wrote, "and I look slightly less like a chimpanzee than usual." He used the money, $2,500, to buy a Jaguar sedan. One of Murray's Yale classmates, Samuel P. Huntington, a fellow of their whimsical political group, the Fathers of the American Revolution, sent his congratulations: "Well done, old boy! FAR scores again. . . . You make all the others of us seem like terrible laggards."

Later that year, Yale University, which had spurned Gell-Mann's application to the graduate physics department more than a decade before, granted him an honorary degree, at age twenty-nine. His old classmate Harold Morowitz, now a professor at Yale, was at the ceremony. Afterward they ran into each other walking across the campus. "Murray, I think they're sorry," Morowitz said.

CHAPTER **9**

THE MAGIC EIGHTBALL

Jeremy Bernstein hadn't intended to be in Paris at the same time as Murray Gell-Mann. It was the 1959–60 academic term and the two young physicists were in France on National Science Foundation fellowships, Gell-Mann primarily at the Collège de France and Bernstein at the Ecole Polytechnique. One day that fall, Bernstein was sitting in an office with his host, the French theorist Louis Michel, discussing a problem involving the weak interaction, when Gell-Mann walked in. Bernstein explained the question that was puzzling them. Murray seemed interested and then left, returning the next day with an answer. *"Messieurs,"* he said in impeccable French, *"le problème est resolu."* Bernstein was impressed. Gell-Mann had found a way to connect the problem with one that he himself had been working on. Then, apparently overnight, he had come up with a general way to solve them both. The year abroad seemed to be getting off to an auspicious start. Finally, Bernstein thought, he was going to get a chance to see how the very best physics was done.

Bernstein had first become aware of Gell-Mann many years earlier while sitting in a math class at Columbia Grammar School. No matter how hard Bernstein or his classmates tried, the teacher, a Mr. Reynolds, was always comparing their work with that of the best student he had ever had: Murray Gell-Mann. Bernstein didn't really get to know Murray back then. Though they were about the same age, Murray was three years ahead in school, and the gap continued to widen. The year Gell-Mann earned his Ph.D. at MIT, Bernstein was just getting his baccalaureate from Harvard. He stayed there to do graduate work in physics, but he had pretty much forgotten about his old classmate until Gell-Mann visited the university in 1954 to give a lecture on his classification scheme for the strange particles. Bernstein was vaguely aware of the work but hadn't followed it closely. He had found it challenging enough to understand the more mundane particles, like protons and pions. When Murray walked into the lecture hall, Bernstein made the connection—the unusual last name, the precocious manner. Later he looked in his old yearbooks and confirmed that this was indeed the boy genius he remembered.

Gell-Mann's lecture had been superb, and Bernstein decided that he had better start paying attention to this new frontier of physics. Several years later, when Gell-Mann's paper on "weak magnetism" suggested possible parallels between electromagnetism and the weak nuclear force, Bernstein took the time to study it in detail. What an exciting idea, he thought—trying to unify these seemingly different phenomena. He and a colleague immediately went to work analyzing the various ways to test Gell-Mann's idea. When their paper was published, Murray liked what he read and sent Bernstein a friendly note. The next time their paths crossed, in Princeton, Bernstein mentioned that he was spending the next academic year in Paris. "That's interesting," Murray said, "so am I." Then he suggested that they might do some work together. Bernstein later remembered the confident way Murray proposed the collaboration: "Stick with me, kid, and I'll put you on Broadway."

After a short visit to London to lecture at Imperial College and meet with Abdus Salam, Gell-Mann prepared to immerse himself in the Parisian lifestyle and, he hoped, get some serious physics done. His plan was to spend much of the year trying to improve the V-A model of the weak force. The theory continued to survive experimental tests involving weak decays of regular particles. But it was still not clear how to expand it to explain the way strange particles disintegrated—why some of the currents in the weak interaction seemed to preserve strangeness, while others changed it. How could both these phenomena be made consistent with the rest of the theory? Part of the problem was that the weak force was not working in isolation. During decay, the strong force also came into play, creating a tangled mess. The challenge, as Gell-Mann put it, was to "cut the Gordian knot" and separate the two effects. In the process, he hoped to learn more about how the strong force itself operated.

At first, he and Margaret and their daughter Lisa were given housing in Plessis-Robinson, a suburb far removed from central Paris. Murray hated it. He wanted to be at the center of things.

Within a few weeks, at his prodding, the school found the family an apartment in the heart of the Left Bank, at 5 Place du President Mithouard, across the street from the church of St. François Xavier, in the chic 7th arrondissement. Just as they were moving in, one of their neighbors, Phillipe de Gaulle, President Charles de Gaulle's brother, died, and the funeral procession began practically at the front door of the Gell-Manns' building—history in the making. The cafés, the historical sites, the French countryside—the perquisites of doing theoretical physics sometimes seemed too good to be true. Murray enjoyed working on his French at least as much as on his physics. And he was just as unforgiving of his occasional lapses, mortified at being wrong. He would come home from a party with Margaret, then spend a sleepless night agonizing over a grammatical error he had made, and agonizing over the time wasted agonizing.

Just about every day, he would meet with Bernstein, Michel, and Maurice Lévy, a theorist of international repute at the Ecole Normale Supérieure, to try to tease apart the tangle of weak and strong forces that seemed to determine how strange particles decayed. The key to understanding the puzzle, they came to believe, was the axial current—the A in V-A. Gell-Mann and Feynman had shown that the vector current was perfectly conserved in the weak decays, just as it was with electromagnetism. But the axial current, it now seemed, was interfered with by the strong force, breaking the symmetry and causing the current to be only partially conserved. V-A could be thought of as a system—an "algebra," Murray called it—of interacting currents. Put this way, the problem was to expand or generalize the algebra to also include the currents involved in the strange decays. And the ultimate goal—still many years off—would be to stretch the algebra so far that it would embrace both the weak and the electromagnetic forces within the same framework, showing that these two phenomena were part of the same thing.

Since many of these ideas sprang from the original V-A paper, Murray asked Feynman to join him and Lévy. But Feynman was unenthusiastic. It wasn't a great time for him. "I am in one of those activity lows," he complained, "where I don't do a damn thing." But he liked the draft Murray sent him and agreed to collaborate from afar, sending back a long list of changes he wanted. When the preprint went out, he was listed as a coauthor. But then he got cold feet, writing to the editor of *Nuovo Cimento*, the journal that planned to publish the paper, that he wanted his name removed in proof. "Listen, I may sound like a fickle woman, but I want out," he pleaded to Murray in a letter. He hadn't worked directly on the theory, he insisted, or had time to check the details. He didn't entirely understand a lot of what Gell-Mann and Lévy were saying. Murray tried to get him to change his mind, but Dick was adamant. He agreed with Gell-Mann that removing his name at the last minute was going to look funny and make people think they had quarreled. "But I find gossip like that passes quickly & is soon forgotten. . . . My only regret is that I have embarrassed you by my dopiness. . . . I won't forget—you are now one up—and you can feel free to do one, or several, absurd & impossible things & I will not be at all perturbed—for I'll remember how I opened your last letter with fear & trembling at what I expected to find (righteous and just anger) but found only friendly patience & lots of physics."

As he tried to follow along on the work, Bernstein sometimes found himself less interested in studying strange particle decays than in observing how Gell-Mann thought about these abstract worlds. He was determined to figure out how his old schoolmate came up with so many creative and ultimately fruitful ideas. For days Gell-Mann would immerse himself in these long streams of thought. Then every so often he would step back and try to record what he had been thinking—a kind of progress report, usually handwritten, that he never bothered to publish. Another thing Bernstein noticed was how much Murray, as irascible as he could be, liked to collaborate. One of Bernstein's colleagues at Harvard had once asked the great Dirac if he had ever been involved in a priority fight. "The really good ideas," Dirac haughtily answered, "are had by only one person." But Murray seemed to thrive on the company of fellow physicists. He liked having someone to bounce his ideas off, to make sure that he wasn't going over the deep end.

Most of the time, what Gell-Mann was talking about seemed so abstract that Bernstein could barely grasp it. Murray played with mathematical objects as effortlessly as a kid with toys. He would say, cryptically, "The currents are commuting like angular momenta," or talk about some new current he had just invented that had "cuter" algebraic properties than the others. Bernstein could more or less follow the logic of the argument, but he was at a loss to understand the overall significance. He got the impression that some great idea was taking shape in the back of Murray's mind but that he couldn't quite bring it into focus. Lately Bernstein had been spending time playing tennis with the great cosmic ray experimenter Louis Leprince-Ringuet. His approach and Gell-Mann's could hardly have been more different. While Leprince-Ringuet, a true Baconian empiricist, believed truths about nature came from pure observation by a mind unprejudiced by theory, Murray seemed to pay little attention to what the experiments were saying. As he and Feynman had learned with V-A, the experiments often turned out to be wrong.

Bernstein enjoyed dining with Murray and Margaret. Sometimes he took walks with Lisa, then three years old. He was impressed with how quick her mind was. In every way, she seemed like a wonderful child. As they strolled along the boulevards, she would identify every car on the street, not just by make but by the year it had been manufactured. It seemed as though another precocious Gell-Mann was in the making. But Bernstein came away from his Paris inter-lude with mixed feelings. Sometimes Murray was so much fun to be with, and other times he was so impossibly difficult. He had a way of making you feel stupid for not knowing something that he found obvious. Bernstein started to worry that Murray must consider him too dense to do good physics. He even took to hiding from him—hanging out in cafés, studying French, and trying to figure out where all this mathematical play was leading. After a few days, he would reemerge. Murray, a little exasperated, would say, "Where have you been? I've discovered millions of things."

Gell-Mann found a more kindred spirit that spring when another young Harvard physicist named Sheldon Glashow rolled into town. For his Ph.D. thesis, Glashow had taken over the idea his teacher Julian Schwinger had proposed about using a Yang-Mills gauge theory to unify the weak interaction and electromagnetism into a single "electroweak" force. He had even published a version of the theory, boldly stating that it could be renormalized. The claim turned out to be foolishly premature, but Gell-Mann thought Glashow was on the right track. His own work was in much the same spirit. In attempting to expand the algebra of V-A, he too was essen-tially trying to find a Yang-Mills framework broad enough to con-nect the weak force and electromagnetism. Maybe they could learn from each other.

Glashow was using his National Science Foundation fellowship to commute back and forth between CERN, the international research center based in Geneva, and the Bohr Institute in Copen-hagen. Gell-Mann diverted him to Paris to give a talk. They hit it off immediately. Like Gell-Mann, the ebullient Glashow—his friends called him Shelly—loved expensive restaurants and sports cars. And as with Murray, the world of theoretical physics had changed his life. The son of a plumber, he had grown up in far northern Manhattan. Now, rushing back and forth between Geneva and Copenhagen in his red TR-3, with a different girlfriend in each city, he fancied himself an international playboy. After his lecture, he and Gell-Mann treated themselves to lunch at a two-star restaurant. Murray insisted that Glashow try a fish course he considered especially good, but Glashow demurred. For all his recently acquired culinary sophistication, he hated seafood: Even the finest, freshest preparation evoked the stench of a fish market in his old neighborhood. Murray made it clear that disliking seafood was not an option. Glashow ordered the dish and had to admit it was very good. Like Bernstein, he found Murray fun to be with, except when he was force-feeding you some fascinating information that you didn't really want to know. And it bothered him that Gell-Mann kept saying nasty things about Schwinger, Glashow's adviser. But Murray liked Glashow's ideas about unification. "What you're doing is good," he said. "But people will be very stupid about it."

Beginning late in 1959, midway through his Paris stay, Gell-Mann decided to switch gears and try applying Yang-Mills algebras to the recalcitrant strong force. Explaining the strong force this way, he hoped, might point the way to uniting all the hadrons, the "stronglies," into a single framework, the kind of periodic table of the particles he had struggled to find before.

The goal, once again, was to put names on the cages in the particle zoo. As with his earlier work in Paris, the key seemed to be this notion of current algebras. In the simplest algebra, the one for electromagnetism, there was a single "operator" called electric charge. Weyl and London had shown that requiring charge to be gauge-invariant gave rise to a great leveling force, electromagnetism, carried by a single "gauge boson"—in this case, the photon. (Though purists may quibble, operators, currents, and the bosons that carry them can be thought of as being roughly analogous.) In their failed attempt to explain the strong force, Yang and Mills had crafted a larger algebra with three operators/currents and hence three bosons, the mediators of the gauge field. Their scheme foundered when they couldn't reconcile the requirement that gauge bosons be massless (so they would travel at light speed) with the fact that pions—presumed to be the carriers of the strong force—were quite massive. But the basic idea seemed sound.

The trick, Murray thought, was to go another step, to find an even larger algebra with more operators and more bosons— enough currents to tie together not just the neutron and proton, as Yang and Mills had attempted, but all the hadrons, the particles that felt the strong force. In Yang and Mills's original effort, the labels "proton" and "neutron" were arbitrary. Flip them randomly throughout the universe, and the strong force would arise to straighten things out. In the scheme Gell-Mann was reaching for, the labels for many different stronglies could be interchanged, and the invisible hand of the strong force would arise to enforce order.

Every day he would have a long lunch with a group from the Collège de France. After sharing several bottles of wine, he would return to his office in the mid-afternoon too fuzzy to fully concentrate. He tried algebras with four, five, six operators, but he still couldn't find the right fit. With each new operator, the calculations became more laborious. Finally he gave up at seven. "The hell with it," he said to himself. He had a hazy glimpse of some kind of structure, but he couldn't pin it down.

He was sure there must be an underlying order. At some very high level of abstraction, he believed, the universe was surely symmetrical, with all the hadrons somehow related. This order was an objective quality of nature, a platonic essence waiting to be identified by inquiring minds. He was annoyed to learn that not everybody thought this way. One day he was talking to a French physicist at the Collège de France about Project Ozma, a precursor of SETI, the Search for Extraterrestrial Intelligence. How would one go about communicating with an alien civilization? Murray suggested that one might start with pulses representing the integers. One beep for 1, two beeps for 2, three beeps for 3. . . . Then, when it had been established that we knew about counting, we could send the atomic numbers for the ninety-two stable elements, leaving out the two that are unstable. The aliens would surely have discovered the same pattern, the periodic table of the elements.

"Wait," the French physicist said, "that is absurd. Those numbers up to ninety-two would mean nothing to such aliens. Why, if they have ninety stable chemical elements as we do, then they must also have the Eiffel Tower and Brigitte Bardot." To the French scientist, the periodic table, and perhaps the whole notion of atoms with positively charged nuclei surrounded by electrons, was just a human construction, our own peculiar way of carving up the cosmos. Other civilizations would create different orders.

Nonsense, Murray thought. There was a real world out there and he was determined to learn its rules. He felt certain that the rules would be elegant. When the early experiments said that V-A was wrong, he and Feynman had persisted with the theory simply because it was beautiful. Their sense of aesthetics had paid off. Murray believed the periodic table of the stronglies he was seeking would be a true reflection of nature's symmetries. But for now he was stuck.

He was tired of particles and needed a break, and he longed to see places even more exotic than Paris. For months he had been pushing for Africa. Margaret was unenthusiastic. She thought it would be too hot. She suggested Iceland or Norway instead. Murray kept pressing. The colonial era was ending, and with African independence he was afraid that wildlife conservation would go completely to hell. The revolutionary regimes would have higher priorities than preserving wild animals for Westerners to hunt or photograph. This might be their last chance to see Africa at its wildest. Anyway, they were more than halfway there. One morning Margaret woke up and said, "If we go to Africa, can I wear a safari jacket?" Murray laughed. He had won her over.

First they stopped off in London, where Gell-Mann was the official Caltech delegate to the 300th-anniversary celebration of the Royal Society, the august scientific organization whose early members had included Isaac Newton. Invited to attend a tercentenary luncheon hosted by the Royal Dutch/Shell Group, Murray prevailed upon a vice-president of the company to help him find a Land Rover to rent in Africa. He wasn't shy about exploiting his growing contacts with the corporate world, no matter how tenuous they were. Then he discovered that the French travel agent had botched their reservations and he had to start again from scratch. Finally the itinerary was fixed and he and Margaret were ready to depart. They hired a baby-sitter to take care of Lisa and flew to the Entebbe airport in Uganda. From there they set off without a guide, just the two of them in the Land Rover with some dictionaries and grammar books. In Kampala they paid a couple of Indian carpenters to build a platform over the back benches in the vehicle. They stored the luggage underneath and slept on top, cooking dinner on their portable Afrigas stove.

In Queen Elizabeth Park, they caught a distant view of a rare African stork called a shoebill. Then they visited a gorilla sanctuary, where Murray composed a handwritten abstract—on the back of a postcard showing a native woman with rings around her neck—of a talk he was to give later that summer at the annual Rochester conference. In the accompanying note he apologized for the lack of secretarial help. He loved being so far from civilization. From Uganda they drove a short way into the Belgian Congo to buy gas. The country was in the throes of revolution, and some locals tried to bully them into driving them all the way to Stanleyville, deep in the interior. They gassed up fast and nervously sped away.

Most of the trip was wonderfully relaxing. Murray had never seen so many different kinds of plants and animals. His bird-watching list quickly grew to include eight hundred species. And he had never heard so many different languages. How strange that a finite variety of subatomic particles could give rise to so much diversity. They headed into Tanganyika and rode the ferry across Lake Victoria from Bukoba to Mwanza—he relished pronouncing the names. Then they crossed the Serengeti Plain, navigating by compass. Sometimes these expeditions made Murray uneasy. Drive too slow and the tsetse flies would come in through the windows and bite you. Drive too fast and you might break an axle on an aardvark burrow. While he drove, Margaret would stand with her head protruding through the game-viewing hatch and sketch pictures of the animals. All her apprehensions about the trip were gone. They saw leopards and lions. One day a giraffe jumped over the car. They ended up in Kenya, visiting two more game parks before turning in the car in Nairobi.

Coming back to civilization was a rude shock. Nairobi seemed to be striving to look like Los Angeles. Soon, Murray feared, the world would be one big suburb, diversity would be erased, and everybody would be speaking English. If only there could be fewer people, he thought, and more wild animals.

After returning to Pasadena, the Gell-Manns, who had been living in a series of rented houses, bought their own place in Altadena, a suburb in the San Gabriel foothills. Murray was impressed by the swimming pool and the glorious old oak tree that stood in the yard. The rambling house, which had been constructed piecemeal over the years, was perched at the edge of a canyon on a remote point of land—four acres, with maybe 3.9 of them almost vertical. A journalist visiting them around that time described what seemed almost a parody of the genteel Southern California life. In his natty tweed jacket, the young theoretician struck the writer as a handsome man who, he ventured, resembled the actor Tony Curtis in horn-rimmed glasses. After a day working at the blackboard with colleagues and dining at the Athenaeum ("Isn't the food here vile," Murray exclaimed), he would drive his Jaguar sedan or Jeep station wagon home to his refuge. "After playing with his blond, precocious daughter Lisa, inspecting the garden, and eating the gourmet dinner his soft-spoken wife has prepared," the journalist reported, "he either settles in a Danish-modern easy chair and reads American or French fiction for a while, or goes up to his den to work on equations."

One of the Gell-Manns' new neighbors was Feynman. Upon their return, Dick announced that he was "catching up" with his colleague. Like Murray, he had acquired an English wife, along with a small brown dog. Next came a house in Altadena. Their rivalry was still mostly playful. The Gell-Manns were happy to see Dick remarried. His first wife, his childhood sweetheart, had died in a tuberculosis sanatorium in New Mexico while he was working on the Manhattan Project. After years of philandering—including a short, disastrous second marriage—he had met his new bride, twenty-four-year-old Gweneth Howarth of Yorkshire, sunbathing at Lake Geneva. Through a strange series of events, the source of some prime Caltech gossip, he had arranged for her to immigrate to California as his housekeeper. Now they were announcing their September marriage at the luxurious Huntington Hotel in Pasadena.

One cold afternoon after the wedding, Murray and Margaret went to visit the new couple. Gell-Mann never forgot his friend's energy and enthusiasm that day—Dick at his best, he would later say. He watched as he playfully crumpled up newspaper pages and threw them into the fireplace. Starting a fire wasn't a mundane task but, like everything, a wonderful game. His spirit seemed almost magical.

Around that time a young physicist named Fred Zachariasen joined the department, moving into the office next to Gell-Mann's. Murray had known Zachariasen's father at the University of Chicago, where he was dean of physical sciences. The younger Zachariasen quickly became one of Murray's favorite collaborators. A visitor to Caltech in the early sixties was struck by the telegraphic intensity of their conversation, Murray leaning back in his chair, feet propped on the desk, as Zachariasen filled the blackboard with equations:

"Q-square omega Q-square . . . can't we call this mu-square? . . . Oh, son of a bitch . . . well, suppose you could find an analog for the form theory. . . . In both cases, d-bar has a desired quality, hasn't it? . . . gamma rho eta eta over gamma rho pi pi. . . . That's it! Bring this here and then factor it out. . . . I understand it better in pictures. . . ."

The dialogue could have been scripted by Beckett. Murray rushes to the board and clacks out his own equations while Zachariasen sits on a table and watches. Line by line, the tension builds. "Gell-Mann stops," the visitor recorded, "walks backward from the board and paces. Then he jumps to the board and writes furiously again, explaining as he does, and gesturing with his other hand. Zachariasen is now on his feet and they are both at the board." And on it goes . . . a typical day at the physics factory.

By now the life Murray was growing accustomed to was becoming more and more expensive to maintain. On top of his and Margaret's needs, he now had his parents to support. Never quite making ends meet, Arthur and Pauline had moved out of the apartment near Columbia Grammar School and into a housing project in the Bronx. Arthur's pension from the bank wasn't enough to live on, so he worked as a tutor. The stress of living soon gave him a heart attack. Several years later, at about the time Murray and Margaret moved to Pasadena, Pauline noticed a lump in her breast. Terrified of what the doctors would tell her, she and Arthur ignored the problem for a year and a half, agonizing only to themselves. When Murray learned what was going on, he was furious. He told them to put everything they owned in storage and fly to Pasadena immediately, where he would ensure that his mother got the best medical treatment possible. He would pay for everything and arrange to ship their belongings later. In their haste, his parents all but abandoned their apartment, bringing only a few mementos and papers. Gone were Murray's childhood coins from Rome and Greece and many of the family's links with the past. The operation was successful, or so it seemed at the time. The cancer had grown slowly, and the surgeon believed he had removed it all.

His parents stayed on in Pasadena. Murray rented them a small apartment downtown, bought them a television, and helped support them for the rest of their lives. They loved Southern California.

They would walk over to Colorado Boulevard to shop or go to the library, where Arthur would find more books to study. He quickly made new friends and took on more pupils to tutor.

Paying for all this required more than a professor's salary, and Murray found additional consulting work. In the late 1950s, his old friend Murph Goldberger recruited him to join an elite group of government advisers. The original idea had come from Murph's wife, Mildred. "Usually some of us . . . go to places like Los Alamos, General Atomic, etc., and for the sake of a few measly dollars work on dull stuff," Murph had written to Murray. "In many cases one has unpleasant or incompatible companions. . . . Mildred proposes that we should get a contract with some agency for summer work on plain theoretical physics, elementary particles, or what have you, set down in a nice place like Cape Cod and work." And that's exactly what they did. The physicists also thought that if they banded together, they could have more influence on strategic policy. The world would be a better place with scientists instead of generals in charge of defense. At first Goldberger and two colleagues, Keith Brueckner and Kenneth Watson, planned to start their own company, Physics Inc., and incorporate in Delaware. But some officials at the Defense Department convinced them instead to form a division of the government's Institute of Defense Analysis. They called it Jason. The letters didn't stand for anything. The name (also Mildred's suggestion) was a reference to Jason and the golden fleece. They would meet each summer and advise the government on antiballistic missile systems, remote detection of Soviet nuclear explosions, and other matters. The membership soon grew to include such luminaries as Hans Bethe, Edward Teller, John Wheeler, Eugene Wigner, and Freeman Dyson, as well as younger physicists like Leon Lederman, Francis Low, Steven Weinberg, and Fred Zachariasen. The symbiosis between physics and the military that sprang from the Manhattan Project was becoming stronger than ever during the cold war.

Gell-Mann enjoyed the distractions of dabbling in this new world of government policy-makers. There was so much more to life than physics. In addition to archaeology, linguistics, and ornithology, he had recently developed an interest in human behavior, amusing himself by devising a simple system to predict the position the right-wing John Birch Society would take on any issue.

He was still immersed in the train of thought that had obsessed him in Paris—coming up with a system of equations that would unite all the hadrons and provide a Yang-Mills field theory of the strong nuclear force. He and Feynman played around with pattern after pattern, trying to get the pieces to fit. But nothing seemed to work. When Glashow came out to Caltech for the 1960–61 academic year, boiling over with ideas, he helped push the search even further, trying to generalize the "Yang-Mills trick," as Murray called it, into field theories embracing both the strong and weak forces. But they weren't making much progress. Murray loved Shelly's enthusiasm but thought he needed to be less impulsive, to write things down and check them more carefully. And Feynman was being as difficult as ever. Zachariasen, whose office was next door, listened to the three scientists argue loudly, shouting at each other as they clacked chalk against slate. Feynman thought Murray was on the wrong track and said he wasn't interested in collaborating on the theory.

Some afternoons Gell-Mann, Glashow, and their other friends would retreat to the Beachcomber, a nearby bar that had an interesting daily special: Your hand was stamped with the time of your arrival, and that determined the cost of the drinks. If you came at 5:30 p.m., it was fifty-three cents a drink for the rest of the evening. Once, to get away from the blackboard, Murray, Shelly, and Fred Zachariasen drove down to Baja California in a four-wheel-drive vehicle stocked with fine wines and an assortment of gourmet foods. Murray seemed as good at identifying cactuses as he was birds. They explored the desert, drinking Mexican rum and driving across the sand. On another expedition, Gell-Mann, Zachariasen, and some friends tried to find an old Spanish mission not far south of the border. After getting lost in a maze of back roads, they pulled over and Murray asked an Indian passerby for directions. The man didn't understand his Spanish, so he rephrased the question in the local Baja dialect. His companions were amazed. With his magic memory, he had been able to absorb enough of this obscure tongue to get them to their destination.

One day in December, while Glashow was away in Massachusetts, Gell-Mann was describing their efforts to a young Caltech mathematician named Richard Block. After listening awhile, Block informed Murray that he had been spending the last few months struggling to reinvent a familiar piece of mathematics: group theory, which had been developed in the late nineteenth century by a Norwegian mathematician named Sophus Lie.

Objects that form a group are related by some kind of symmetry.

Consider, for example, an equilateral triangle. You can rotate it 120 degrees, 240 degrees, 360 degrees, and the shape is preserved. These rotations can be thought of as a group. Or consider an object shaped like a book. It can be rotated on three different axes: two axes (x and y) run vertically and horizontally, crossing at the center of the cover, while the third axis (z) is perpendicular to the two. All the possible rotations are related by specific rules. First flip the book 180 degrees, end over end, on its x-axis, then rotate it to the right 90 degrees on the z-axis. Note the final position. Now start again but perform the operations in the reverse order: Rotate the book 90 degrees to the right, then flip it end over end. The book winds up in a different position. These two operations—flipping end over end and rotating to the right—do not commute. A × B is different from B × A.

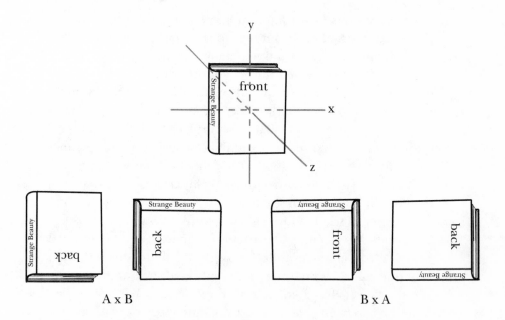

Groups of Rotations.

Around the turn of the century, a French mathematician named Elie-Joseph Cartan had catalogued every possible kind of these Lie groups. For example, a simple group called U(1) represents the continuous symmetry of a circle: when rotated around its center, it looks the same. The group—the name means "unitary one-dimensional"—is said to have one "operator," whose job is to carry out this single allowed rotation. U(1) can also be used to describe electromagnetism. Here the "operator" is the photon, and the "rotation" is the effect electromagnetism wields on a particle, transforming the shape of its quantum wave.

The three possible operations/rotations involved in flipping a book can be captured by SU(2), for "special unitary group in two dimensions." As it happened, SU(2) could also be used to describe the isospin symmetry—the group of abstract ways in which a nucleon can be "rotated" in isospin space to get a neutron or proton, or a pion to get the negative, positive, and neutral versions. These rotations were what Gell-Mann had been calling currents. These groups were what he had been calling algebras.

He couldn't believe how much time he had wasted. He had been struggling in the dark, while all these algebras, these groups—these possible classification schemes—had been studied and tabulated decades ago. All he would have had to do at the Collège de France was go to the library and look them up. Instead he struggled to re-create them from scratch. In fact, he soon learned, one of the regular lunch companions with whom he had been polishing off those bottles of wine was a leading expert on Lie algebras. Murray himself had even learned about Lie groups back at Yale and consulted books on group theory as a graduate student. But all the material had been presented in such an abstract manner that he had never really grasped what it was about, how it might relate to the real world. In 1951, he now remembered, he even sat in on a series of masterly lectures on group theory given by the physicist Giulio Racah in Princeton. But Murray had been so mesmerized by the speaker's beautiful Florentine accent that he never focused on the substance of the talk. Maybe he had a mental block against the whole idea of thinking in terms of groups and symmetries. Pais, who was a friend of Racah, had been an enthusiastic promoter of the subject; it formed the very basis of his vision of uniting particles into families associated with new quantum numbers.

Now Gell-Mann wished he had paid closer attention. It all seemed so straightforward. The algebra Glashow had come up with for unifying the weak force and electromagnetism was a combination of two groups—what Cartan called SU(2) × U(1). In Paris, as Murray struggled to expand the algebra of the isospin doublet, SU(2), to embrace all the hadrons, he had been playing with a hierarchy of more complex groups, with four, five, six, seven rotations. He now realized that they had simply been combinations of the simpler groups U(1) and SU(2). No wonder they hadn't led to any interesting new revelations. What he needed was a new, higher symmetry with novel properties. The next one in Cartan's catalogue was called SU(3), a group that can have eight operators—eight ways to rotate its members to convert one into the other.

Because of the cumbersome way he had been doing the calculations in Paris, Murray had lost the will to try an algebra so complex and inclusive. He had gone all the way up to seven and stopped. Maybe SU(3) would provide the scaffolding for the periodic table of the stronglies. One day in January he sat down and began working out the details. The ideas flowed onto the page with an ease Murray had never known before. As in his earlier attempt at classifying the hadrons, the aim was to take the smaller pieces of the puzzle—the neutron-proton doublet, the pion triplet, and so on—and combine them into higher symmetries. The smaller structures would be fragments of a larger, all-embracing whole.

Murray quickly learned that there were several different manifestations of the SU(3) symmetry—what the mathematicians called representations. One had three members, one had eight, one had ten, and one had twenty-seven. All were different ways in which SU(3) revealed itself in the world. Whichever representation one used, the result was an abstract polygon; rotate it according to the rules of SU(3) and it would give rise to all the various members of the cluster. For now, Murray concentrated on the representation that had eight members. The number of spin-½ baryons was also eight. Maybe he could arrange them according to SU(3). The members included the familiar proton-neutron doublet and six strange particles: a triplet of sigmas (positive, neutral, and negative), a doublet of xis, and a single lambda.

These particles could be distinguished by their different charges, −1, 0 and +1, and by their different strangenesses, 0, 1, −1, −2. So suppose you drew a graph plotting charge against strangeness. The doublets, triplets, and singlets would line up into a neat, hexagonally shaped array.

As he worked out the equations, Murray marveled at the beauty of the architecture, the pattern unfolding in his mind. Using the eight operations allowed by SU(3), each particle could be converted into any of the others. Or, to use a different metaphor, there were eight "currents" through which the members of the group could flow from one into the other, assuming new identities by changing their quantum numbers. Each of the operations/currents presumably involved some kind of hypothetical boson as the carrier. If one is limited to the flat plane of a piece of paper, it is hard to do justice to the elegance of the symmetry. SU(3) is best described as an eight-dimensional mathematical object that can be rotated in a multidimensional space, manifesting itself with each twist as eight different particles. Like the familiar toy called the Magic Eightball, each face displays a different message.

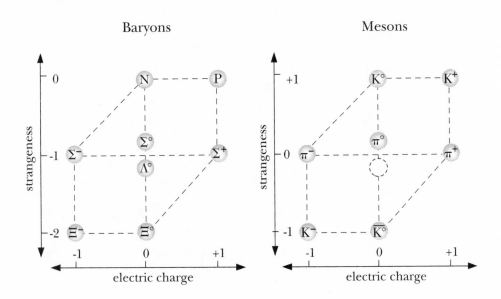

The Eightfold Way. When certain hadrons, the particles that obey the strong force, are arranged according to their strangeness and electrical charge, they line up into hexagonal patterns. In the first hexagon, baryons with spin ½ are arranged with the neutron and proton in the upper row, three sigmas in the middle row, and two xis in the bottom row. Notice that the center position is occupied by two particles, the chargeless version of the sigma and the chargeless lambda.

In the second hexagon, mesons with spin 0 are arranged with two kaons in the upper row, three pions in the middle row, and two kaons in the bottom row. Here one of the two particles in the center position is blank. Gell-Mann predicted that the gap would be filled by a particle he called the eta.

It took imagination to make the model work. Each of the eight particles in the family had a different mass. What was so symmetrical about that? But these, Gell-Mann realized, could be dismissed as extraneous details. One had to peer through the haze of distractions to behold the pattern lying underneath. In an ideal world, the masses would be identical, but something interacted in different ways with the particles, breaking the symmetry and mucking things up.

Murray constructed similar eightballs for all the mesons with zero spin: the kaons, pions, and antikaons. These also formed a hexagonal array. Since there were only seven known particles in this category, Murray confidently predicted an eighth: a singlet later called the eta. And he proposed a third octet consisting of mesons with spin –1. Here he was really sticking his neck out. Most of these particles still seemed to exist only in the theorists' minds. But he was sure that they would be found by the experimenters.

The ease with which the pieces were clicking together was exhilarating. Within days the whole scheme was in place. Murray called it, with a touch of irony, the Eightfold Way. Not just because there were eight possible rotations and eight particles in each representation, but because of a saying of the Buddha about the eight ways to achieve nirvana: "Now this, O monks, is noble truth that leads to the cessation of pain; this is the noble Eightfold Way: namely, right views, right intention, right speech, right action, right living, right effort, right mindfulness, right concentration."

Murray himself was in nirvana. For years all these different particles had come raining down on earth or ricocheting from the blasts of accelerators. Where there had been confusion, now there was hope of order. This sense of well-being extended to his inner world. He was seeing a psychotherapist he really liked, learning to control his anxieties and free himself from the paralysis of indecision. Now he felt so good about his new theory that he resolved to be a better person, more considerate of others, more tolerant. He started talking about what he called "the new me." The Buddha would have been pleased.

But these noble feelings didn't last long. By January 20, he was confidently passing around a preliminary preprint of the theory. Over the next couple of months, he worked on a revised version, but his characteristic uneasiness soon returned. What if the idea turned out to be crackpot? When Glashow, who had just returned from his vacation back east, heard about the paper, he thought his name should be on it too, but he didn't press the matter. He was less preoccupied with the strong force than with his attempt at electroweak unification. Maybe they could join forces and go for an even greater unification. Glashow and Gell-Mann struggled to combine their two schemes, to use group theory to describe both the electroweak and strong forces, showing how they worked together on the subatomic realm. First they demonstrated with some rigor what now seemed almost obvious: that every Yang-Mills gauge field could indeed be described by a Lie group—U(1) for electromagnetism, SU(2) for the isospin symmetry, SU(3) for the Eightfold Way. But they just couldn't get the strong and weak forces to mesh. They clashed intolerably, as Murray later put it. When they wrote up the work half a year later, they lamented, in print, over how ugly the mechanism was. "The model we have discussed is not seriously put forward as a physical theory," they concluded, "but it is a good illustration of the ideas involved in the gauge method." On this front, they had failed.

Why couldn't he get the Eightfold Way to fit with Glashow's scheme for the electroweak force? In the back of his mind, all the old doubts were resurfacing. Maybe he and Glashow couldn't get their models to jibe because there was something basically wrong with field theory. Gell-Mann was also troubled by rumors of some experimental data that seemed to contradict the Eightfold Way. According to some results at Berkeley, the sigma and lambda particles had opposite parity. If this was true, then they wouldn't fit into the same octet, where he had put them.

He was fretting over this discrepancy during a visit to Berkeley to talk to some experimenters. He arrived late in the morning, and one of his colleagues handed him a cup of coffee. Murray looked down at the slogan that had been printed on the cup to taunt him: "Sigma-lambda odd, sigma-lambda odd" was emblazoned around the rim. Just what he needed. In March, the final version of the Eightfold Way preprint came out, but he was almost ready to disown the idea. About a week and a half later, he sent *Physical Review* a paper tentatively titled "Theory of Strong Interaction Symmetry"— a kind of state-of-his-mind report as of spring 1961—but he barely mentioned the Eightfold Way.

His confidence reached its lowest point in April, when he and Margaret went to the University of Illinois to visit David and Suzy Pines. At a party, he ran into William Fowler, a Caltech nuclear physicist who was also visiting at the university. Fowler had led many of the crucial cloud chamber experiments in the early 1950s that had tested the theorists' notions about strange particles. His team had demonstrated the reality of associated production, the linchpin of Gell-Mann's strangeness scheme. Still, he and Fowler didn't get along. Murray often made it clear that compared with particle physics, he considered nuclear physics an intellectually inferior pursuit. What could be more boring than limiting your sights to the nuclei of atoms when all the real action was deeper inside? Fowler, a jovial, blustering sort, enjoyed antagonizing his colleagues. (He had once told a graduate student, George Zweig, "You're not smart enough to be a theorist. You should do something else.") Most people laughed off Fowler's good-natured insults. But Murray wasn't so cheerful about taking what he so gleefully dished out. Now, with several drinks under his belt, Fowler walked up to Murray and began deriding the Eightfold Way and the whole enterprise of particle physics. Why didn't Murray spend his time working on something worthwhile instead of this dead-end field? Margaret and the Pineses steeled themselves for a blistering counterattack. But Murray didn't have it in him. Anyway, he was really trying to be nice to people. He timidly replied that he personally found Fowler's work quite interesting and was sorry that Fowler didn't feel the same way about his own. Margaret and David and Suzy looked on in wide-eyed disbelief. "What's happening, what's the matter with you?" they asked later. "Why didn't you bite his head off?" Murray didn't know. The feelings of gloom deepened. That was the end of the "new me." He began reverting to his old ways.

In June he withdrew the paper from *Physical Review* to rewrite it. The experiments with the sigma and lambda particles, which had seemed to contradict the octet idea, were starting to seem doubtful. The model was so beautiful it just had to be true. He changed the title of the paper to "Symmetries of Baryons and Mesons" and at the very end of the paper put back the Eightfold Way. Then he flew off to India to teach at a physics summer school in Bangalore. Monsoon season wasn't the best time to explore the remote areas of India, but he wanted to see as many different kinds of exotic people and native wildlife as he could. Even with all these distractions, he couldn't stop worrying about being wrong. In his lectures, he hardly mentioned his classification scheme.

When he fell into one of these dark moods, his only solace came from Margaret. "My dearest love," she wrote to him in India, describing a lonely evening sitting in the backyard of the house in Altadena. Lisa was asleep and there was almost nothing to break the silence but the chirping of a cricket, the insistent ticking of a clock. "Outside the garden is cool and sweet-smelling, the pool is bone dry and shining white (it looks good that way when you are used to it). . . . Right now I should like to climb the stairs to your study and watch you biting your pipe and writing equations. But after all, that is a nice way to miss someone, as long as you are sure he is coming back."

Early the same summer, Gell-Mann helped Keith Brueckner, one of his Jason colleagues, organize a conference at the University of California campus in La Jolla. They invited Oppenheimer and took him to lunch at the Hotel del Coronado, an imposing nineteenth-century Victorian retreat across the bay from San Diego. Murray resisted the urge to order the wine, deferring to Oppenheimer, who asked the vintage of one of the Bordeaux. "Sir, all our wines are at least five years old," the waiter said. Murray thought it was hilarious; Southern California could be so gauche. At the conference, he spoke about the Eightfold Way, Goldberger about dispersion relations. Abdus Salam came from Imperial College in London, and there was a rising young South African theorist named Stanley Mandelstam. In the midst of all this, Murray was a little shocked when one of the speakers, a Berkeley theorist named Geoffrey Chew, took the floor and delivered a manifesto. Field theory, he declared, was no good. At least for explaining the strong force, physicists needed a new theory based not on abstract fields but on the concrete realities of the S-matrix, the long-sought table of numbers that would describe all the possible ways one particle could scatter off another. Field theory no good? It sounded to Murray like Lev Landau talking.

Chew believed that of all the possible S-matrices that might be constructed, physicists could prove that only one was mathematically self-consistent. All the others would contain some kind of internal flaw. This one true S-matrix would be no less than the blueprint of the cosmos. The hadrons would have the properties they did—charge, mass, and so forth—because these were the only values that satisfied the S-matrix. Thus it would be shown, as Einstein once had hoped, that God had no choice in making the universe. There was only one possible way that things could be.

And there was more. In Chew's magical S-matrix, the hadrons

would be united by something he called "nuclear democracy." Physicists were used to thinking of some particles, most notably the neutron and proton, as fundamental, with the more exotic baryons and mesons as somehow secondary, a kind of nuclear aristocracy. There had even been attempts, none successful, to show that the secondary particles were somehow made from the fundamental particles. Chew argued that physics must abandon this prejudice. Just because the proton and neutron were discovered first didn't mean they were privileged. They were just the stablest, lowest-energy states of hundreds of possible hadrons. None was any more fundamental than each of the many possible ways a planet could orbit around the sun. The ultimate goal, he declared, must be to show that there are no fundamental particles, that all particles can be thought of as being composed from one another. It was a dizzying idea. Suppose, for example, that a proton and an antiproton were bound together by a meson shuttling back and forth between them. Now what if the meson were, in turn, made from the proton and the antiproton? More generally, particle A would be made from B and C, while B was made from A and C, and so forth. Each particle would, as Chew put it, pull itself up by its own bootstraps.

Gell-Mann liked the idea of nuclear democracy (though he later pointed out that it should more precisely be referred to as "hadronic egalitarianism"—after all, the particles weren't voting). He even spoke favorably of the bootstrap idea in his own La Jolla talk, speculating that it might somehow underlie the Eightfold Way. (The next year, he nominated Chew for the Heineman Prize.) But he was put off by this full frontal attack on field theory. He had always admired Chew for his seeming modesty and lack of pomposity. Back in the Chicago days, when Chew was at Urbana, Gell-Mann and Goldberger had tried to interest him in dispersion relations. Murray had always felt that Chew was resisting the idea. Now he had apparently not only accepted dispersion theory, he was recasting it in this bombastic new form.

When it came to the Eightfold Way, Gell-Mann was still suffering from stage fright. He would wake up in the middle of the night, sure he was making a fool of himself. Late that summer he withdrew the *Physical Review* paper again, thoroughly revising it. "Most of the predictions remain the same," he wrote to the journal's editor, S. A. Goudsmit, "but they are no longer tied to a particular and doubtful model." And so it went, his mood swinging back and forth.

But ultimately the experimental evidence trickling in pulled him back toward the scheme. Experimenters were starting to find the particles that fit the third octet he had proposed in his Caltech preprint. Before 1961 was over, they had discovered the eta, the missing eighth member of the second octet. The symmetry was impossible to ignore.

After almost a year of self-torture, he settled on the final draft. Right up front, he put a caveat: He was using fields in his theory, knowing full well that some physicists, like Chew, were arguing "that the apparatus of field theory may be a misleading encumbrance." He wanted his readers to know that he was presenting his ideas this way simply as a mathematical convenience; they would stand whether or not the framework itself survived. He was keeping his options open. Having dispensed with the disclaimers, he forged ahead with his ideas about symmetries, groups, and algebras, ending with a detailed account of the Eightfold Way. (He manipulated the format so that it would appear in Section VIII.) With all the delays, the article wasn't published until February 1962. It was far from a bold presentation. But he had staked his claim. And as it turned out, not a week too soon.

CHAPTER 10

HOLY TRINITY

Murray Gell-Mann was not the only physicist thinking about using group theory to make sense of the particle zoo. Several months before Gell-Mann's epiphany at Caltech, a late-blooming physics student in London was putting together a scheme that turned out to be all but identical to the Eightfold Way. Had Murray known, he might have worked a little faster. His unknown rival was not one of the usual suspects—a member of the close-knit fraternity of leading particle physicists one might expect to see at a Rochester conference. In the world of physics, Yuval Ne'eman, an Israeli army colonel and engineer, was unknown. After spending more than a decade fighting for Israeli independence and rising nearly to the top of the new nation's military establishment, he had suddenly gone back to school to pursue his first love, theoretical physics. Now he was making up for lost time.

Ne'eman had been born in Tel Aviv, four years before Gell-Mann, and he was proud to belong to what Israelis called an "old settlement" family—the equivalent, he would say, of having come over on the *Mayflower*. His ancestors were from Lithuania, members of a rabbinical school that had championed rationalism over the more mystical approach to God taught by Hassidic Jews. Ne'eman took his heritage to heart. As a boy in Palestine, he learned everything he could about science. And like Murray, he was driven to classify things—minerals, languages, the Chinese dynasties, the kings of Europe. He was in love with all kinds of charts. Early on, he became preoccupied with Linnaeus's classification scheme for the flora and fauna and Mendeleev's periodic table of the elements. He was also enchanted by more abstract orders. After learning about the wonders of imaginary numbers, involving the square root of −1, he invented his own mathematical system, an invisible toy he lovingly turned over in his mind. When a cousin gave him a copy of Sir Arthur Eddington's popular physics-and-philosophy book, *The Nature of the Physical World,* he read it again and again. Like Murray, he was a prodigy, graduating from high school at age fifteen.

But there the similarities ended. For all his love of the abstract, Ne'eman had spent his early years immersed in the concrete realities of war. After high school, when Murray was biding his time before college on FDR's campaign, Ne'eman signed up with the Israeli underground, the Haganah. When he wasn't fighting for the Resistance, he was studying the very practical subject that Arthur Gell-Mann had tried to steer Murray into: engineering. After Ne'eman got his degree, he joined the military, commanding an infantry battalion in the 1948 war and serving as acting head of the Secret Service.

It wasn't until 1957, when he was thirty-two years old, that he was able to start pursuing physics. He asked his boss, Moshe Dayan, then the Defense chief of staff, for a two-year leave of absence. Not wanting to lose Ne'eman completely to science, Dayan appointed him to the post of London defense attaché, suggesting that he use his spare time, so-called, to work on a Ph.D. At first Ne'eman thought he wanted to study general relativity and cosmology at Kings College. But he dreaded the long commute from Kensington, the royal district of London where the Israeli embassy was located. Instead Ne'eman went practically next door to Imperial College, walking into Abdus Salam's office in full uniform and presenting his letters of recommendation, including one from Dayan. Salam found this introduction a little weird, but he let the young man join the department.

Before long Ne'eman was teaching himself group theory (while negotiating for the purchase of two submarines and fifty Centurion tanks for the Israeli government) and trying to find a scheme that could be used to classify the hadrons as neatly as the plants and animals and elements and kings. Salam was dubious of such a grand plan. "You are embarking on a highly speculative search," he warned. One by one, Ne'eman tried out various groups to see if he could make the particles fit. One of his favorites was shaped like the Star of David, but the design turned out to lack the universality he sought. By November 1960, a few weeks before Gell-Mann learned about Lie groups from the young mathematician Richard Block and started crafting the Eightfold Way, Ne'eman told Salam that he was convinced the hadrons could be arranged according to the pattern called the SU(3) octet.

Salam approved of the idea, and Ne'eman was so relieved that he immediately called his wife to tell her the good news—perhaps his plan to become a physicist was a good gamble after all. But he was disappointed when Salam told him that he was not the first to recognize the potential of SU(3). A Japanese team in Nagoya, led by the maverick physicist Shoichi Sakata, had been studying the same group. Salam had been thinking about how the Sakata model might be cast in the form of a gauge theory. He proposed that he and Ne'eman collaborate. Publishing with the great Salam would be a coup. But after a few weeks of hearing nothing, Ne'eman feared that his paper had become lost in a pile on his boss's desk. With too much to do, Salam agreed that Ne'eman should publish his classification scheme by himself. The wait for the department typing pool seemed interminable, so Ne'eman asked his secretary at the embassy to prepare the paper. Finally, in February, while Murray was revising his preliminary preprint of the Eightfold Way, Ne'eman sent "Derivation of Strong Interactions from a Gauge Invariance" to the journal *Nuclear Physics* and waited for the editor's response.

The paper came back like a boomerang, landing on Salam's desk with a thud. Didn't his students know that submissions had to be double-spaced? And there were other typographical infelicities as well. At about the same time, the early draft of Gell-Mann's preprint made its way to the department mailbox. "Look at that!" Salam said to Ne'eman. "Gell-Mann is suggesting your model." He thought the presentation was magnificent, and he loved the name "Eightfold Way." He admitted, several days later, that he hadn't really taken Ne'eman's idea all that seriously until he saw that Murray was working on the same thing.

As a courtesy, Ne'eman immediately wrote to Gell-Mann to request his own copy of the preprint, pointing out that he had sent a similar theory to *Nuclear Physics* six weeks before. He had caught a glimpse of Murray when he had come through London in 1959 to give a talk at Imperial College on his way to Paris. Ne'eman, who had learned about the lecture at the last moment, quickly threw a jacket over his uniform and rushed to the lecture hall. Afterward he listened from the periphery while one of his colleagues asked Gell-Mann some questions. Later both were at a party at the British physicist Thomas Kibble's home, but they didn't meet. Ne'eman's only real impression of Gell-Mann was that he seemed very young and very smart. He later read one of his Paris papers and was deeply impressed. In fact, he felt more encouraged than disappointed that

Murray had scooped him. It was comforting to know that he was in such august company. Ne'eman had his paper retyped, including a mention of the Gell-Mann report, and resubmitted it. Because of the typing problem, Ne'eman had lost his slight head start, and Gell-Mann had beat him into print. But because of Murray's months of agonizing, Ne'eman's paper was actually the first to be published in a journal. It was another of those occasions, so common in particle physics, when the same idea crystallized almost simultaneously in two people's minds.

One of the great enduring mysteries is why mathematics seems to mirror so perfectly what goes on in the real world. The physicist Eugene Wigner puzzled over this in an essay called "The Unreasonable Effectiveness of Mathematics in the Natural Sciences." Why indeed should particles obediently line up according to the rules of group theory? Murray himself didn't much care for this kind of philosophical question. But it intrigued him that when there existed two alternative explanations for a physical phenomenon—like weak nuclear decay—the more mathematically elegant one often turned out to be correct.

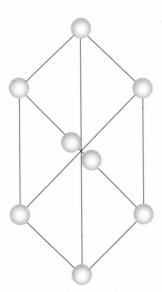

Fundamental triplets. The Eightfold Way is made up of a mosaic of simple triangular building blocks.

In the scheme devised by Gell-Mann and Ne'eman, particles seemed to arrange themselves in octets. But what about the other

SU(3) representations—those with three, ten, or twenty-seven members? Especially intriguing was the simplest three-member arrangement, called the fundamental representation. All the more intricate patterns—the eights, tens, and twenty-sevens—could be formed by combining these triplets like tiles repeated in a mosaic. The triplet, in fact, could be thought of as a kind of building block for the larger patterns.

If SU(3) octets indeed described the hidden architecture of the subatomic world, the very logic of the mathematics seemed to demand that these patterns be erected atop a foundation that had something to do with the number 3. But what could this possibly mean? Could the particles that belonged to the Eightfold Way be built somehow from triads of other particles? Since the mid-1950s, Sakata's group in Japan had been fiddling with an SU(3) scheme in which the hadrons were ultimately made from combinations of protons, neutrons, and lambdas. In the scheme, these particles were the fundamental building blocks of matter.

But many physicists considered the Sakata idea suspect. Were protons and neutrons really fundamental, or did they just seem so because they were discovered first? As for the lambda, it was apparently thrown in simply as a means of endowing certain particles with strangeness, like a sprinkling of fairy dust. It all seemed a little contrived. Anyway, the whole idea of fundamental particles went against the ever-more-popular doctrine of the bootstrap, the nuclear egalitarianism in which everything was recursively made from everything else. Even worse, many of the Japanese physicists seemed to be driven less by experiment than by their passion for Marxism, believing that Friedrich Engels's doctrine of dialectical materialism required that all particles be reducible to solid little subunits.

Still, whether one was driven by Marxism or by a Platonist belief that the world aspired to imitate mathematical forms, the appeal of a universe made from SU(3) triads was hard to resist. The idea of Sakata-like triplets had even been an influence on Gell-Mann, part of the reason he played down octets in the first unpublished version of his *Physical Review* paper (what came to be called "Symmetries of Baryons and Mesons"). He had even wondered aloud back at the La Jolla conference whether an octet or a triplet—if either—would turn out to be the correct configuration for making sense of the hadrons (the "stronglies"). But maybe it wasn't a matter of either-or. Could the two ideas somehow be combined?

As far back as his Eightfold Way preprint, Gell-Mann had demonstrated that the octets could indeed be built up from triplets of what he called "fictitious" leptons, "which may have nothing to do with real leptons but help to fix the physical ideas in a rather graphic way." They were, in other words, a pedagogical device that made it easier to explain the octets. "The particles we consider here for mathematical purpose do not necessarily have anything to do with real leptons," he had written. Then, hedging as usual, he added: "but there are some suggestive parallels." The notion that hadrons, usually assumed to be indivisible, might be made of smaller units as basic as leptons was a radical notion. And Murray was too cautious to go so far. He just left a hint of the idea. It was all so confusing. Were octets fundamental or were triplets? And triplets of what? And did there have to be fundamental building blocks at all?

Ne'eman, who considered the Sakata group's dialectical materialism to be about as scientific as Hassidic mysticism, was also thinking about triplets. For months he had been fascinated by how neatly they could be combined into octets. Maybe the power of the triplet had something to do with the fact that space had three dimensions. "Could a baryon be made of three leptons?" he had asked Salam. Salam just laughed. Nothing, he said, was too crazy so long as it worked.

After finishing his paper on SU(3), Ne'eman returned to Israel for a while and hammered out a triplet scheme with a colleague, Haim Goldberg. Goldberg had done his thesis under the guru of group theory, Giulio Racah, the man Murray had listened to (but not heard) in Princeton so long ago. In February 1962 Ne'eman and Goldberg sent out a preprint describing their idea. Then, in a seminar at the Weizmann Institute in May, Ne'eman made what would turn out to be a prophetic remark: Group theory almost seemed to imply that baryons are really made of fundamental triplets. But surely this didn't mean that the baryons could be pried apart to reveal three smaller subparticles of some kind. For one thing, Ne'eman and Goldberg realized, such hypothetical building blocks would have to be fractionally charged. A proton had a charge of $+1$. If it was made of three other particles they would have $\frac{1}{3}$ charges. It was an article of faith that the smallest unit of charge was -1 or $+1$. No one had ever seen charge come in fragments. It was hard enough trying to sell the Eightfold Way. Arguing that it was erected on a scaffolding of fractionally charged subparticles would be a public relations disaster. SU(3) triplets, they decided for the time being, must be little more than a curious mathematical accounting device.

.   .   .

The same month that Ne'eman and Goldberg mailed off their speculative paper about triplets, Murray's mother died. The doctors had been wrong; the surgeon had not succeeded in removing all the malignant tissue. It spread, eventually reaching her brain, and she passed away in a nursing home (a "dying home," Murray called it) after a long, long illness. Arthur was despondent. If only they had sought treatment immediately, when Pauline first found the lump, maybe she would still be alive. He tried to fill the emptiness by listening to the radio, talking to his friends, and walking around Pasadena. But the loss left him dumbfounded.

That summer, Murray brought his family to Geneva, where he was spending a few months at CERN, the European particle accelerator center. Arthur wrote from California, still trying to make sense of what had happened. "The shock of her death has stupefied and dulled my mental faculties to such an extent that I was unable at first to realize the magnitude of my loss," he wrote, going on to describe an "agonizing emptiness" that had caused him to call into question his lifelong faith in the ability of scientists and philosophers to understand the world.

"What is Life and what is Death?" he asked, drawing on all his old rhetorical powers. "Why are we born, why do we suffer? Of what significance is all human activity when we contemplate man's puny position among the flaming stars and whirling galaxies? Science may have given us a better understanding of that grand illusion which we call the universe, but it has deprived our life of any meaning. The deeper we delve into the mysteries of nature, the more we realize its utter futility and worthlessness. How much better off is the crawling ant which is happily unaware of the many dangers threatening its existence, and unconscious of the ruthlessness, injustice, and demoniacal cruelty of its creator."

In Geneva there were plenty of distractions to help Murray and Margaret get their minds off Pauline's death. They could walk through the Jardin Botanique with five-year-old Lisa, identifying the plants and flowers, or admire the architecture in the old part of the city; they could stroll along the shore of Lake Geneva, gazing up at the Jet d'Eau, the world's largest fountain, spewing water almost five hundred feet into the sky.

They were renting an apartment on a cross street just off the main road to CERN, so Murray could easily get there by public transport when Margaret was using the car. Weisskopf was now serving as director-general of CERN, and Gell-Mann enjoyed seeing his old teacher again. But most of his time was spent across the street from his apartment in a café, writing in his notebooks, occasionally ordering something to eat or drink, enjoying the European tolerance for intellectual loitering. With all the excitement in the air, he felt that physics was on the verge of a revolution akin to the discovery of quantum mechanics in the 1920s ("except, of course," he had written to Harold Morowitz, "for a few orders of magnitude in the number of people and the amount of money involved").

It was not so much the Eightfold Way that had him excited as it was Chew's bootstrap and a related bit of esoterica he had been working on with Fred Zachariasen, called Regge poles. In the late 1950s, the Italian physicist Tullio Regge had proposed that mesons and baryons could be arranged in meaningful patterns by graphing their energy against their spin. If energy was plotted on one axis and spin on the other, the particles would line up along "Regge trajectories." Since spin was quantized—it had to come in either integral or half-integral amounts—particles could appear only at certain points along the trajectories called Regge poles. Recent experiments seemed to bear this out. The result was another possible classification scheme, an alternative to the Eightfold Way. Gell-Mann manipulated the equations, trying to predict the outcome of accelerator experiments while avoiding the nonphysical answers (negative probabilities and such) that the physicists called "ghosts."

In July the Rochester conference was held at CERN, and Murray, who was scheduled to talk about the Regge approach (how to weed "sense solutions" from "nonsense solutions"), finally got a chance to meet Yuval Ne'eman. High on the conference agenda were two groups of newly discovered particles, so fleeting, they didn't even linger long enough to leave a track in a bubble chamber. Their existence had to be inferred from their decay products. Physicists called such short-lived creations resonances. A quartet of resonances called deltas had been discovered by Fermi's group back in the early fifties. More recently experimenters had used their new accelerators to create a triplet of closely related resonances called sigma-stars and a doublet called xi-stars. It wasn't clear where these new particles fit into the big picture.

On the first day of the conference, Ne'eman was thinking about these latest interlopers as he boarded the shuttle bus from his hotel to CERN. His eye was immediately caught by an attractive young

Israeli woman. *"Shalom,"* he said, sitting down next to her and her husband. Sula and Gerson Goldhaber, who were experimenters at Berkeley, had heard about this Israeli colonel who had independently discovered the Eightfold Way in his spare time. They were eager to hear firsthand about SU(3) and talk about the puzzling new resonances. In exchange, they told Ne'eman about a maddening new experimental result. They had been trying to find a certain resonance that some theorists said must arise when kaons were scattered off nucleons. But the harder they looked, the more convinced they had become that the theorists were wrong. The resonance didn't exist. Ne'eman mentally filed away this unexpected observation, thinking it might turn out to be an important clue.

Over the next few days, he thought about the new resonances and the one that was missing. Slowly the pieces started to fit together. He had already played with SU(3) triplets and with SU(3) octets. Ne'eman realized that the new resonances—baryons that each had a spin of ¾—would fit neatly into a ten-member manifestation of SU(3), called a decuplet. He scribbled his new theory on a piece of notebook paper, attached it to some of his recently published work, and gave them to the Goldhabers.

A diagram of the decuplet version of the Eightfold Way—a kind of tenfold way—could be drawn as a pyramid tipped onto its side:

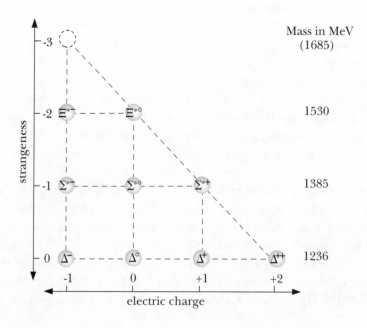

The four deltas were in the bottom row. The next higher row was occupied by the new triplet of sigma-stars, and the third row by the new doublet of xi-stars. While the particles in the bottom row had a strangeness of 0, those in the next row had a strangeness of −1, and those in the next row a strangeness of −2. An interesting pattern. The mass of the particles also changed in an orderly manner from row to row of the chart. Physicists often measured mass by how much energy it contained (according to the formula $E = mc^2$). The mass difference between each level was roughly 150 million electron-volts (MeV).

The very top row, the apex of the pyramid, was empty. No known particle fit into this gap. In his note to the Goldhabers, Ne'eman predicted—at this point, it was simple arithmetic—that the unoccupied point must be filled with another spin-$3/2$ particle with a strangeness of −3 and a mass of about 1,685 electron-volts. He also explained how the missing resonance—the one the Goldhabers had failed to find—ruled out the possibility of fitting the particles into a larger, twenty-seven-member representation. All signs were pointing to a decuplet—a new manifestation of the Eightfold Way.

On Tuesday, July 10, at a session discussing the new particles, Ne'eman was waiting for a chance to spring his prediction on his colleagues. If the predicted particle could be found, it would be the strongest evidence yet that the Eightfold Way was real. At an opportune moment, Ne'eman raised his hand. The rapporteur looked his way and said, "Professor Gell-Mann." Murray, sitting behind Ne'eman, had been scribbling down calculations about the masses of the new particles. Realizing that they would fit into a decuplet, he had jumped up and waved his hand, seizing the spotlight as eagerly as he had back in elementary school. Then he walked up to the podium and began describing the arrangement— the four deltas, the three sigmas, the two xis. He explained that the Goldhaber gap would rule out a twenty-seven-member cluster. And then he predicted the existence of the triply strange tenth particle (a possibility he had actually raised as long ago as the 1955 Pisa conference). The name he had suggested back then, omega-minus, seemed particularly appropriate now: It would be the final particle in the decuplet. He was so excited by this realization that at first he called the new particle the omega-zero; then he corrected himself, pointing out that it would be negatively charged.

The announcement barely caused a ripple in the room. The discussion quickly moved on to other topics. But once again Ne'eman had been scooped, just barely, by Gell-Mann. Goldhaber leaned over and asked, "Did you tell him about your prediction?" Ne'eman said no; Gell-Mann had come up with it on his own. Walking back to his seat, Murray stopped to say hello to Goldhaber and another experimenter, Yehuda Eisenberg. Gell-Mann said that the decuplet scheme might provide an explanation for a particle Eisenberg had observed years earlier in cosmic rays. Eisenberg might have already seen an omega-minus! Then Gell-Mann spotted Ne'eman's name tag, introduced himself, and asked him and the others to join him for lunch in the CERN cafeteria. While they were eating, two experimenters from Brookhaven, Nicholas Samios and Jack Leitner, walked up to the table. Struck by the elegance of the Eightfold Way, Samios thought that testing it would be an experimenter's dream. He knew it would take considerable time and a lot of money, but the answer would be clear-cut: The omega-minus either existed or it did not. The investment was risky, but Gell-Mann's batting average had been very high. He asked Murray what kind of pattern the experimenters should look for, how the omega-minus would likely decay. Gell-Mann jotted down some calculations on a napkin along with a note to the director of Brookhaven, Maurice Goldhaber (Gerson's brother), urging him to let Samios search for the particle. Then he continued to talk to Ne'eman. Gell-Mann had heard that he was an Israeli army colonel and felt sure he would be a fascinating person to know. Though they lived on opposite sides of the Atlantic, they had been dwelling in the same mental spaces. They hit it off immediately, and Murray invited him to come to Caltech the following autumn to work for a couple of years.

Neither Ne'eman nor Gell-Mann seemed especially concerned about having to share credit for the Eightfold Way. It was far from clear at that early stage that $SU(3)$ octets were anything worth fighting over. Many of their colleagues suspected that the two physicists were simply seeing pictures in the clouds. Freeman Dyson insisted that the theory couldn't be right. If you slammed, say, two protons together, the likelihood that a certain particle is produced in the debris is called its cross-section. Dyson was bothered that some of the strange particles, which were supposed to be members of the very same octet, had wildly different cross-sections. "You don't understand," Murray told him. "The cross-sections are just details." If you stepped back and looked past such distractions, the order came shining through. Dyson wasn't convinced.

In many people's minds, the Eightfold Way had another count against it. It was a field theory, and field theory was in disrepute. The idea of particles interacting by exchanging some kind of virtual bosons—the supposed carriers of the field—still wasn't working for the strong force. For all the efforts of the theorists, the equations, with their ever-mounting infinities, remained unsolvable. All of this continued to feed suspicions that understanding the strong force, and classifying the particles under its sway, could only be done with the S-matrix.

Gell-Mann himself remained uncomfortably on the fence. Field theory seemed too powerful to discard, yet he didn't want to seem like a reactionary at a time when the S-matrix (or dispersion relations, as he still preferred to call the theory) was the avant-garde. He found that if he drastically oversimplified the nature of the strong force, he could come up with toy field theories that could be used to make calculations. He knew the toy theories were wrong, but they could be used as devices to abstract general principles about particle behavior. He hoped these principles would hold true whether or not field theory was eventually vindicated. After all, everyone knew that the earth revolved around the sun, and not vice versa. But navigators found it easier to calculate the position of the stars using the old-fashioned geocentric model.

In the end, Gell-Mann still hoped that the S-matrix would turn out to be compatible with field theory, not a replacement for it. The same summer as the Geneva conference, he and Murph Goldberger met in Berkeley for a Jason gathering and started talking about how the two approaches might be reconciled. Chew and his followers had become increasingly excited about Regge poles as a means of classifying particles. Goldberger and Gell-Mann were also enthusiastic about the theory. But while Chew's group had embraced Regge's trajectories as an integral part of the S-matrix, Murray and Murph suspected that the phenomenon could also be explained with old-fashioned quantum field theory. If field theory failed to account for the Regge phenomenon, it might indeed be in trouble. Gell-Mann went back to Caltech and worked on the matter with Zachariasen. And Goldberger started discussing it with Francis Low.

Goldberger and Gell-Mann had a chance to get back to the problem several months later when they were invited as visiting professors to MIT, where Low was now on the faculty. In January 1963 the Gell-Mann family moved temporarily to Cambridge and rented Julian Schwinger's house while he was away on leave. After some intense bouts of calculation, the three scientists were able to show that Reggeism was indeed consistent with field theory. Whatever one thought about Chew's doctrine, the fact that particles lined up along Regge trajectories could not be used to show that the S-matrix was truer than quantum field theory—more evidence that these two seemingly immiscible theories were just different perspectives on the same thing.

It was like old times with Murray and his two old colleagues back together again. One day, for no particular reason, they all decided to collaborate on the best gourmet dinner imaginable, potluck-style. Murph and his wife dug a fire pit in their driveway and, with the advice of a friend, turned a garbage can into a meat smoker, cooking a turkey suspended on a coat hanger over the fire. Margaret prepared a filet mignon. A theoretical chemist at MIT was in charge of the champagne. (He wouldn't let Murray transport it in the old car he had borrowed for his visit because the suspension was bad. All the jostling would break the champagne's long-chain molecules.) Everyone, including the Lows and some other couples, converged at Schwinger's house for the feast. It was a black-tie affair with a menu printed for the occasion. After he returned, Schwinger had it framed.

Gell-Mann, Goldberger, and Low made for a powerful combination of talent, and the three friends wished they could work together all the time. During the stay in Cambridge, they began discussing the possibility of moving en masse to Berkeley, forming the core of a new institute or department. Gell-Mann was personally offered a corner office with a view of San Francisco Bay. Geoffrey Chew, Sheldon Glashow, and Steven Weinberg were already part of the formidable cast at Berkeley. When Chew heard the news, he was overwhelmed at what he thought of as a "coup to end all coups" but also a bit apprehensive. Would the other theorists feel that Gell-Mann, Goldberger, and Low were getting special treatment? And would the three newcomers agree to work for $20,000 for a nine-month appointment? "Murray particularly worries me," Chew wrote to the three scientists, "because his talents are so exceptional and he has a right to expect some unusual recognition. If Murray gets his Nobel Prize soon this problem may be solved. If he has to wait for another five or six years, he may make the rest of us uncomfortable. How about it, Murray?" And complicating matters, Chew wrote, was the fact that Abraham Pais had a part-time position at the Berkeley radiation lab. "If we start a new Institute is Pais 'in' or 'out'? Again I call on Murray!"

There were other potential problems. Murray wasn't the only one who felt that Chew's single-minded pursuit of the bootstrap sometimes bordered on zealousness. After hearing one of Chew's recent sermons, Goldberger started calling him "the Billy Graham of physics." He added: "After his talk I nearly declared for Christ. Since I had already changed from Jewish to Regge-ish, it was the only thing I could think of."

The worries turned out to be irrelevant. The dream institute never came to pass. But the idea of forming a new research center remained at the back of Murray's mind. Not necessarily a center focusing on physics but one that would bring together people from many different fields to solve the world's problems.

In late March of 1963, near the end of his leave, Gell-Mann traveled to New York to give a series of three lectures at Columbia on the strong interaction, the Eightfold Way, and the bootstrap. The day before the second talk, the physicist Robert Serber took him to lunch at the faculty club. Walking up Broadway with Murray and a group of other physicists, Serber was eager to discuss the Eightfold Way. A couple of weeks earlier, another physicist had given a background talk to get the Columbia scientists up to speed on $SU(3)$. Serber, who had learned group theory as a tool for analyzing the energy levels of atoms and molecules, immediately wondered why $SU(3)$ gave rise to particles arranged in eights and tens but not in threes. He played around with the idea and satisfied himself that octets and decuplets could be built from triplets, implying that there was some kind of trinity of fundamental subparticles from which the baryons were made. (He also noticed that, in addition, mesons would be made not from triplets but from pairs of these hypothetical subparticles, combining one with its antimatter equivalent.) Sitting at a table in the faculty club, he asked Gell-Mann why he hadn't entertained this idea.

T. D. Lee, who was sitting with them, said it was a terrible notion.

Murray started scribbling calculations on a napkin, showing Serber that (as Ne'eman had realized the year before) the fundamental particles making up the triplets would have to have charges of $-\frac{1}{3}$ or $\frac{2}{3}$. Everyone agreed that this was a serious problem.

But Serber's question started Murray's mental wheels turning. He was vaguely aware of Ne'eman and Goldberg's rather abstract treatment of $SU(3)$ triplets, but it hadn't sparked a fire. Like the "pedagogical" triplets he had mentioned in his paper on the Eightfold Way, the idea seemed little more than an empty mathematical formalism. At MIT he had been mulling over a scheme in which baryons were made from *four* fundamental objects, which allowed him to avoid, in a rather contrived way, the seeming absurdity of fractional charges. Now his thoughts went careening down a different avenue. Fractional charges seemed impossible because no one had ever seen them. But that didn't necessarily mean they didn't exist. What if the subparticles were somehow hidden away inside the hadrons? Maybe they could be thought of as having infinite mass; then no amount of energy could spring them loose. Crazy as it seemed, this idea had an added attraction. One could believe in triplets without forsaking the bootstrap. If Chew was right—if there was indeed nuclear egalitarianism—then particles like neutrons, protons, and so forth were ultimately all made from one another, nodes in a self-supporting network. Having tinier objects (or whatever they were) within the particles wouldn't necessarily derail the theory—so long as these pieces were forever unobservable.

In his next lecture, Gell-Mann mentioned Serber's question about fractionally charged triplets. During coffee afterward, he and some other physicists mulled over these hypothetical "quirks" of physics, and in the course of the conversation Murray started calling them "kworks," a nonsense sound that was just supposed to mean "funny little things."

On their way back to California, the Gell-Manns stopped at Woods Hole, Massachusetts, on the southwestern tip of Cape Cod, for a Jason gathering. There were soon four of them. A son, Nicholas, was born at a Boston hospital in July. The family set up housekeeping at the Miramar Guest House, a grand oceanside manor with a private beach and a two-story living room. By now, Jason was paying him $200 a day for these sessions. Particle-beam weapons, long-range sonar, radar-jamming devices—Jason was at the center of the development of high-tech warfare. Murray was chairman of a group studying the calculus of disarmament and the likely effects on the arms race of building antiballistic missile systems.

After Murray returned to Caltech for the fall semester, he and Ne'eman, who had arrived from Israel, quickly began discussing the idea of fractionally charged subparticles. Under Gell-Mann's tutelage, Ne'eman was learning that there was more to doing physics than coming up with original theories. He would arrive at work with a barely formed idea, then watch as Gell-Mann worked out the implications, showing how it fit in with the rest of the picture physicists were drawing. Murray, he quickly learned, seemed to have an almost innate sense of what a theory should look like, how the final product should be packaged. It had never occurred to Ne'eman to give the SU(3) scheme a catchy name like the Eightfold Way or to call the missing particle in the decuplet something like the omega-minus. He just left it as a symbol in an equation. And he never would have come up with a funny sound like "kwork." Discussing his nascent idea of fractionally charged triplets with Murray, Ne'eman realized that what he had left as a mathematical abstraction, Gell-Mann was solidifying into a theory about the physical world.

The two physicists also found that they shared a love for languages and history. But not all their interests overlapped. Ne'eman couldn't imagine why anyone would want to get up early in the morning to watch birds. And Ne'eman's devotion to Israel left Murray cold. In fact, Ne'eman later noticed that his friend had developed a strange mental block. He would stubbornly refuse to say the word "Israel," referring to "Palestine" instead. He didn't even like to be considered Jewish. Some of his colleagues were offended by this odd behavior. Others were more charitable. Glashow figured that his friend simply refused to be categorized. Religion had, after all, played no part in Murray's life. He was a man of science. Back in Vienna, his father had considered himself a citizen of the world, and so did Murray. Ne'eman was also willing to overlook this idiosyncrasy. Gell-Mann, at least, was someone who, like himself, was deeply interested in world affairs.

Feynman was a different story, constantly getting on both their nerves. The three were sharing a meal one day when Ne'eman announced that he was being called back to Israel for the inauguration of a new campus of Tel Aviv University. In the course of the conversation, Ne'eman mentioned the many contributions Jews had made to science. "Jews and science?" Feynman asked, as though he was genuinely puzzled. He wasn't aware of any great Jewish contributions to science. Now, Hungarians were important, Feynman said, naming Szilard, Teller, Wigner, and others. Murray looked at him in mild disbelief. "Don't you know, Dick, that all these Hungarians are Jewish?" When he ventured even a few inches outside physics, Feynman seemed to Ne'eman like an ignoramus.

It was about this time that Murray was paging through a copy of Joyce's bottomless novel *Finnegans Wake*, whose dense network of puns and allusions had become the obsession of so many scholars, when his eyes alighted upon the lines he was about to make famous:

> —*Three quarks for Muster Mark!*
> *Sure he hasn't got much of a bark.*
> *And sure any he has it's all beside the mark.*

Perfect. It took triplets of quarks to make the octets and decuplet of the Eightfold Way. Murray was a little bothered that Joyce's "quark" rhymed with "bark." He had it in his head that it was supposed to be pronounced "kwork." But you could read just about anything into *Finnegans Wake*. Putting on his hat as amateur Joycean, he noted that "Three quarks for Muster Mark!" could also be taken as a call at a tavern for a round of drinks (the novel, after all, is about the drunken dream of a pub owner, H. C. Earwicker). "Three *quarts* for Muster Mark!" That would be in accord with Murray's preferred pronunciation. Literary exegesis could be as inventive as particle physics. He was also amused that "quark" in German refers to a kind of soft, smelly cheese and can be slang for "nonsense."

One day he tried to explain quarks to his old teacher Viki Weisskopf during a phone call to CERN. "Please, Murray," Viki said, "let us be serious; this is an international call." About that time, Murray showed a draft of a paper on quarks to Stanley Mandelstam, who worked with Chew at Berkeley. He exclaimed that Murray was trying to start a counterrevolution against the idea of the bootstrap by proposing that there were fundamental particles. Gell-Mann pointed out, as he would many times in the coming years, that nuclear democracy was safe if the fundamental particles were unobservable. He wasn't quite sure yet what that meant. But it seemed to be the only way to get the idea to fly.

By the end of the year, Gell-Mann was putting the finishing touches on "A Schematic Model of Baryons and Mesons." Elegant and succinct, just two pages in length, the paper, like so many of his others, was a model of understatement. The implications and details of the quark scheme are implicit in the equations, not baldly stated, almost as though he were hiding the radical implications.

He opened the paper by paying obligatory homage to the boot-strap. And though he described his new idea using the language of field theory (pretending that the quarks were held together by some kind of hypothetical bosons), he made it clear that this was just a mathematical convenience. He wasn't making a statement about how the world worked but simply providing a model, a device for making calculations. Everything, thus far, was politically correct. Then, just to get it over with, he described his uglier model involving four integrally charged subunits. But if one were willing to accept the idea of fractional charges, he ventured, a more elegant scheme was at hand. And he proceeded to show the wonderful things one could do with quarks, even citing James Joyce in the footnotes. Suppose quarks come in three varieties, u, d, and s (standing for "up," "down," and "strange," though he didn't say so in the paper). The up quark has a charge of $\frac{2}{3}$; the down quark and the strange quark each have a charge of $-\frac{1}{3}$. Combine two up quarks and the down quark ($\frac{2}{3} + \frac{2}{3} - \frac{1}{3}$), and you get a proton charge of $+1$. Combine two downs and an up ($-\frac{1}{3} -\frac{1}{3} +\frac{2}{3}$), and you get a neutron with zero charge. Since the proton has more up than down quarks, it has an up isospin, and vice versa for the down-spinning neutron. Strange particles have strange quarks inside them, the total being indicated by their strangeness number. (The triply strange omega-minus, for example, would contain three of them.) When particles decay, Gell-Mann speculated, the weak force changes the value of one of their quarks, and therefore their identity—a neutron (ddu) becomes a proton (duu).

If nothing else, this was an extremely clever accounting device—and maybe more. "It is fun to speculate," Gell-Mann wrote cagily toward the end of the paper, "about the way quarks would behave if they were physical particles of finite mass (instead of purely mathematical entities as they would be in the limit of infinite mass). . . . A search for stable quarks . . . would help to reassure us of the non-existence of real quarks."

He would spend the rest of his life trying to explain those enig-matic lines. What could it possibly mean to say that everything is ultimately made of "purely mathematical" particles that are perma-nently bound together and can never be observed in isolation? Did quarks really exist? This was the annoying kind of philosophical question that Arthur might ask. For Murray, quarks were not tiny little objects so much as they were patterns, symmetries underlying nature. As the ancient Pythagoreans put it, All is made from number. Still, just as the discovery of protons, neutrons, and electrons—the components of atoms—explained the symmetries of Mendeleev's periodic table of the elements, quarks, whatever they were, explained why the Eightfold Way worked so well.

Dreading the thought of dealing with *Physical Review Letters*—he imagined trying to get the word "quark" past the editorial establishment—he sent the paper to a new CERN publication called *Physics Letters*. The editor had been bugging Murray to submit a paper and was happy to accept even something as crazy as a discussion of fractionally charged particles. Received on January 4, 1964, it was published with lightning speed on February 1.

Few physicists were likely to embrace something as outlandish as quarks while the Eightfold Way was still up in the air. Across the country at Brookhaven, Samios and his colleagues were in the midst of a heroic and maddening effort to test the theory by seeing if the unknown particle it predicted, the omega-minus, really existed. Things weren't going well. Using their gargantuan accelerator, they sent protons spinning around and around a circular track, a full half-mile across, then slammed them into a hard target. The debris from the minute collisions was then funneled into a pipe hundreds of feet long and filtered by magnets until nothing should be left but a pure beam of the strange particles called kaons. If all went according to plan, one or two of the kaons would break down into the omega-minus, leaving the trail Gell-Mann had described on the napkin back at CERN.

The experiment was as difficult and excruciating as Samios had predicted. It took months of tinkering just to tune the machine to spit out a steady stream of kaons. Then thousands of photographs a day of the disintegrating particles were analyzed with the aid of Long Island housewives trained to measure the angles and curves and help spot the telltale signature of an omega-minus. After weeks of this tedium and some 50,000 pictures, there was no sign that the particle existed. Ne'eman was sitting by a swimming pool at a conference in Coral Gables, Florida, in January 1964, when Maurice Goldhaber told him the bad news. It was a little shocking to him

that such a beautiful theory could be wrong. Returning to Pasadena, Ne'eman told Gell-Mann. Murray, who was occupied with
planning a trip to Japan later that year, suggested that the two theorists might have to jump off Mount Fuji in disgrace. Ne'eman, going
along with the joke, said he could always go back to the Israeli army.

Then, on the very last day of the month, in the 97,025th picture,
Samios saw something funny. Working the night shift, he was staring at a bubble chamber photograph when, through the thicket of
lines and spirals, he thought he perceived a track of a kaon suddenly interrupted in mid-flight as though it had struck another particle. And a foot away, on the other side of the picture, was a track
like a V—a shape generally interpreted as a neutral particle decaying into positive and negative remnants, each pulled in different
directions by the bubble chamber's magnetic field. Could this
invisible neutral particle be distant debris from the collision of the
original kaon? If so, then what happened in between? Could an
omega-minus have made a fleeting appearance?

The next morning, Samios showed the picture to some other
physicists, and they peeled through the possible layers of meaning.
Was that the mark of a gamma ray—two spirals curling in opposite
directions as a photon decayed into an electron and a positron?
And, look, over there—was that *another* gamma ray? They measured
the lines and angles, calculated the energy involved in the various
interactions, and derived the particles' masses. They finally decided
that they had seen the very reaction Gell-Mann had advised them
to watch for: A negative kaon had collided with a proton and produced three particles. One of them, a positive kaon, was easy to
see. The second, a neutral kaon, couldn't actually be seen. Chargeless particles leave no tracks. One had to infer their existence from
their breakdown products. So far, nothing interesting had occurred.
But then there was a third particle: From its tiny track, one could
make a case that, for the briefest instant, there existed a negatively
charged particle with a mass of about 1,682 million electron-volts,
plus or minus 12 (Murray had predicted 1,685), and a strangeness
of 3. An omega-minus.

According to the story the physicists slowly pieced together,
the omega-minus then quickly decayed into two more particles:
a negative pion and an invisible neutral xi. This second particle
then decayed, in turn, into still more invisible particles: a neutral
lambda and a neutral pion, which disintegrated into the two gamma-
ray photons. The gammas finally decayed into the two electron-

positron coils, the lambda into a proton and a pi minus—the very V that had first grabbed Samios's attention. At first he and his colleagues were stunned, looking at one another numbly. Then they started to smile. Someone did a little dance. They were the only ones on earth who knew that the particle existed. The theorists might think they knew, but here was the first physical evidence. The empty slot in the Eightfold Way had been filled.

Obviously, as with reading Joyce, a great deal of interpretation was involved. The omega-minus was not something anyone would likely notice if not already primed to see it. But the physicists' minds were now as perfectly tuned as the magnets on the accelerator, and before long similar events were found at both Brookhaven and CERN.

With all the excitement, Samios forgot to call Gell-Mann right away. When he finally remembered, Murray had to pretend to be surprised. He and Ne'eman had heard rumors about the experiment and had called Brookhaven themselves for details. No one at the lab had thought to call Ne'eman directly. Samios later sent photographs with an apologetic note: "Please excuse the oversight, but you knew it existed before we did!"

Not everyone wanted to be convinced. Frank Yang didn't believe that finding the omega-minus necessarily vindicated the Eightfold Way. There might be another explanation. For months, Feynman, maybe a little miffed at having blown his chance to collaborate on the theory, would periodically burst into a room where Gell-Mann and Ne'eman were working. "Have you heard the news? The omega-minus has a spin of one-half!" That would invalidate the whole decuplet scheme, which was made of particles with a spin of $3/2$.

But that was wishful thinking on Feynman's part. As far as most physicists were concerned, the Eightfold Way was a smashing success. In his original periodic table, Mendeleev had left gaps that were filled in later as new elements were found. Soon people were talking about Gell-Mann as the new Mendeleev. (If they were scrupulous with their citations, they conferred the honor on Ne'eman as well.) Coincidentally, Gell-Mann's quark paper was published the day after the omega-minus was found. Robert Serber kicked himself for not publishing the idea himself. (Murray acknowledged their talk in a footnote.) But the idea was too strange and new to cause much of a sensation. Quarks were going to take a while to win acceptance, even from their inventor.

CHAPTER 11

ACES AND
QUARKS

Gell-Mann was once asked, when he was barely in his thirties, why theoretical particle physics was a young man's game. In biology and medicine, so many of the great breakthroughs were made by older, distinguished researchers with many years of experience. In particle physics just about every important new theory was coming from young upstarts who, like Murray, were still only a short way into their careers. The pattern began with the quantum revolution in the early part of the century. Planck himself was already forty-two when he came up with the notion that energy was emitted in packets called quanta. But the astonishing strides Heisenberg, Pauli, and Dirac made in working out the surprising implications were accomplished when they were in their early twenties.

The reason for such precociousness, Gell-Mann speculated, was the very different nature of physics compared with, say, biology and medicine. The so-called life sciences consisted of a huge repository of facts, many of them unconnected, that had been piling up for decades. It took years of study and rote memorization for an aspiring scientist to master what was already known. By the time a researcher was ready to make an original contribution, he was probably well advanced in his career. Watson and Crick's stunning revelation of the double-helical structure of DNA was obviously an exception. But in particle physics, sudden breakthroughs by inexperienced minds happened all the time. The young science was not yet weighed down with learning. By the time a student was working on his Ph.D. dissertation, he was already at the cutting edge of the field. "Being a physician requires immense knowledge, vast libraries, and much learning and experience," Gell-Mann told his interviewer. "We need only a handful of books, a few copies of the

Physical Review, and off we go. . . . But we use a lot of deduction, induction, and reasoning. Once you understand, you can throw away your books and reason out what you've forgotten—except the latest data, which you don't understand at all!" In particle physics, a handful of basic concepts were elaborated on using a handful of powerful mathematical tools, many of them at least half a century old. Agility of mind was more important than wisdom and experience, and so younger brains had an edge.

And there was another factor that made particle physics different from some other sciences: There was a richer vein to mine. In the early 1960s, with the field just barely beginning, most of the big discoveries were yet to be made. Combine that with the pack mentality that exists in all disciplines—at any one time, there tends to be only a few problems that people are working on—and one can go a long way toward explaining why so many of the important discoveries occurred to two people almost simultaneously. Why one of the discoverers would go on to get most of the glory was more difficult to explain.

George Zweig began to wonder whether "elementary" particles like baryons and mesons could be made from tinier pieces while he was finishing his doctorate at Caltech. He came to the question by a rather roundabout route. Originally Zweig thought he wanted to be an experimenter, not a theorist. In the early 1960s, he was running back and forth between Pasadena and the Lawrence Berkeley National Laboratory in northern California, site of an accelerator called the Bevatron, trying to see if decaying kaons violated something called time-reversal symmetry, the idea that a reaction is equivalent whether it runs forward or backward in time. Since kaons were strange particles, they were barred from decaying through the strong force. It was now well established that when the weak force came into play, wearing the particles down, it violated parity, the symmetry called P. A reaction and its mirror image were not the same. And the weak force violated the C symmetry, charge conjugation. Switch the particles with their antiparticles and, again, the two reactions would diverge. Zweig wanted to see whether the weak force violated time symmetry, called T, as well.

It was while pondering this question that he became acquainted with Gell-Mann. He occasionally wandered into Murray's office—he thought of him as the house theorist—to discuss what was showing up in the kaon experiment. As a second-year graduate student, Zweig had heard Gell-Mann's first seminar on the Eightfold Way and been blown away. It was a bravura performance, and Zweig came away convinced Murray was right, even before the discovery of the omega-minus. And he was amazed by Gell-Mann's output. In 1961 and 1962, he generated a paper almost every two months.

Zweig got along reasonably well with his rather intense teacher— he figured that he, a student, didn't pose much of a threat to Gell-Mann's fragile ego—and he found their discussions on the abstractions of physics immensely stimulating. Graduate students, in particular, found Gell-Mann cordial and sometimes even kind. They were amused by his natty dress—he lectured in a houndstooth jacket during the hottest months of summer—and his ability to pick up enough of an obscure language to convince the innocent that he spoke it like a native. Once, after visiting the Yucatan and learning a few words from some Indians, a student tried to test Gell-Mann's claim that he knew Mayan. Without missing a beat, Murray said the sentence the student was speaking was Lower Mayan. He only knew Upper Mayan.

But those who had to compete with Gell-Mann didn't always find him so charming. By this time the abrasive Gell-Mann style was becoming legendary. Some physicists, like Leon Lederman, could barely stand to talk to him anymore. If Lederman wanted to discuss electroweak unification and made the mistake of mentioning W particles (the hypothetical carriers of the weak force), Murray would interrupt: "Uxyls. They're called uxyls." It was impossible to carry on a conversation. Instead, Lederman joined the longer line of visitors waiting outside Feynman's door. For all his brashness, at least Dick didn't make him feel stupid about asking questions.

More and more, Gell-Mann seemed to have an almost obsessive need to put people down, especially potential competitors. During talks by visiting physicists, he would pointedly spread out a newspaper, ruffling the pages in exaggerated boredom. It was difficult to carry on a conversation with someone who was constantly correcting your pronunciation or rephrasing what you had just said, trying to wrest control over every situation. Feynman, it was true, could be just as bad. He would announce in the middle of a seminar that the speaker's work epitomized everything that was wrong with physics and then stride melodramatically out of the room. Science for him was a kind of performance art.

Steven Weinberg got the Caltech treatment when he came down from Berkeley in the early sixties to give his first seminar there. He had reason to be apprehensive. He had gotten off to a bad start with

Gell-Mann several years earlier at a physics conference in Gatlinburg, Tennessee. As a graduate student at Princeton, Weinberg had read Murray's paper on weak magnetism, the same one that had so impressed Jeremy Bernstein. Weinberg was amazed at how Gell-Mann had started the paper with some very sophisticated assumptions about fundamental theory and ended up with a detailed practical analysis of nuclear physics that one might expect only from an experimenter. Most physicists focused either on abstract mathematical theory or on understanding what was actually going on in the experiments, the phenomenology. Gell-Mann clearly was doing both, and the paper set a standard for Weinberg of just what a theorist could do. Riding from the Gatlinburg airport with a group of physicists, Weinberg asked one of them what he thought of something Gell-Mann had just published. "Why don't you ask him yourself?" was the reply. Murray was there in the car. When Gell-Mann gave a talk on weak magnetism at the conference, Weinberg was ready with a pointed question. In his analysis, Gell-Mann had implicitly made an assumption about something called second-class currents. Shouldn't this be made explicit? Murray thought about Weinberg's question for about ten seconds and said, simply, "No." The room was silent. Weinberg paused a few seconds to regain his composure and said, "Yes." The audience broke out laughing.

A few years later, at the Rochester meeting in Geneva, Gell-Mann got back at Weinberg. After Murray's talk on Regge poles, Weinberg raised his hand and asked about what he considered a mathematical difficulty in the equations: "Does the vanishing of the denominator $2J + 1$ at $J = -\frac{1}{2}$ have anything to do with the difficulty of analytically continuing from $J > -\frac{1}{2}$ to $J < -\frac{1}{2}$?"

"Probably," Murray answered, letting the word hang in the air. He thought Weinberg was a creative theorist, though perhaps a little too preoccupied with mathematical rigor. Some problems were just too complex to solve exactly, at least with the current level of understanding. But one could find hints of what nature was doing by playing around heuristically—trying this idea and that idea, relying on rules of thumb that seemed to work, even if you couldn't explain exactly why. When you were working at the edge of the unknown, the last thing you needed was to get hung up on details. Gell-Mann thought Weinberg didn't fully appreciate this rough-and-ready style.

When he made his Caltech debut, both Feynman and Gell-Mann needled him mercilessly. Weinberg, feeling a little traumatized, thought Feynman was just playing cat-and-mouse: He loved making people squirm. Murray's aggressiveness at least seemed motivated by a genuine interest in the physics; he pounced on any idea he thought was leading in the wrong direction. But it was possible to stand up to Gell-Mann and, if you were right, even win him over. Not so with Feynman, Weinberg felt.

Gell-Mann and Weinberg developed a grudging, prickly respect for each other. But from then on, Weinberg felt gun-shy about lecturing at the school. Physicists started to talk about a Caltech style of doing physics, more brutal even than Oppenheimer's relentless grilling or Pauli's acerbic wit. Ideas were attacked with a viciousness that some found shocking, others exhilarating. If your theory stood up to this kind of scrutiny, then maybe it would survive the supreme arbiter, nature itself. Particle physics is so intensely competitive that even close collaborators like Yang and Lee found their friendship could not survive. (They bitterly parted ways several years after their triumph over parity.) But no school was more combative than Caltech. Visiting speakers learned to prepare themselves for what might develop into a no-holds-barred fight if either Murray or Dick was in the audience. On the increasingly rare occasions when both of them were at the same seminar, spectators would be treated (or subjected) to a tournament of wills, all the more exciting because it was being held between the men whom many considered the two smartest people in the field. Murray would make a sarcastic crack about something Feynman had just said. Dick would fire back. But while Feynman usually laughed off the absurd exchanges, Gell-Mann would get angrier and angrier. Dick didn't let anything get under his skin; Murray let everything get under his. The two were so different and so competitive that it was becoming harder for them to be friends.

Though put off by this kind of intellectual pugilism, Zweig was finding the idea of being a theorist rather than an experimenter more appealing all the time. After two and a half years of building equipment, tweaking the adjustments, and poring over preliminary data, he concluded that the experiment on time symmetry was going nowhere. The apparatus was simply not detecting enough kaon decays to bother with detailed analysis. In the summer of 1962, he and his wife, a physicist named Erica Jen, took a vacation to the Yucatan Peninsula. Getting away from science was just what Zweig needed. After a couple of months of relaxation, he decided to write his thesis on a theoretical problem. He returned that fall and told Gell-Mann, who was getting ready to leave in a few months for the sabbatical at MIT. "Why don't you go see Dick?" he said. Zweig considered Feynman unapproachable and somewhat terrifying. But Murray offered to call and pave the way.

"Murray says you're O.K.," Feynman said when Zweig walked into his office. "And if Murray says you're O.K., you must be O.K." He agreed to become Zweig's adviser. Like so many young physicists, Zweig was immediately pulled in by Feynman's charisma. By the time Murray left for the East Coast in January, Zweig was deep into his thesis problem. He graduated that spring, and by the time Gell-Mann returned from sabbatical, all fired up about fractionally charged triplets, Zweig was off to Geneva to work as a postdoctoral researcher at CERN—ships passing in the night.

And so it was with an unencumbered mind that Zweig began to consider novel ways to explain a perplexing phenomenon he had heard about while putting the finishing touches on his dissertation. Experimenters at Brookhaven had recently discovered that the strong force caused a particle called the phi-meson to decay, against all expectations, into a kaon and an antikaon, ignoring the far likelier, more energetically favorable decay path: into a rho and a pi-meson. Feynman had taught him that in strong interactions anything that can possibly happen must happen unless barred by a conservation law. (Or as Murray put it, echoing Orwell, anything not forbidden is compulsory—the totalitarian doctrine.) Zweig found the phi's behavior bizarre and was excited by the possibility that some unknown physical process was involved. What if the mesons were composed of smaller pieces? A particle would decay when these subparticles came unglued. If the phi-meson was made of certain pieces and if the rho and pi—the particles the phi was *supposed* to decay into—were made of different pieces, that would explain everything. There would be no way to take the subparticles that made up the phi and reassemble them to make pis and rhos. You couldn't get from point A to B any more than you could add two even numbers and come up with one that was odd. The phi would have to decay by a different route.

After settling in at CERN that summer, Zweig began to consider the implications of this radical suggestion. He knew the notion that hadrons were composed of smaller units was way out on the fringe. About the closest thing was Sakata's model, in which hadrons were made from various combinations of protons, neutrons, and lambdas. Like most physicists, Zweig thought Sakata was wrong—not

because there was something abhorrent about fundamental particles (as devotees of the bootstrap insisted) but because the Sakata
triplets couldn't be used to build up the patterns of the Eightfold
Way. Zweig played around with $SU(3)$ and saw that the octets and
decuplet *could* be built up from triplets if they consisted of some
heretofore unknown particles that were, of all things, fractionally
charged. At first he was bothered by the idea. He wanted the
constituents—by now he was calling them "aces"—to correspond
somehow to the simple leptons like the electron and neutrino,
which had integral or zero charge. As physics now stood, there were
four of these lightweight leptons—the electron, the muon, and the
neutrinos associated with each of them. But there were scores of
heavier hadrons: baryons like the neutron and proton and mesons
like the pion. How much neater the universe would be if this
embarrassing abundance of hadrons could be shown to consist of
combinations of tiny aces. Then everything would ultimately consist
of a handful of integrally charged particles, leptons and aces. But
he couldn't make the idea work.

If he accepted, however, the idea of fractional charges, like ⅓
and ⅔, the pieces all seemed to fit together. Mesons and baryons
were built from three different subparticles represented in Zweig's
scribblings by a circle, triangle, and square—the equivalent of what
Murray (unbeknownst to Zweig) was calling the up, down, and
strange quarks. (Zweig also suggested the existence of a fourth
ace—as many as there are in a deck of cards—that would be found
inside more exotic particles. Then there would be four leptons—
the electron, the muon, and their neutrinos—and four aces. A
pleasing symmetry. But for now, the fourth ace's place in the grand
scheme was unclear.)

In Zweig's emerging system, baryons were made of triplets (he
called them "treys") of circles, squares, and triangles connected by
"strings." The proton was made from two circles and a triangle (two
ups and a down, in Gell-Mann's terminology) and the neutron from
two triangles and a circle (two downs and an up). Mesons, as in
Murray's scheme, were made from an ace and an antiace, a "deuce."
Zweig was on a roll. He would excitedly jump up in the midst of a
calculation, pace back and forth, then rush back to his desk to see if
what was unfolding still really seemed true. He was seeing the world
through new eyeglasses. Weak decay occurred when one of the aces
in the neutron, for example, changed into a different kind of ace. It
was not the whole neutron but one of its components that decayed.

That fall, while Gell-Mann was talking over the triplet idea with Ne'eman at Caltech, Zweig began writing his own report. It appeared on January 17, 1964, as a CERN preprint. Then, two weeks later, he picked up *Physics Letters* and felt a sick, sinking feeling as he read Gell-Mann's paper on quarks. Zweig had figured he had all the time in the world to work out the idea completely and savor the results. Now he rushed to expand his theory in a second CERN document, which was distributed a month later.

After he had recovered from the shock, Zweig consoled himself with the realization that he and Gell-Mann had approached the idea of SU(3) triplets in very different ways. While Gell-Mann's paper was a mere two pages (with only the latter part devoted to the quark model), Zweig's second preprint, counting all the charts and diagrams, was some eighty pages long. In his own paper, Gell-Mann was tentative, even timid. Quarks were "mathematical," not "real," more like patterns than things, part of the abstract symmetries of the Eightfold Way. The members of the octets and decuplet flowed from one into the other along the intertwined paths—the currents—described by the group of transformations called SU(3). Quarks provided an especially elegant and compact way of describing the choreography. Gell-Mann called his theory a "schematic" model of baryons and mesons. His rarefied "current quarks," as they came to be known, seemed more like symbols than things—mathematical cogs in the abstract calculational scheme he called current algebra. Zweig's aces were something that could be snapped together like Tinkertoys.

While in the Gell-Mann paper the ramifications of the quark hypothesis were left largely implicit—there was a great deal to be unpacked from those few paragraphs—Zweig laid out the story with a panoramic sweep. "We will work with these aces as fundamental units from which all mesons and baryons are to be constructed," he asserted. He speculated on what might clamp the aces together to form particles: probably the strong force but maybe a yet-undiscovered interaction stronger than the strong force. He ticked off the questions that would have to be answered: "Are aces particles? If so, what are their interactions? Do aces bind to form only deuces and treys? What is the particle (or particles) that is responsible for binding the aces? . . . And more generally, why does so simple a model yield such a good approximation to nature?" This was what seemed to bother him the most: The model was almost too good to be true. "In view of the extremely crude manner in which we have approached the problem, the results we have obtained seem somewhat miraculous."

Zweig conceded that aces might be no more than a tool that "supplies a crude qualitative understanding of certain features pertaining to mesons and baryons. In a sense, it could be a rather elaborate mnemonic device." And then he concluded with what he believed in his heart was true: "There is also the outside chance that the model is a closer approximation to nature than we may think, and that fractionally charged aces abound within us."

These were the words of an iconoclast, a romantic, someone happy to take on an establishment convinced that there were no fundamental particles, that everything was made from everything else according to the egalitarianism of the bootstrap. Gell-Mann couched things so that he could have both quarks and the bootstrap. "Mathematical" quarks, permanently trapped and unobservable, gave rise to "real" particles you could see. And these "real" particles formed the nodes of the self-generating net that Chew was talking about. The explanation begged the very question of what is meant by "real." Zweig, on the other hand, seemed hopeful that experiments might soon turn up tiny particles with fractional charges. The Experiments Committee at CERN was also intrigued by the possibility, and before long physicists were scanning bubble chamber photographs for evidence of aces.

In the meantime, Zweig had to stake his own claim by getting the CERN report published in a refereed journal. Gell-Mann had been so sure *Physical Review Letters* would pummel him with editorial quibbles that he had published the quark paper in the less prestigious *Physics Letters*. The head of the theory group at CERN told Zweig that he was expected to submit his paper there too. It was, after all, the CERN journal. But Zweig, ever the maverick, refused to go along. He wanted his brainchild to debut in physics' premier showcase. He appealed to Weisskopf, as head of CERN, and he gave his approval to submit the paper to the *Physical Review*. And so Zweig threw himself to the lions. If Gell-Mann had been right that someone of his stature would have trouble publishing quarks, then a young postdoc like Zweig barely had a prayer. The referees hounded him with so many questions and objections that he gave up. The report wasn't properly published until sixteen years later, when it appeared in a historical collection of papers tracing the history of the quark.

In August, eager for a chance to spread the word about aces,

Zweig traveled to the western coast of Sicily to lecture at a physics summer school held in the medieval town of Erice. Perched on a cliff overlooking the Tyrrhenian Sea, the village of old stone buildings and cobblestone streets had become a favorite summer haven for physicists, who gathered at the Ettore Majorana Centre for Scientific Culture. Housed in two former monasteries and a historic palazzo, the institution was named for a Sicilian theoretical physicist who had disappeared in 1938 on a ferry from Palermo to the mainland. Feynman sat in on the lectures, and Zweig was dismayed when his ideas barely made a dent. His adviser seemed completely unimpressed with the argument. He didn't exactly say that it was wrong. He just ignored it.

Pleased with the success of the Eightfold Way in predicting the omega-minus and still unsure what to make of quarks, Gell-Mann was happy to have other things besides physics to think about. Sometimes he wondered about the series of flukes and coincidences that had led him into this field. Ever since brother Ben had introduced him to the wonders of the natural world, he had wanted to study its glorious diversity and try to find its underlying taxonomies. But he didn't particularly care which part of the picture— linguistics, archaeology, ornithology, physics—he concentrated on. He loved trying to figure out how the world of subatomic particles worked, yet there were so many other things he found just as interesting. He remained fascinated by psychology and wondered whether it could be made into a science that was as solid as physics. For all the ambiguities of quantum mechanics, particle behavior was clear-cut. These tiny shards of energy and mathematics followed precise statistical laws. But people, made from particles, were much harder to fathom. Was there a way to devise theories to explain the unconscious drives and neuroses that compel people to behave the way they do? Could one mathematically predict how a human would come apart under certain circumstances, as one could predict how particles are likely to decay? He longed for a way to approach these kinds of matters more precisely, as a kind of algebra of unconscious drives. Part of his motivation was to understand some of his own exasperating qualities. Why did he have this need to treat some people so badly, and why did he find writing so difficult—even a simple letter to a friend? "Dear Murray," Murph Goldberger had written to him in 1962, "Is one to come to the conclusion that you are trying to be like Feynman and never answer letters, or is it J. Schwinger you're modeling yourself after? Another explanation is that you are too stupid to answer the questions I keep raising."

Gell-Mann had tried to interest the psychoanalysts he knew in the possibilities of a rigorous science of the mind. But so much of their craft seemed to him like folklore. The issue that interested them was not how to bring psychoanalysis into the domain of science, but just the opposite: how to explain psychoanalytically why scientists are driven to understand the world through the formulation and testing of hypotheses. The psychologists he talked to were also unenthusiastic. Behaviorism was still all the rage. According to this doctrine, psychology must deal only with observable phenomena. One could talk about stimuli and responses, but what happened in between the ears, involving mushy concepts like mind and mental processes, was unknowable and outside the realm of science. The argument was not so different from the one followers of the S-matrix used to denigrate field theory: Science should concentrate on measuring the quantum numbers of particles before and after a collision, and forget about the virtual processes that happened in between. Gell-Mann found both attitudes ridiculously doctrinaire.

Everywhere he turned with his new enthusiasm, he hit a wall. For years he had been trying to convince Caltech to expand its narrow interests by starting a department of behavioral sciences. True, these subjects hardly deserved to be called sciences yet, but that was just the point. The school had an opportunity to get in on the ground floor, to develop ideas as basic as those the first quantum theorists wrestled with in the 1920s. But the school wasn't interested. Sometimes, he complained, the place seemed like a barber college, a trade school that focused on science and engineering to the exclusion of everything else. Sometimes he was tempted to jump ship. When Harvard made him another offer, he eagerly accepted. But then he began to get cold feet. Murray called Murph Goldberger's house and told his wife Mildred that he was going back east. But the more he talked, the less sure he sounded. Finally Mildred said, "Murray, you don't have to go. They can kill you, but they can't eat you!" He decided to renege on the deal. Margaret was unhappy about the reversal. She really wanted to live in Cambridge, which was so much more like England. Murray wasn't sure why he was unable to take the leap (more evidence of the need for a science of psychology). "I became frightened about a number of things, especially how to care for my father at a great distance," he wrote to Weisskopf, who was still trying to get him back. And he didn't care to work so closely with Schwinger. What Murray had really wanted all along was a joint appointment at Harvard and MIT. That way he wouldn't be tied down to a single institution. He would be free to roam. MIT still seemed a little grubby to him, but it wasn't a barber college. His alma mater was amenable to the idea, but Harvard wouldn't agree. After listening to his constant complaints about Caltech, his colleagues found his decision to turn down Harvard baffling. Some of them joked that the school had met all his demands except one—changing the name of the school to the Gell-Mann Institute.

Caltech wasn't exactly holding Murray down. By now he was traveling everywhere, using his position as one of the most sought-after people in particle physics to see as much of the world as possible, adding new species to his tally of rare birds, marveling at the profusion of languages and life. "I like diversity, and I like the natural history behind diversity," he told a journalist around that time. "Why are there so many different tongues, so many different birds, even so many different human neuroses? It's interesting to find the pattern behind them." In 1964, as his quark theory rippled through the infosphere, he embarked on three trips—to India, Japan, and the Soviet Union. This constant globe-trotting was becoming a routine part of his life.

Early in the year, he flew to Udaipur, India, to meet with foreign scientists about arms control, which he continued to study for Jason. The meeting was the latest in a series launched in the mid-1950s by the British philosopher Bertrand Russell, with the encouragement of such scientific luminaries as Albert Einstein, Max Born, Hideki Yukawa, and Linus Pauling. The first of these International Conferences on Science and World Affairs, in 1957, had been underwritten by a Cleveland industrialist, Cyrus Eaton, who insisted that it be held in Pugwash, Nova Scotia, where he was born. And so the high-profile meetings became known as the Pugwash conferences. Some of the better-known participants had included Hans Bethe, I. I. Rabi, Leo Szilard, Victor Weisskopf, Eugene Wigner, and, from the realm of geopolitics, Henry Kissinger. Now Gell-Mann was joining the American delegation at Udaipur, a beautiful sixteenth-century city of marble palaces and temples surrounding three lakes—the Venice of the East, some called it. This was the kind of exotic locale that Murray loved. At the hotel, a converted palace built on an artificial island, barefoot turbaned waiters served breakfast while birds—Alexandrine parakeets, Murray informed his colleagues—flew overhead.

During the meeting Gell-Mann and two other American scientists, Jack Ruina and Carl Kaysen, argued against using antiballistic missiles, or ABMs, to protect cities from nuclear strike. At first sight, the notion of shielding innocent civilians with a nuclear umbrella seemed unassailable. Who could possibly object? But on closer scrutiny, it seemed that ABMs might undermine the doctrine of nuclear deterrence known as MAD: mutual assured destruction. MAD was based on the idea that the Soviets would be forever deterred from launching a first strike because they knew they would suffer a devastating counterattack. The United States would likewise be deterred. But a country that had protected its cities with ABMs might feel more confident about launching a first volley.

When Ruina tried to make the case to the Russians, Mikhail Millionshchikov, vice-chairman at the All Union Academy of Sciences, was so baffled that he insisted the interpreters must be fouling up the translation. That night Ruina wrote up the ideas, and Gell-Mann added his thoughts at breakfast. Then they gave their paper to Millionshchikov. The problem hadn't been the language barrier. The Russian said that the Soviet people would never stand for a pledge not to use ABMs. They wanted to know what their government was doing to protect *them* from the American threat. A government that opposed ABMs would never be reelected. Murray tried to keep a straight face. He sympathized with the Russian scientists. They were interested in the same thing, both political and scientific, as he was. But the romanticized view he once had held of the Soviet government had faded. (Later that year at a scientific conference in the Soviet Union, Gell-Mann was surprised when Millionshchikov sought him out and asked if he remembered their breakfast in India. "Of course! You said our ideas were crazy," Gell-Mann replied. "Well," the Russian said. "They're not so crazy." Then he got up and left the room. Eight years later the Antiballistic Missile Treaty was signed.)

The trip to India was quickly followed that spring by a visit to the Research Institute for Fundamental Physics at Kyoto University, where Gell-Mann had been invited by the great Yukawa, inventor of the pion. The center, modeled after the Bohr Institute in Copenhagen, had been established in 1953 in honor of Yukawa, after he had become the first Japanese to win a Nobel Prize. The ancient imperial city, with its rolling hills studded with the peaked roofs of Buddhist temples, had long been a magnet for seekers after truth. One could stroll along the Path of Philosophy, going from one shrine to the next. Margaret came along on the trip, and she and Murray stayed in a traditional Japanese *ryokan*. Yukawa was away at the time, but he made sure Gell-Mann was set up in a nice office.

Kyoto was also a stronghold of followers of Sakata, who was based in the north at Nagoya. His group's continuing attempt to build all the hadrons from triplets of protons, neutrons, and lambdas, in accordance with dialectical materialism, was seeming more and more like an obsession. If you wanted to have triplets, Murray asked, why not use quarks instead? The Kyoto physicists wouldn't hear of it. After all, no one had ever seen fractional charges. Well, if there *were* such things as quarks, Murray replied, they were probably permanently trapped inside the hadrons, unobservable. Nonsense, his Japanese colleagues said. According to Engels, matter must be made from material substance, something you could grab, at least with a particle detector. Quarks were too abstract, an example of bourgeois idealism. Of course one didn't have to be a Marxist to find something objectionable about building blocks that were, in principle, unobservable. The idea was just as repulsive to many devotees of the S-matrix.

Gell-Mann left Kyoto convinced that the Japanese physicists' perceptions were addled by dogmatism. The basic building blocks had to be something you could explain to peasants and fishermen, to the masses. His irritation was still brewing when he was invited to deliver a public lecture in Tokyo sponsored by the newspaper *Yomiuri Shimbun*. (He was amused to learn that the name literally meant "read/sell.") He used the occasion to condemn the doctrinaire attitude he had encountered in Kyoto. The talk didn't go over well. Murray stopped after half an hour or so and asked for questions. There were none. He worried that he had made the lecture too brief. His writer's block was still a problem. The sponsors seemed to feel that for what they had paid him, he should have talked for an hour or more. Whatever its length, it was not likely that a lecture by an American attacking Japanese physicists in their own country would be met with a standing ovation. But at least, he felt, he had stood up for the principle that Weisskopf and Fermi had taught him: The most elegant formalisms were nothing if they clashed with what was happening in the real world.

Gell-Mann had barely resettled in Pasadena when, several months later, he set off on his third international trek of the year, flying to the Soviet Union for another Rochester conference. The location was Dubna, on the Volga River north of Moscow, a center of research laboratories sometimes called one of Russia's "Science Cities."

The visitors were housed in a dreary hotel with bad plumbing. On the plaza outside, a physicist named Val Fitch introduced some of his hosts to an advanced American technology: the Frisbee. As he hurled the plastic disk back and forth, Fitch had reason to celebrate. He and another experimenter, James Cronin, had just published an astonishing discovery: The weak forced violated yet another symmetry. The earlier shock about violation of parity (P) and charge conjugation (C) had been softened by the belief that the weak force respected the combined symmetry, CP. Replacing particles with antiparticles and reflecting the reaction in a symmetry mirror was supposed to guarantee an identical outcome. In an exceedingly delicate experiment, Cronin and Fitch showed this wasn't always so. Very, very occasionally, the double symmetry would be violated. Once again the unruly weak force was refusing to play by the rules. For subtle reasons, CP violation is taken as equivalent to saying that time symmetry, T, is broken—just what Zweig had tried unsuccessfully to show in graduate school. (Even after Cronin and Fitch's finding, the weak force couldn't be dismissed as a complete renegade. Take all three symmetries together, it was believed, and beauty could be restored. CPT would always be conserved. Replace matter with antimatter, left with right, and past with future, and any two experiments would come out the same.)

This was the first Rochester conference since the unveiling of fractionally charged triplets, and though many scientists were careful to cite both Zweig and Gell-Mann, everyone was already calling the subparticles—or symbols, whatever they were—quarks, not aces. And though most people knew Ne'eman had also thought up SU(3) octets, the Eightfold Way was identified in most people's minds with Gell-Mann, as was strangeness, though it had been found simultaneously by Kazuhiko Nishijima. Gell-Mann had learned the power of naming. He had not just discovered a series of abstract concepts that helped make sense of the subatomic realm, he had bestowed the names that anchored them in people's minds. But his growing fame was not just a matter of his forthright style. It was hard to think of a recent theorist who had come up with so many powerful new ideas, who cut so agilely to the heart of so many phenomena to find order where there had been confusion. Who else had discovered three ideas as powerful as Murray's triumvirate—strangeness, the Eightfold Way, and quarks? Well, maybe not quarks. The idea was still going over, as Gell-Mann later put it, "like a lead balloon."

In a speech that was one of the highlights of the conference, Salam declared that the successes of the Eightfold Way were "more impressive than one has any right to expect." But when it came to triplets he (and just about everyone but Zweig) was far more cautious. In addition to the "revolutionary quark model," he described what he called "conservative" triplet schemes, with whole instead of fractional charges. It was hard to deny that the mathematics of group theory led one to conclude that underlying the Eightfold Way was a fundamental representation involving some kind of triplet. But the CERN experimenters looking for fractionally charged quarks had found nothing. Nor had a team at Brookhaven. Trying to save the idea, theorists devised elaborately complex models in which hadrons were made not of one but two or three different kinds of $SU(3)$ triplets, cleverly woven to yield "quarks" with whole-number charges. Schwinger had recently come up with a complicated model (spurning the nuclear democrats, he called it a "field theory of matter") in which all hadrons were built from triplets of fermions as well as triplets of bosons. In his paper, which appeared in *Physical Review* during the Dubna conference, he relegated Gell-Mann's triplets to a footnote: "He introduces particles of fractional charge which can be detected, presumably, only by their 'palpitant piping, chirrup, croak, and quark.' " (Apparently the *Physical Review* could be more accommodating to flights of linguistic fancy than Murray gave it credit for.) T. D. Lee had his own triplet model. And Sheldon Glashow was at Dubna lobbying for a theory (done with a colleague, James Bjorken) in which there were four quarks—up, down, strange, and (carrying on the Gell-Mann tradition of whimsical nomenclature) charm. "Some people," Salam said wryly, "do not know when to stop." Like Zweig in his aces preprint, they were attracted by the notion of a symmetrical world with four kinds of leptons and four kinds of quarks.

For doubters, these elaborations were reminiscent of Ptolemy, the ancient astronomer who tinkered endlessly with his cosmological theory to preserve the fiction that the earth was at the center of the universe. When the heavenly lights failed to move in perfect circles around the earth, he proposed that their orbits included curlicues called epicycles. And when observation and theory still wouldn't mesh, he added epicycles to the epicycles.

Salam ended his talk with a standing joke begun by Victor Weiss-kopf at the Rochester gathering at CERN two years before. Weiss-kopf had summarized the confusion of particle physics by showing a cartoon in which two very British archaeologists wearing pith helmets are looking down at a tiny pyramid, no more than several inches high, which they have just excavated. One says, puffing on his pipe, "This could be the discovery of the century. Depending of course on how far down it goes." Salam summed up the work on the strong force with another pyramid cartoon. Two Egyptian leaders are watching their slaves add layer after layer to an experimental new structure—a pyramid that is standing upside down, balancing precariously on its tip. One says to the other: "I hope this structure holds till the next conference." But Salam held out a ray of hope: Triplets, "at their most exciting, may be a new form of matter. It is a prospect before which the imagination reels."

Gell-Mann carefully titled his own talk "Possible Triplets in the Eightfold Way." "Field theories involving triplets may be very useful for abstracting mathematical relations and symmetry principles," he said, "but they are no doubt premature (at best) as real descriptions [of a] hadronic system." Just that spring, he had tried using his new inventions this way, as a calculational tool, while remaining agnostic as to their reality. For convenience, he imagined that particles were made from quarks held together by the shuttling of a single boson—one of those virtual force-carrying particles that are the hallmarks of a field theory. Using this fiction, he constructed a model of the strong interactions "which may or may not have anything to do with reality." The theory was obviously wrong for a number of reasons. But it seemed like a useful approximation, a good mathematical tool. After he had used the toy theory to abstract some general principles about interacting currents— new rules for his growing current algebra—he threw the field model away. He was using quarks somewhat as he had used the pedagogical triplets in the Eightfold Way paper. Telegdi told him the technique was like one used by the great French chef Escoffier: Cook a pheasant between two slices of veal, then throw away the veal.

What quarks "really" were in this sense was difficult to say. As Gell-Mann ambiguously described them, they seemed more physical than mnemonic devices—but less so than, say, an electron. But what could that mean? It was an issue on which he was not willing to be pinned down. After all, quantum mechanics had shown the peril of trying to apply everyday, ordinary notions like "particle" or "wave" to the subatomic realm. Human language, the outgrowth of a long chain of historical accidents, was too crude for describing the underpinnings of reality. Murray struggled for the words that would imperfectly describe quarks, calling them not only "mathematical" but even "fictitious." One colleague interpreted Murray's maddening ambiguity like this: "If quarks are not found, remember I never said they would be; if they are found, remember I thought of them first." But the difficulty cut much deeper, rooted in the confusion that besets every scientist: To what extent were quarks or fields or even neutrinos and electrons out there in the universe, and to what extent were they inventions of the human mind? When pressed, Gell-Mann sounded like a Platonist talking about the mathematical harmony of the spheres. But he was growing increasingly impatient with discussions over the meaning behind the mathematics. Feynman liked to claim that he had orders from his doctor forbidding him to discuss metaphysical matters. Gell-Mann went him one better and actually obtained a written prescription from a doctor in an extension class he taught at UCLA. Philosophy was bad for his health.

After the Dubna conference, Gell-Mann and his Caltech colleague Fred Zachariasen boarded an old Aeroflot turboprop and headed, via the Nepalese city of Kathmandu, for a physics conference in Sapporo, on the Japanese island of Hokkaido. As Murray watched the terrain from the window, he followed their route on the map and was surprised when the plane flew over the western fringe of China. Relations with the Chinese were so bad that he had expected the pilot to take a more southerly route. He thought nervously about what might happen if they were forced to make an emergency landing. But it was thrilling to see this strange new part of the world. He identified the remote oasis of Kashgar and watched as the plane flew over the rugged Karakoram Range before dropping down into the Delhi Plain.

That night in Delhi, Murray bought a copy of a book describing the Devanagari alphabet, used in Hindi, Sanskrit, and Nepali. Then they flew on to Kathmandu. On the plane Zachariasen watched in amazement as Murray opened the book, scanned the characters for several minutes, then wrote them down from memory. To Zachariasen all the squiggles looked more or less identical. He was amazed when, after a few trials, Gell-Mann had mastered the entire alphabet. This photographic memory, Zachariasen thought, was part of the key to Murray's genius. He was so smart that he would have been a good physicist in any case. But with his incredible ability to remember every experimental result and the detail of every paper he had read or written, he was unbeatable. To be sure, Zachariasen sometimes found these great mnemonic feats annoying. Once the Gell-Manns and the Zachariasens went camping together in the coastal jungles between Mazatlán and Acapulco. Murray, of course, insisted on reciting the name of every bird they encountered. Zachariasen had once liked bird-watching, but after a few trips with Murray, he came to hate it, perversely insisting that all birds looked to him like ducks. Still, he was impressed with how Murray combined what could have been just a cheap parlor trick with such mathematical verve and creativity.

By the time they landed in Kathmandu, Murray could translate the shop signs and speak a few words of Nepali. They got permits to trek to a high mountain lake but dropped the plan because it was monsoon season. Anyway, Murray learned, it was a religious holiday and Brahmins from all over the country would be flocking to the same place. He didn't like to feel crowded. They flew on to Calcutta, Bangkok, Tokyo, Sapporo, and finally Los Angeles, circumnavigating the globe.

As though the idea of fractional charges weren't bad enough, physicists were slowly realizing that if quarks existed, they would also violate the revered law of physics known as Pauli's exclusion principle. Pauli's rule was supposed to apply to all the particles that make up matter: protons, neutrons, electrons, mesons—the so-called fermions. Unlike the bosons, these particles were said to have "Fermi-Dirac statistics," meaning they cannot be in the same quantum state at the same time. Otherwise there would be nothing to keep all the electrons circling an atom from collapsing into the lowest orbit; instead of the variety of atoms, each clothed with a different veil of electrons, matter would be a uniform, indistinguishable blob. Since quarks were supposed to make up fermions, they presumably would be fermions themselves. But what then was one to make of the delta particle, which was made from three up quarks, or the omega-minus, which was made from three strange quarks? Putting two u, d, or s quarks together in the same particle was allowable—one could be spinning up, the other down. But adding a third should be impossible, a flagrant violation of all that the Pauli principle held dear. (Even worse, the omega-minus and delta each had a total spin of $3/2$, meaning that all three quarks—$1/2 + 1/2 + 1/2$—had to be spinning the same way.)

Gell-Mann and Ne'eman had briefly considered the problem back when Murray was writing up his original quark paper. They had heard about a new kind of statistics that might provide a loophole to keep the quarks from violating Pauli's law. But when they went to the Caltech library to look up the paper, they couldn't follow the argument. (Ne'eman later blamed this on an error in the scientist's calculations.) Murray had decided to forget about the statistics problem and write up quarks anyway. Maybe if some physical principle, like infinite mass, kept them from coming out of the hadrons, the weird statistics wouldn't matter. What you can't see won't hurt you.

Others took the problem more seriously. Shortly after Gell-Mann's and Zweig's original papers, a physicist named Oscar Greenberg proposed that the quark's odd behavior could be explained with something he called parastatistics. Fermions obeyed Fermi statistics, bosons obeyed something called Bose-Einstein statistics (allowing them to crowd into the same state), but quarks obeyed "parastatistics of rank three," providing a loophole to the Pauli principle. Unfortunately the theory also allowed for exotic cases in which baryons could obey Bose-Einstein statistics, clustering together like photons in a beam of light.

At the University of Chicago, Yoichiro Nambu (with the help of a Syracuse University graduate student named Moo-Young Han) came up with his own solution—another example of what Salam called "conservative" quark theories, in which two or three overlapping SU(3) triplets conspired to give quarks integral charge. In the Nambu scheme, quarks not only had "flavor" (up, down, or strange) but also something called charm. (The name didn't stick since Glashow had already appropriated it for an entirely different use, as the name of the fourth quark he and Bjorken had invented.) In the delta particle, for example, the three up quarks would have different charm numbers. Thus they would not be identical and the Pauli principle would be saved. There were wheels within wheels within wheels; more epicycles, the skeptics grumbled—those who bothered to read the paper at all.

Gell-Mann liked to say that a theorist should be judged by the number of correct ideas minus twice the number of mistakes he had published. With strangeness, V-A, and the Eightfold Way under his belt, he had no trouble weathering all the doubts about quarks. Zweig was much more vulnerable. Close to an appointment at a major university, he was blocked at a faculty meeting when a senior physicist declared that aces must be the work of a charlatan. Caltech, familiar with the high quality of Zweig's work, soon hired him. But he found little refuge there from the criticism. His mentor, Feynman, still hated the quark/ace theory. And Murray quickly made it clear that he didn't have much use for Zweig's insistence that hadrons were actually pieced together from solid little quarks somewhat in the way that atomic nuclei were made from protons and neutrons. He considered Zweig an outstanding, imaginative physicist, and he admired his originality and breadth. But he dismissed Zweig's version of quarks as the "concrete block model." It was for "blockheads," he said. While the discovery of the electron had eventually given rise to a bustling electronics industry, Gell-Mann doubted that anything like a "quarkonics industry," as he put it, would ever emerge. Quarks were something more subtle—so subtle that he could barely describe them, even to himself. Distancing himself from Zweig's more direct approach, he continued to draw a distinction between his own more abstract "current quarks" and Zweig's less reputable "constituent quarks."

For all his caginess, though, and despite his obsessive fear of being wrong, Gell-Mann couldn't quite let go of such a compelling idea: that the subatomic world was woven according to the deepest symmetries of SU(3), with triplets somehow forming the Eightfold Way. We live in a world where beauty, simplicity, and elegance are "a chief criterion for the selection of a correct hypothesis," he marveled during a lecture. That, he said, "makes the adventure of working in our field particularly rewarding." He spoke of hopes that, once the strong force and the weak force had been tamed, a unifying principle would eventually be found to unite all the forces of nature. Electromagnetism, the weak force, and the strong force would at last be brought into a single framework. He didn't know what form it would take—a field or a bootstrap theory. The latter, he said, looked more promising for now, but field theory might still have a future. After summarizing physics' recent successes and its persistent puzzles, he closed with a hint of what might yet come: "Finally, there is the really exciting prospect of total surprises, things completely outside our experience, which our present-day theoretical language is inadequate to describe. . . . The strain has been accumulating for fifteen years; the shock should come fairly soon."

CHAPTER 12

THE SWEDISH PRIZE

The sun rises eight hours later in Pasadena than in Stockholm, and so it was still dark on the October morning in 1965 when Richard Feynman received a call from a reporter informing him that he had won a Nobel Prize. He, Julian Schwinger, and Sin-itiro Tomonaga were finally to be honored for their work so many years before in tidying up the theory of quantum electrodynamics—getting rid of the infinities and turning it into a precision device for describing the choreography of electrons and photons. The first call was followed by another and then another. Around daylight, photographers began arriving at the door, unaware that QED explained the very process by which images left their mark on film.

Over the years the ritual surrounding one's first day as a winner of science's most prestigious prize had become as predictable as the ceremony in Stockholm that followed a couple of months later, presided over by the Swedish king: the telephone ringing in the dark, the good-natured grumbling about having one's sleep interrupted, the public agonizing over the questionable effects of such sudden fame on one's scientific output, the (sometimes uncharacteristically) modest assessments of the recipient's own work. Feynman followed the script to the letter. It wasn't entirely an act. In these days when the doubts about quantum field theory were at their deepest, he seemed genuinely concerned about the value of his work on renormalization, wondering aloud if he and his colleagues had simply found a way of taking the infinities and "sweeping them under the rug." It was "a dippy process," he would complain, "hocus-pocus."

Except for the brief moment when he had vainly convinced himself that he alone had discovered the V-A theory of the weak force, Feynman still hadn't experienced the emotional rush of finding something completely new about nature. True, his Feynman diagrams had revolutionized the way people did physics. But he was still a little disappointed that the work he would be best remembered for involved simply repackaging what others had discovered—the great truths of quantum mechanics found by Planck, Heisenberg, Schrödinger, Bohr. Still, he didn't feel moved to decline the prize, which carried with it a check for $55,000. He spent his third of the money on a beach house in Mexico. Some of his students draped a banner across the dome of the prominently located Throop Hall: "Win Big, RF." At Caltech, home to ten previous laureates, jousting for a Nobel Prize was as exciting as football.

Gell-Mann had trouble working up much enthusiasm for his friend's success. Never mind that Feynman was a decade older than Murray or that the work being honored had been done in the late 1940s, when Gell-Mann was still a teenager. Murray seemed dejected, despondent, like someone whose father had just died. It wasn't easy always having to share the spotlight with Feynman. Gell-Mann could have moved to just about any physics department in the world and been the brightest star. At Caltech, he always had to wonder if he was really as smart as Dick. On the scientific grapevine, Gell-Mann's name was regularly mentioned as a likely Nobelist. Every year he wasn't chosen seemed like a blow. He took to dismissively calling the honor "the Swedish prize," unwilling or unable to utter its real name.

If Murray was going to get his own Nobel, it now seemed, it probably wasn't going to be for quarks—at least not for "real" ones. No matter how fine a net the experimenters cast, they still couldn't snare one of these beasts. But they weren't quite ready to abandon hope. When a predicted particle fails to leave its mark in a particle detector, all is not lost. Theorists can take the failed experiments as evidence that the particle is too massive to be produced by the limited energy of existing accelerators. Then they can wait for the next generation of more powerful machinery. Or they can turn their sights to outer space, where the natural acceleration of cosmic rays provides astronomically higher energies than can be created on earth. And so the old art of gathering stardust was revived. Cosmic ray physicists, happy to have something new to do, sent up balloons and ascended mountaintops, but still there were no quarks. It was possible that naked quarks were just very, very rare; that if one waited long enough—billions of years, perhaps—one or two would come wafting by. But perhaps there was a quicker way to search. Since cosmic rays had been showering the earth since creation, a few fractionally charged particles might be sitting around somewhere.

So the search moved to the bottom of the oceans, the polar ice caps, the dust floating in the air. And after all this sifting, there still were no signs of fractionally charged particles. By the summer of 1966, almost twenty experiments had failed to turn up a single one. One scientist ground up oysters, hoping they might have sucked up quarks along with the other muck on the ocean floor. He would call up Murray, sometimes late at night, with the negative reports. Around that time, an Australian physicist reported that he was "99 percent sure" he had discovered a quark dislodged from oxygen and nitrogen atoms smashed by cosmic rays. But he too turned out to have failed.

There was some very oblique evidence that quarks might exist. The reasoning went like this: Baryons were supposedly made of three quarks and mesons of two quarks (a quark and an antiquark). Thus the probability that a baryon would scatter off another baryon compared with the probability that a meson would scatter off a baryon should be 3:2. This figure was approximately borne out by experiment. But that was hardly enough to clinch the case.

Throughout the search, Gell-Mann remained evasive. In a funny way, *not* finding quarks bolstered his argument that they were "mathematical," either an abstract accounting device or some philosophically maddening entity somehow trapped forever within the baryons and mesons. Whether quarks could be found in the laboratory or in outer space was, he kept insisting, irrelevant—as long as they helped make sense of the symmetries. But it was never really clear just what Gell-Mann was trying to say. Were quarks really "there" (in some loose sense of this seemingly simple word) but hidden by an insurmountable physical barrier? Or did they only exist as ideals in the platonic phantom zone?

In most physicists' minds, the whole idea of subparticles—whether real or mathematical—remained suspect, and Gell-Mann took great pains to convince others that his wild idea was not to be taken as an important part of his oeuvre. Because of his earlier work and the sheer force of his personality, he was well on his way to becoming the preeminent person in the field. He didn't want to risk seeming too radical in his pronouncements. A scientist should be judged by the number of times he was right minus twice the number of times he was wrong—his motto never seemed more true.

In the summer of 1966, Gell-Mann was asked, during a discussion at the Ettore Majorana summer school in Sicily, whether he agreed that "the best feature about quarks is their name." "Yes," Murray replied. "The whole idea, as far as I introduced it (and I still think it is right), is that they are a useful mathematical creation in order to express the commutation rules of the currents and the approximate symmetry properties of the particle states. Maybe they are real things but probably not." Then he went on to explain his ideas about quarks being somehow permanently trapped and unobservable. The quark, he agreed with his questioner, was like the imaginary prey in Lewis Carroll's "Hunting of the Snark." It could not be found: "For the Snark *was* a Boojum, you see."

When the Rochester conference was held shortly afterward in Berkeley, stronghold of S-matrix theory and the bootstrap, Gell-Mann was given the honor of delivering the opening address: a sweeping overview of the state of particle physics since the very first Rochester meeting in the early 1950s. Rumor had it that the organizers had originally thought of having five or six scientists divide up the task. When Murray's name kept coming up as the best person for each topic, it was decided that he might as well just give the entire presentation. During his hour-and-a-half marathon, Gell-Mann tried to make the best of the confusing times. "In some respects, it's rather humbling to think about how little progress we've made in the last fifteen years," he lamented, "but if we actually look at the data accumulated and the theoretical analysis, it's clear that we are much further on our way toward understanding the particles." He diplomatically suggested that theorists stop wasting their time arguing the merits of the S-matrix and field theory. "It would be better if all the efforts that we expend on the discussions . . . were devoted to arguing for a higher-energy accelerator so that we can do more experiments over the next generation and really learn more about the basic structure of matter." He was as convinced as ever that advocates of field theory and the S-matrix were just arguing over terminology, that the two approaches opened different windows on the same thing.

Then he launched into a demanding, detailed explication of the arcana of Regge poles and current algebra, ending with a reminder of how much in physics was still unknown. Physicists had yet to find the "photons" of the weak force, the bosons that acted as its carriers. And that was the least of it. Some recent theories seemed to require that the universe contain magnetic monopoles, tiny half-magnets that absurdly had only one end. Everything seemed up in the air. Gell-Mann facetiously predicted that someone at the conference might announce a single new particle that filled all the needed roles—and that acted as an $SU(3)$ triplet obeying parastatistics. "For this particle I suggest the name 'chimeron,' " he said.

He ended on a note of solace. "I think that it's not discouraging, but rather that it's marvelous not to know all of these fundamental things. We still have problems to work on."

About quarks he was as dismissive as ever, saying they were "probably fictitious." "The idea that mesons and baryons are made primarily of quarks is difficult to believe," he said, "since we know that, in the sense of dispersion theory, they are mostly, if not entirely, made up out of one another. The probability that a meson consists of a real quark pair rather than two mesons or a baryon and antibaryon must be quite small." Quarks, he said, repeating what was becoming a familiar theme, are probably "mathematical entities that arise when we construct representations of current algebra." But buried in the talk was a prediction that would turn out to be amazingly prescient. If quarks were indeed an element of the universe, then one had to explain the apparent rarity of their occurrence (so far, zero) in the measurable world. Either they would have to be very massive (as he had suggested in his original quark paper) or, more likely, he now proposed, lightweight and bound permanently together by some kind of very high or even infinite barrier. The grip holding them together would be unbreakable. This idea would linger in the air for years before its power was realized.

That spring, during a lecture at Cambridge University, Gell-Mann was surprised to learn that the great Dirac actually liked quarks. Murray couldn't understand it. Quarks had so many annoying properties. Then he realized that Dirac (like Zweig and Glashow) was intrigued by the fact that both quarks and leptons were fermions—particles with spins of ½. If hadrons were indeed made from quarks, then all of creation would consist, at its bottom level, of spin-½ fermions. If only there was some way to reconcile the quark's wispy ontological status.

Around this time, Gell-Mann, this thirty-seven-year-old elder statesman, was given the highest honor of academia, an endowed chair. He was named the first Robert Andrews Millikan Professor of Physics, in honor of the great Caltech experimenter who had measured the charge of the electron and named cosmic rays. His salary was raised to $35,000. In the fall of 1966, asked to speak at a Caltech seventy-fifth-anniversary conference, Gell-Mann seized another opportunity to belittle the idea of quarks. With his smooth, measured delivery, resonant voice, and precise sense of timing, he was developing into a virtuoso lecturer. In giving a layman's overview of particle physics, he started with some favorite set pieces. First there was John Updike's famous *New Yorker* poem about neutrinos, "Cosmic Gall," found taped to the doors of so many physics professors' offices:

> *Neutrinos, they are very small.*
> > *They have no charge and have no mass*
> *And scarcely interact at all.*
> *The earth is just a silly ball*
> > *To them, through which they simply pass,*
> *Like dustmaids down a drafty hall*
> > *Or photons through a sheet of glass. . . .*

(Updike had actually written that neutrinos "do not interact at all," but in the name of "scientific license," Murray edited him.)

> > *They snub the most exquisite gas,*
> *Ignore the most substantial wall,*
> > *Cold-shoulder steel and sounding brass,*
> *Insult the stallion in his stall,*
> > *And, scorning barriers of class,*
> *Infiltrate you and me! Like tall*
> *And painless guillotines, they fall*
> > *Down through our heads into the grass.*
> *At night, they enter at Nepal*
> > *And pierce the lover and his lass*
> *From underneath the bed—you call*
> *It wonderful; I call it crass.*

The audience laughed appreciatively, and Murray continued with another *New Yorker* poem, "Perils of Modern Living" by Harold P. Furth, inspired by a statement Edward Teller had made about the possible existence of mirror-image antimatter worlds:

Well up beyond the tropostrata
There is a region stark and stellar
Where, on a streak of anti-matter,
Lived Dr. Edward Anti-Teller.

Remote from Fusion's origin,
He lived unguessed and unawares
With all his antikith and kin,
And kept macassars on his chairs.

One morning, idling by the sea,
He spied a tin of monstrous girth
That bore three letters: A.E.C.
Out stepped a visitor from Earth.

Then, shouting gladly o'er the sands,
Met two who in their alien ways
Were like as lentils. Their right hands
Clasped, and the rest was gamma rays.

The performance brought down the house. Then, with the audience in the palm of his hand, he went on to describe more cutting-edge physics: the beautiful symmetries of the bootstrap theory, with all the hadrons made up of one another, and a "far crazier picture." Quarks. He placed a transparency on the overhead projector— "Hadrons are made of quarks and anti-quarks," it read—and removed it so quickly that the startled audience laughed at his abruptness. Were quarks really to be such a flash in the pan?

He seemed to enjoy himself as he described his theory, never extracting his tongue entirely from his cheek. "Now what *is* a quark. . . . One possible derivation of the name—scholars are already disputing this, some assuming it comes from the German word for rotten cottage cheese—is from the heading of a page in *Finnegans Wake*. . . ." He filled in the wonderfully simple details of the scheme, marveling at the elegance, then ended with the inevitable question: Do these shards of mathematics actually exist?

"We must face the possibility—and this is a possibility that we're quite willing to face—that quarks are not real," he said. "Actually that is just as well; mathematical quarks are even easier to work with than real ones, because certain restrictions imposed by the reality of the particles can be dispensed with." There was more laughter, and

Murray smiled impishly. It was hard to tell when he was being serious and when he was being coy. The ideas of the bootstrap and
mathematical quarks would probably both turn out to be right and
equivalent, he predicted—two different ways of saying the same
thing. "It is also possible, of course, that they are equivalent and
both wrong—or inequivalent and both wrong. However, if it turns
out that they are equivalent and one is right and the other is wrong,
we will probably be in trouble."

The following summer, when he gave the closing lecture at
another session of the Erice summer school, he seemed even more
unsure. Quarks might well turn out to be "purely illusory," he said,
"a passing phase in our description, which will go away after a while,
when we will learn how to use the bootstrap methods and solve our
equations without using quarks."

In the fall of 1967, Gell-Mann and Feynman—"the hottest properties in theoretical physics today," one colleague called them—
were featured in a joint profile in the *New York Times Magazine*.
The quark, Gell-Mann said, splitting the usual philosophical hairs,
is likely to turn out to be nothing but "a useful mathematical
figment."

And that is where things stood when the first indirect signs of
quarks began to appear in a new particle accelerator that had just
been built in the Northern California hills around Stanford University. By now the method of exploring the insides of nuclei was
almost routine: Accelerate hadrons around circular tracks, faster
and faster with each revolution, and then send them crashing into
other hadrons. New physics might be found in the debris. That,
anyway, was how it was done at Brookhaven, CERN, Chicago, and
other laboratories. Since just after World War II, physicists at Stanford had been working on a different method—accelerating electrons down a linear track several feet long and then firing them into
hadrons. Since electrons don't feel the strong force, they could
penetrate inside the protons and give clues of what, if anything, was
inside.

The first of these machines, the Mark I, had a gun barrel twelve
feet in length, long enough to accelerate the electrons to an energy
of six million electron-volts. In the 1950s, the physicist Robert Hofs-
tadter used a longer, more powerful version, the Mark III, to bounce

electrons off protons and show from the angle of the recoils that the larger particles had size: Electrons might be thought of as dimensionless mathematical points; protons were more like blobs. The work earned him a Nobel Prize. By 1960 the Mark III, now three hundred feet long, reached energies of a billion electron-volts. By 1966, after years of congressional wrangling, the Mark III was followed by a very expensive machine nicknamed "the Monster": The two-mile-long Stanford Linear Accelerator could speed electrons to energies of twenty billion electron-volts, pushing them deeper into the nucleons. Half a century earlier, Ernest Rutherford and his assistants had fired alpha particles at gold atoms, showing from the recoils that there was a hard little mass inside: the nucleus. At the Stanford Linear Accelerator Center (soon shortened to SLAC), experimenters hoped to use the electron bullets to see whether there was something inside the proton—if it had not only size but an internal structure. Who knew? Maybe they would even find quarks. Most, of course, were doubtful about the existence of any such structure. The proton might be bigger than a point, but it was probably mushy inside—a homogeneous smear of positive charge.

The first experiments seemed to uphold this prediction. When electrons bounced off protons at fairly low energies, the two recoiled elastically like billiard balls—just what one would expect if the proton was a featureless blob. As one physicist put it, "The peach didn't seem to have a pit." Many physicists then decided that there was no reason to look further: Clearly there was nothing, not quarks or anything else, inside. A more dramatic test, however, would be to see what happened when the electrons struck the proton hard enough to excite it to a higher energy state, even bursting it apart. The results of this "deep-inelastic scattering" would be extremely hard to interpret, but they might say some interesting new things about structure.

Looking for a hard problem to crack, James Bjorken (the young theorist who had collaborated with Glashow on the quixotic paper about charmed quarks) stepped in. If the proton was indeed nothing more than a homogeneous mass of charge, Bjorken predicted, the high-speed electrons would tend to hit the mushy ball and glance off at narrow angles. But if there were hard, pointlike objects inside, the electrons would often strike them and glance off at wide angles.

He also devised a way to test for quarks by applying a more abstract idea, one that his colleagues had to struggle to understand. Drawing on Gell-Mann's esoteric current algebra—the veal-and-pheasant mathematical cookery—Bjorken worked with two "structure functions" describing what would happen during deep-inelastic scattering. If there was nothing inside the proton, then these functions would rapidly get smaller as the particle was hit harder and harder by the electrons. But if there was an internal architecture—consisting, say, of quarks—then the functions would decrease far more slowly, becoming almost flat. Bjorken also showed that if the data were graphed in a certain way, the curve would exhibit a property called scaling: No matter how hard the bullets struck, the shape of the graph would be the same. According to Bjorken's interpretation, scaling would be another sign that protons, against most expectations, were complex, compound objects with something hard inside.

When Bjorken described some of his ideas at a conference at Stanford in 1967, T. D. Lee, who was sitting in the audience, asked if he really meant to imply that quarks might exist. Bjorken backed off. "I would also like to disassociate myself a little bit from this as a test of the quark model." But he was being too timid. That fall, experiments at SLAC by Jerome Friedman, Henry Kendall, and Richard Taylor found the effects that Bjorken had anticipated: Many more electrons than expected were bouncing away at wide angles—strong evidence of pointlike masses inside the proton. And they found this obscure phenomenon called scaling. But the experimenters didn't really understand how Bjorken got from his assumptions to the predictions. Why do point charges imply scaling? It was all very hard to make sense of.

Confronted by the anathema of fundamental subparticles, the bootstrappers quickly sought to explain away the data. When the electron bullets interacted with the proton target, the oppositely charged particles would exchange a virtual photon. Nothing new there, just basic QED. But then, the theorists ventured, the virtual photon might somehow turn into a hadron and react strongly with the proton. With this theoretical contraption, one could explain the experimental results without assuming there was anything inside the proton but electromagnetic mush. That such a seemingly ad hoc explanation was taken seriously is a mark of just how unpopular the idea of subparticles really was.

In the summer of 1968, in the midst of the confusion, Richard

Feynman descended on SLAC. He had been out of the loop in recent years, trying to understand quantum gravity and superconductivity, making pronouncements about what was wrong with science education, and studiously ignoring just about anything to do with the strong force, which he considered a hopeless mess. Making sense of the accelerator experiments, he complained, was like studying the craftsmanship of pocket watches by smashing two together and examining the pieces. Sometimes he felt embarrassed by how little he knew about the recent attempts, all futile, to explain the strong force. But he saw little reason to follow the esoterica of dispersion theory, Regge poles, or the bootstrap. "I've always taken the attitude that I have to explain the regularities of nature," he said. "I don't have to explain the methods of my friends."

Recently, though, Feynman had been slowly wading back into the controversy. He agreed to give a talk on "Current Algebras and Strong Interactions" at a sixtieth-birthday celebration at Cornell for his old teacher Hans Bethe. "One of the reasons why I haven't done anything much with the strongly interacting particles," Feynman said, "is that I thought that there wasn't enough information to get a good idea. My good colleague, Professor Gell-Mann, is perpetually proving me incorrect." He even allowed that there were signs of an imminent breakthrough: "We suddenly hear the noises of the crackling of the breaking of the nut." But he couldn't resist taking a dig at the Eightfold Way and the faddish popularity of group theory. "Physics always seems to consist of (I can remember at the conferences) some kind of mumbling in which you repeat yourself, you know like simple baby talk, like boo-boo. I remember the session in which I found everyone muttering $Y(5)$, $Y(5)$, and it's $SU(3)$, $SU(3)$ and so on and on, and there are quark-quark and quark-antiquark." The jargon, he thought, was like the chanting of an exotic tribe.

But maybe the jabber was worth listening to sometimes. After trying to back out of an agreement to speak at a conference Robert Marshak was organizing on particles and fields (he no longer felt he had anything of importance to say), Feynman decided to attend after all. And for the first time in years, he was teaching a course in particle physics. One day around this time, Zweig bumped into him at Caltech and they went to lunch. Feynman excitedly described his new syllabus. "Did I leave anything out?" he asked. "What about quarks?" Zweig replied. Slouched over the cafeteria table, Feynman pushed himself up with his arms. "All right," he said. "I'll look into it."

Though he was dubious about the idea, Feynman had been playing with a model in which hadrons had something inside them. He was agnostic about whether these parts were fractionally charged or came in threes. He called them "partons"—the quanta of some unknown field inside the hadrons. If one took this as a simplifying assumption, maybe it would be easier to understand the strong force. When two protons approached each other at near the speed of light, he figured, relativistic effects would flatten them in the direction they were moving. They would look to each other not like spheres but like pancakes. Suppose there was a swarm of pointlike particles buzzing around inside them. Relativity would also slow their clocks, freezing these tiny objects to a standstill. All the complications of the collision—of the strong interaction—could be reduced to two flat layers of stationary particles coming together. Most of the tiny particles, the berries in the pancakes, would pass each other by, but occasionally two of them would collide and rebound.

Sitting at a picnic table on the grounds at SLAC, Feynman would expound on the idea, excitedly slapping his hands together like hadronic flapjacks. Bjorken was off mountain-climbing, but another researcher told Feynman about the weird scaling phenomenon. Elatedly he realized that the SLAC experiments might provide an even simpler test of his idea. Instead of slapping two pancakes together, you could pin a single pancake to your target and shoot an electron at it. To the electron speeding down the track, the proton would appear like one of Feynman's pancakes studded with partons. The negatively charged bullet would recoil off one of them and you could measure the result. That evening he went to a topless bar to flirt with the waitresses and think, then returned to his motel. Just as he was falling asleep, he had a flash that brought it all together. He saw how to show mathematically that scaling was entirely consistent with his notion of partons.

With his simple picture, he could explain the different results of the various scattering experiments. At low energies, the electron bullets were not moving rapidly enough to be faced with significant relativistic effects: The partons would appear to be darting around inside the proton so crazily that it would be all but impossible to single one out. The proton would indeed act like an undifferentiated mass, a peach without a pit. But at the higher energies of the deep-inelastic scattering experiments, an electron would see single, frozen partons.

When Bjorken returned, Feynman confronted him with his discovery. "*Of course* you must know this. *Of course* you must know that." Bjorken knew about some of the things Feynman was saying. Others were news to him. And Bjorken knew things that Feynman did not. Overall the young researcher was struck that Feynman had conjured up such a vivid way of describing what he had done with the abstractions of current algebra. Feynman was, once again, getting worked up about an idea that somebody else had already discovered. Yet again he had taken something maddeningly abstract and made it solid and understandable. As with QED, he had found a better, more intuitive way. Because of Feynman's enthusiasm, partons soon became all the rage at SLAC. Some people thought of them as quarks, others as something else—perhaps virtual pions and nucleons. Maybe with more experiments, their identity would become clear.

The rest of the physics world was more cautious. Near the end of the summer, SLAC's director, Wolfgang Panofsky, circumspectly described the Stanford experiments at the Rochester conference in Vienna. The sessions were held in the ornate halls of the Hofburg Palace, and the grand marble walls caused the sound to ricochet and the ricochets to ricochet like swarms of virtual particles. Maybe it was the acoustics, or maybe it was Panofsky's careful, understated description of the potential bombshell. In any case, the words didn't make much of an impression. There were still so many questions to answer. If the electrons were truly being scattered by pointlike particles—partons, quarks, whatever you wanted to call them—then why weren't these fragments knocked free? Despite the indirect evidence, it was still hard to get around the fact that individual quarks had not been found anywhere but in the equations of group theory.

Gell-Mann was too distracted to follow the details of the developments at SLAC. As the evidence emerged, he was a continent away, in Princeton, where he had gone with his family in the fall of 1967 to spend a year at the Institute for Advanced Study and try to decide what to do with his life. As much as he loved the mountains and deserts of Southern California, he and Margaret were tired of Pasadena. The sky there was becoming darker with smog every year, and the streets were suffering from the same urban problems that plagued so many cities in the sixties. A headline in the *Los Angeles*

Times said it all: "Pasadena's Crown City Image Tarnished: White Flight, Urban Blight, School Problems." Gell-Mann had lost hope of convincing Caltech that the social sciences were worth pursuing. And Feynman was becoming more and more of a pain. Recently he had been spending time in a topless bar near his house, drinking Seven-Ups, doodling equations on the paper placemats and sketching the dancers. When the club was raided he went to court to testify on its behalf. ("That," Gell-Mann told Telegdi, "is Dick's view of civil rights.")

When Murray heard that Feynman was running around talking about things inside of nucleons called, of all things, "partons," his disgust only increased. Gell-Mann was still ambivalent about quarks. But with the new evidence emerging at SLAC, it now seemed that Feynman was trying to lay claim to his idea. Partons! Half Latin, half Greek, it was an unbelievably ugly word. Murray started calling them "put-ons." And he was revolted by this idea of flapping pancakes. As far as he was concerned, Bjorken had explained everything more elegantly with his beloved current algebra. Never mind that hardly anyone could understand it. While Feynman tended to think in pictures, Gell-Mann thought more in terms of abstract relationships. Nature was a language, and he was intent on figuring out its grammar and its syntax. Why resort to messy metaphors?

For a man with so many frustrations, Princeton seemed at first to offer a welcome change. In addition to its coterie of resident physicists and mathematicians, the Institute had a School of Historical Studies, formed from the merger of the School of Economics and Politics and the School of Humanistic Studies. This was no glorified trade school (as he still saw Caltech). Even so, Gell-Mann was a little put off by the lack of a cooperative spirit at the Institute. He did his best thinking when he was arguing with colleagues at the blackboard. At the Institute, people tended to hole up in their offices and contemplate in silence. There was actually very little of the crossdisciplinary interaction that he longed for. And he soon learned that the Institute was a place of Machiavellian intrigue, where the mathematicians put down the sociologists and warred with the new director, Carl Kaysen (a mere economist), who had succeeded Oppenheimer.

Still, Gell-Mann enjoyed the first-class treatment given to him and his family. When they arrived for their year in residence, they were provided with an idyllic home on Mercer Road, an extension of Mercer Street, famous now as Einstein's old address. The house bordered on the Institute's five-hundred-acre woods, and Murray enjoyed walking or bicycling along the footpaths on his way to work or to Nick's nursery school. Murph Goldberger was just across town at Princeton University. New York City was a short train ride away. When Oppenheimer retired in 1966 (he died the following year), there had been talk of offering Gell-Mann the directorship. Now, when the Institute offered him a full-time faculty position, he seriously considered moving there.

But as he had done with Harvard, he turned down the offer, troubled again by his inability to leave a town and a school that he had come to find so wearisome. Part of the problem was money. He didn't feel he would get paid enough by the Institute. At Caltech, summer pay was built into his contract, giving him a leisurely interlude for travel. Recently he and his family had been spending much of each summer in Aspen, where he would hang out with his friends at the Aspen Center for Physics. At the Institute for Advanced Study, he would only get summer pay if he cut a separate deal for a summer appointment, and the last thing he wanted to do was spend the hottest months in New Jersey. But there was always some excuse. When Cambridge University sounded him out about succeeding Paul Dirac as Lucasian Professor of Mathematics, a chair once held by Isaac Newton, he turned them down as well.

During that year, Francis Low came down to Princeton from MIT. He, Goldberger, Gell-Mann, and another Institute visitor, Norman Kroll, tried unsuccessfully to put together a field theory for the elusive weak force, complete with Murray's uxyls as the carriers. One physicist later called the result, published in 1969, "an absolutely disgusting model." Murray dismissed it as "a piece of trash," insisting that most of the ideas were not his own. Ultimately, it was not a productive year, and he later felt that the paper had damaged all their reputations. "The mountains labored and produced a ridiculous mouse," he would say, paraphrasing a favorite Latin proverb.

When springtime came, the Gell-Manns reluctantly packed up the car and drove back to California. On the way, they heard that Martin Luther King, Jr., had been assassinated. (They detoured around Kansas City to avoid the riots.) Weeks after reaching Pasadena, Murray turned on the television and learned that Robert F. Kennedy had been shot to death. He was shaken by the news. He hadn't been crazy about Kennedy, but he had come to think of him as the best possible hope. Kennedy had just announced the members of his science advisory team. Gell-Mann was to be one of them.

He had looked forward to using the platform to promote his ideas on conservation and defense policy.

As the 1960s drew to a close, campuses across the country were erupting in demonstrations against the Vietnam War, where the number of American troops rapidly approached and then surpassed half a million. Students turning their backs on the status quo seemed to divide into two amorphous groups. The more militant took to the streets and occupied university buildings, hoping their anger would spread until the entire American system of higher education was shut down. Others fled the cities and university towns for rural communes. Instead of trying to bring down the system, they would ignore it. For all the disenchanted, big science was the enemy, a collaborator with the military-industrial war machine—the "technocracy," Theodore Roszak called it. His *Making of a Counterculture* was as ubiquitous on dormitory bookshelves as *The Whole Earth Catalog*, with its advice for turning back the clock to a nineteenth-century pastoral existence, made a little more comfortable by solar energy and wind power. Science must be returned to the people—that was the protestors' rallying cry. Groups like Jason, where some of the country's best physicists held secret meetings to advise the Pentagon, were condemned as traitors to the human cause.

In fact, a more restrained version of the same debate was going on within Jason itself. Many of the Jasonites considered themselves liberals and idealistically expected to sway the generals with pure reason. They liked to think that their opposition to a large-scale ABM program had helped restrain the arms race. Some of them, like Steven Weinberg and Freeman Dyson, forcefully objected to any research connected with the Vietnam War. Others, including Gell-Mann, thought they could end the war by helping to design an "electronic barrier" to be erected across a remote section of the Ho Chi Minh trail, the main supply route from North to South Vietnam. Convoys crossing it would trip the cybernetic sensors and be hunted down by aircraft and destroyed. The physicists hoped the barrier would allow the United States to stop bombing and bring both sides to the conference table. And they felt betrayed when Lyndon Johnson continued to pursue the war full force.

To some of his Jason colleagues, Gell-Mann seemed far less interested in fighting the war than in understanding its sociology. He didn't see why Jason shouldn't be able to get together a group of experts in many different fields and figure out a solution. Briefers from the Defense Department got the kind of treatment Murray would give to visiting physicists lecturing at Caltech. He regaled them with lessons from history, urging them to read a book by the eminent scholar Bernard Fall, *Street Without Joy,* so they could really understand the French experience. Fall himself came to a Jason meeting and asked some tough questions: "Does anyone know what the main source of fuel is in Saigon?" Murray, always the star student, correctly replied, "Charcoal." Gell-Mann may even have had some effect in broadening the generals' outlook. When Dyson later had occasion to visit the Pentagon, it seemed to him that every general had a copy of the book Gell-Mann had recommended.

As far back as his days at Yale, Murray's political sentiments had been pulled in two directions: the young Ivy League leftist who liked to hang out with the Social Register crowd. Whether the issue was field theory versus the bootstrap, the ontology of quarks, or American defense policy, it was hard for him to take sides. He was more opportunist than ideologue. When Lee DuBridge, the president of Caltech, pulled some strings and got Murray appointed to Richard Nixon's presidential transition team, Gell-Mann, a Democrat at heart, had no objections to serving the Republicans. Like his work for Jason, the appointment was a chance to spread his views about how the world should be run. He met with Nelson Rockefeller and Henry Kissinger to discuss Vietnam. Soon he was named to Nixon's science advisory committee. Not that he cared much for this president. Back in the late fifties, when Nixon was in his second term as vice-president, Gell-Mann had used Pavlovian conditioning to train the family dog, an African hound called a basenji, to grimace whenever he said "Nixon." The dog's name was Ruwenzori ("Rui," for short), after a mountain range on the border between Uganda and the Congo. "Rui," Murray would say. And when the dog, with its floppy skin several sizes too large, looked his way, he would shout "Nixon!" while blowing cigarette smoke in the animal's eyes. He showed off the trick at a dog show, introducing the animal as the only registered member of both the American Kennel Association and the Democratic party. Now Gell-Mann was among the handful of scientists being briefed by the Nixon administration on the war.

Unsure if they wanted to stay in Southern California, the Gell-Manns had sold their house in Altadena before departing for Princeton. Much as he loved the rambling old place, he had come to feel that it was not "serious" enough for the kind of life he aspired to. It had been perfect for a young couple but, for reasons he couldn't quite articulate, it no longer seemed appropriate. Now the family had to find a new place to live. He and Margaret had purchased a lot in the foothills, thinking they might build there someday. But that seemed too permanent a decision. After a month in temporary digs, they rented a massive, completely serious two-story house on Armada Drive overlooking the Rose Bowl and the Arroyo Seco, the long dry riverbed that runs along the western edge of Pasadena. The six-thousand-square-foot house had seven baths (counting the two cabanas), four bedrooms, and a grand entranceway.

There was a certain cachet to living by the Arroyo. From the late nineteenth to the early twentieth century, this area was home to the bohemian Arroyo Culture, a group of artists, craftsmen, and writers who turned their back on Pasadena's upper crust, embracing instead the Spanish and Indian cultures of the Southwest. The Arroyoans rejected the manicured lawns and rosebushes of Pasadena for life on the edge of the wilderness. Or so they liked to think. Over the years, the wilderness had slowly given way to suburban neighborhoods; the Arroyo was lined with concrete and turned into a flood-control ditch. The noisy Rose Bowl, brimming with football fans, had become a more imposing presence. Still, the Arroyo remained a good place for walking, riding, or, when it became fashionable, jogging. Like Central Park, it was a place where a person could get away from the city. As one strolled down its winding course, the houses above disappeared in the foliage, with only an occasional bridge arching overhead as a reminder that the wildest thing here was the automobile.

Dark and imposing, with an anonymous, almost institutional facade, the Gell-Manns' modern new residence was far from anyone's conception of an Arroyo bungalow. It struck some visitors as more of a museum than a house, a place to keep Murray's growing collections of antiquities. But for now it was home, and the Gell-Manns dabbled in what was left of the Arroyo tradition. On the first floor of the house, Murray gradually accumulated a gallery of ancient American Indian pottery, set in sand to avoid earthquake damage. Margaret installed a loom and practiced the craft of weaving. With Gweneth Feynman, she also joined the Arroyo Singers, an amateur group that gave performances of baroque and Renaissance music. Another member was Judy Macready, wife of an old Yale classmate of Murray's, Paul Macready, who lived across the street. Macready, an expert glider pilot who had studied at Caltech, later became famous for his Gossamer Condor, the first human-powered aircraft. He was just starting a new company to pursue experimental aviation and other interests, and asked Gell-Mann to be a director. Macready also shared Murray's fascination with human psychology, and the two became friends.

Murray's daughter, Lisa, kept a horse at a stable across town in Eaton Canyon and charmed her teachers at Pasadena's Polytechnic, the prestigious private school associated with Caltech. She missed Altadena, with its mix of different people, and the students at her old school. The new neighborhood seemed so middle-class and homogeneous. But like her mother, she rarely voiced her opinion. She was the perfect student, quiet and intelligent, with an irresistible smile. Nick, some six years younger, had a wilder streak. He was more interested in sports than studying—not really a chip off the old block. When the family decided to get a new dog, Murray did all the research—reading, going to dog shows, talking to other dog owners about the relative merits of the various breeds. It was as though he were choosing the best wine or luxury car to buy. Nick had it in his mind that he wanted a Siberian husky. His will prevailed. But Murray decided that the dog had to be given a name in the language spoken by the Chukchi people of far northeastern Siberia. On a visit to Washington, he used his influence as a presidential science adviser to borrow books on the language from the Library of Congress, and he began learning the rudiments. Using a Russian-English dictionary to decipher the Chukchi-Russian dictionary, he came up with the name Taengychyn, meaning "the best"—Tengi for short. The family also had Burmese cats. (Murray's favorite, U Nu, named for the Burmese prime minister, was run over and killed on Mercer Road while they were in Princeton.) He and Margaret even joined the Cat Fanciers Association. Whatever was worth doing was worth doing well, at whatever the cost.

Murray still wished he could work up the nerve to get out of Pasadena. He liked the Bay Area. An appointment at Berkeley or Stanford would be nice. He regretted all over again turning down Harvard and the Institute for Advanced Study. Margaret was still longing to return to the East Coast, but she wasn't one to impose her wishes. Her parents had quarreled constantly, and she refused to argue about anything. Murray was pretty sure she would cheerfully pick up and go wherever he wanted. But that was too easy. He wanted to debate the matter, just as he would with colleagues working together on a physics problem. He wanted to consider the pros and cons of every possibility, dismissing the weaker arguments one by one until the right view—preferably the one he was pushing—prevailed. He wanted to be right. Margaret refused to play the game, and they stayed locked in a state of inertia. They talked about escaping Pasadena by building a log cabin with all the amenities on their lot in the foothills. But that never happened. Margaret was afraid she would be lonely up there away from all their friends. A few years later, when the landlord threatened to raise the rent on the Armada house, Murray took the path of least resistance and reluctantly bought the place.

In the spring of 1969, Margaret's brother died of colon cancer. He was the last of her immediate family, and she was left with the sad task of dealing with the solicitors in England about his estate, which included a collection of antique Broadwood pianos. The Gell-Manns also spent more of their time taking care of Murray's father. He had never fully recovered from Pauline's death. Finally succumbing to diabetes, pulmonary edema, and congestive heart failure, he died in August, never knowing that a few months later his son would win a Nobel Prize.

Gell-Mann had reason to believe that 1969 might be his year. It had gotten off to an auspicious start when he won the $10,000 Research Corporation Award for predicting the omega-minus. Previous winners had included physicists as renowned as Werner Heisenberg and Ernest O. Lawrence. Chien-Shiung Wu had received the award for her experiment overthrowing parity; Watson and Crick, for deciphering DNA. A few months earlier Gell-Mann received what may have been a hint about the Nobel Prize at a meeting of the Ramsey panel, a government advisory group assembled to discuss a powerful new accelerator to be built in Illinois at what would become known as Fermilab. The scientists had been trying to coordinate their calendars. Would Murray be available in early December? He said he thought he would. "Are you sure?" T. D. Lee enigmatically asked. The Nobel Prizes, announced in October, were awarded in Stockholm on December 10, the anniversary of Alfred Nobel's death.

Lee may not have known any more than anyone else. As Murph Goldberger later wrote in a tribute to his friend's triumph: "Standard physics cocktail party conversation for the past six years in late October was always, 'I wonder if Murray will get it this year.' " Anticipation grew when Gell-Mann canceled a trip to England, insisting that he had an earache. A few days earlier, another Caltech scientist, Max Delbrück, had learned that he had won a share of the Nobel Prize in physiology or medicine for his work on viral genetics. A film crew had flown in from Sweden to interview him, and some people in the physics department noticed that it hadn't left yet. Gell-Mann was unaware of this, but he went to bed on October 29 knowing that he just might be awakened by the long-awaited call. At best, he thought he might share the prize.

When the phone rang at 3:30 a.m., the day before Halloween, he sleepily heard someone on the other end, a journalist, talking about Sweden and what sounded like "elementary articles." But if he understood correctly through his grogginess, he was getting the whole thing—375,000 Swedish kronor, or about $72,800. As the calls kept coming, Margaret decided to get up and make breakfast. Lisa buried her head under the pillow and tried to go back to sleep. Nick, sensing the excitement, put on the pirate costume he had bought to go trick-or-treating. The next day a wire-service photo at the top of page one of the *Los Angeles Times* showed a vignette of life in the Gell-Mann kitchen: Nick sitting at the breakfast table wearing his false mustache, black eyepatch, and skull-and-crossbones hat; Murray in a sweater and black-framed glasses; Margaret solicitously pouring hot water into his teacup. Lisa was still asleep.

Caltech held a press conference at 9 a.m. (A TV reporter jokingly called Murray "Dr. Quark.") Robert Bacher, the school provost and former head of the physics department, declared that "Dr. Gell-Mann has contributed probably more than anyone toward bringing order out of chaos in high-energy physics."

Feynman was particularly generous in his praise: "This event marks the public recognition of what we have known for a long time, that Murray Gell-Mann is the leading theoretical physicist of today. The development during the last twenty years of our knowledge of fundamental physics contains not one fruitful idea that does not carry his name." He also mentioned Murray's interests outside physics. "If further confirmation is needed that some scientists can be as sensitive and as active toward human problems as any humanist," he pronounced, a little grandiosely, "we are proud to exhibit Gell-Mann."

The praise kept coming full blast. The *New York Times* said this "youngest son of Austrian immigrants" was "a possible successor to the mantle of Albert Einstein"—a cliché science writers were finding irresistible. The campus newspaper, the *California Tech,* rushed out a special edition: "Extra! Gell-Mann a Nobel." Local businesses were happy to chip in by taking out special ads. "Congratulations! Dr. Murray Gell-Mann from Campus Barber Shop. (Three Barbers to Serve You.)" Later that day Murray had a case of champagne brought in for his theory group. After work, he and Margaret celebrated at the Beachcomber with the Zachariasens.

In the interviews that followed the announcement, Gell-Mann did the usual worrying about whether the prize meant the end of his productive career. Asked what he would do with the money, he said he wanted to buy a plot of land and protect it. He had seen a place he liked in Aspen. Murray was becoming more and more involved with conservation. He and Margaret had joined the Sierra Club, and he was talking with friends about starting an institute to apply rigorous quantitative science to city and rural environmental problems. Whenever he was asked about quarks, he played down the idea. "The quark is just a notion so far. It is a useful notion, but actual quarks may not exist at all." He said whether or not they existed was "immaterial."

In fact, the official Nobel citation didn't say anything about quarks. The prize was being given "for his contributions and discoveries concerning the classification of elementary particles and their interactions." It was a kind of lifetime achievement award, for someone barely forty. That he was apparently being recognized for so many contributions helped explain why the judges hadn't included any of the codiscoverers in the award: Nishijima for strangeness, Ne'eman for the Eightfold Way, or Zweig for quarks.

Murray was sad that his father had died too soon to hear the news. But he was delighted to learn that his brother Ben, who had done so much to interest him in science, would be at the ceremony. A group of Ben's friends in Carbondale, where he was still working for the local paper, raised the money to send him to Stockholm.

That December, Murray and Margaret set off for Europe. First they went to London. Then, with Murray decked out in a cashmere overcoat and a fur hat, they flew with an Imperial College professor, Derek Barton, one of the two winners in chemistry, to Göteborg on the west coast of Sweden to lecture at Chalmers University. From there they went on to Stockholm, checking in at the elegant Grand

Hotel. The view was magnificent, overlooking the harbor and the medieval stone skyline of Gamla Stan, the Old Town, which sits on one of the city's fourteen islands. The rest of the week was a blur of limousine rides and speeches as the laureates were taken from celebration to celebration. (Murray skipped a talk he was supposed to give at the old university town of Uppsala and went bird-watching with Ben, spotting a rare black woodpecker and coming across an old Viking tomb marked with runestones.)

The climax of the week, Nobel Day, was the social event of the season. Tickets to the white-tie awards ceremony and the dinner and ball afterward are highly coveted even by those who have no idea what the scientists are being honored for. In his will endowing the prizes, Alfred Nobel had put physics at the top of the list. And so, when the ceremony took place on the evening of December 10 in Stockholm's ornate Concert Hall, the physics prize was the first to be awarded. Men in white tie and tails and women in evening gowns filled the floor of the auditorium, overflowing into the balcony ("Penguin Mountain," it is sometimes called). The Stockholm Philharmonic Orchestra played a scherzo from the Pastoral Suite by Lars-Erik Larsson. Then, with the members of the Royal Swedish Academy of Sciences and other dignitaries arrayed on the stage behind him, and the royal family sitting to one side, Professor Ivar Waller, a member of the Nobel Committee for Physics, began the solemn presentation speech.

"Elementary particle physics, which is now so vigorous, was still in its infancy when Murray Gell-Mann in 1953 . . ." For the next ten minutes, he described Gell-Mann's work with strangeness and the Eightfold Way, even alluding to the obscurities of current algebra, finally ending with a polite mention of quarks. "It has not yet been possible to find individual quarks," he said, "although they have been eagerly looked for. Gell-Mann's idea is none the less of great heuristic value." He called him the person who had "during more than a decade been considered as the leading scientist in this field."

Then came the point in the presentation when the speaker dramatically pauses and turns from the audience. He faces the new laureate, addressing him directly.

"Professor Gell-Mann: You have given fundamental contributions to our knowledge of mesons and baryons and their interactions. You have developed new algebraic methods which have led to a far-reaching classification of these particles according to their symmetry properties. The methods introduced by you are among the most powerful tools for further research in particle physics.

"On behalf of the Royal Swedish Academy of Sciences, I congratulate you on your successful work and ask you to receive your Nobel Prize from the hands of His Majesty the King."

It is an electrifying moment. With a fanfare of trumpets, the new laureate bows to the king, who bestows the leather-bound diploma and the twenty-three-karat gold medal. Continuing with the custom, he bows to the Royal Academy, then to the audience, finally sitting down amid thunderous applause. When Gell-Mann had a chance to examine the medal, he saw that engraved on the front was the stern visage of Alfred Nobel. On the back was the Goddess of Nature, holding in her arms the cornucopia of knowledge as the Genius of Science lifts the veil from her eyes. Arching overhead was an inscription from Virgil's *Aeneid: Inventas vitam juvat excoluisse per artes:* "And they who bettered life on earth by newfound mastery." Or, translated more literally: "Inventions enhance life, which is beautified through art." On the very bottom was the name of the winner: Murray Gell-Mann.

After physics came, in the order of importance Nobel had given them, the prizes for chemistry, medicine, and literature (the winner that year was Samuel Beckett), and finally, given for the first time that year, the memorial prize in economics. (The Nobel Peace Prize is presented in Oslo, Norway.) Gell-Mann thought it appropriate that the winners of the economics prize were seated separately from the other winners. Economics could hardly be considered a science.

When the ceremony was over, the laureates and their families were packed into limousines and chauffeured across the river for the banquet at Sweden's architectural gem, the Town Hall. To the accompaniment of a chamber ensemble, the laureates filed into the elegant Golden Hall, with its twenty-four-karat mosaics. In further keeping with the preeminence Nobel had given physics, Margaret, who had borrowed a mink coat from Gweneth Feynman, entered the hall on the arm of the king and sat by his side during the long dinner. Murray was wedged between the princess and the queen. After the toasts, course after course of lavish dishes was announced with musical fanfares, the army of waiters striding in formation to serve the food. Champagne (Krug Brut Reserve) was followed by an appetizer of avocado stuffed with salmon and then, for the main course, a roast filet of beef with truffles (Périgord-style), served with a 1959 Bordeaux from Château Potensac. (But alas, no pheasant cooked between slices of veal.) Then, as the orange sorbet and coffee and liqueurs were served, the after-dinner speeches began. In a long toast to the laureates, Gell-Mann was praised for having made sense of the "bewildering mess" of subatomic particles: "To laymen you appear as the maid entering the children's playroom to clean up and bring order where an overall smashing game is going on. Thus we are now prepared for a continuation of this smashing business, the game of dividing the indivisible." When it was Murray's turn to speak, he walked to the podium, accompanied by another sounding of trumpets, and began:

"Your Majesty, Your Royal Highnesses, Your Excellencies, Ladies and Gentlemen. As a theoretical physicist, I feel at once proud and humble at the thought of the illustrious figures that have preceded me here to receive the greatest of all honors in science, the Nobel prize. I think also of my colleagues in elementary particle theory in many lands, and feel that in some measure I am here as a representative of our small, informal, international fraternity.

"We are driven by the usual insatiable curiosity of the scientist, and our work is a delightful game. I am frequently astonished that it so often results in correct predictions of experimental results."

A murmur of laughter rippled through the hall.

"How can it be that writing down a few simple and elegant formulae, like short poems governed by strict rules such as those of the sonnet or the waka, can predict universal regularities of nature? Perhaps we see equations as simple because they are easily expressed in terms of mathematical notation already invented at an earlier stage of development of the science, and thus what appears to us as elegance of description really reflects the interconnectedness of nature's laws at different levels.

"For me, the study of these laws is inseparable from a love of nature in all its manifestations. The beauty of the basic laws of natural science, as revealed in the study of particles and of the cosmos, is allied for me to the litheness of a merganser diving in a pure Swedish lake, or the grace of a dolphin leaving shining trails at night in the waters of the Gulf of California, or the loveliness of the ladies assembled at this banquet."

With its pedantic references to the waka, an ancient form of Japanese poetry predating the haiku, and the merganser, the speech was already vintage Gell-Mann. Then, to the astonishment of the audience, he broke into Swedish. The king, who had nodded off, woke with a start.

"Detta lands folk har visat kärlek och hänsyn till skönheten som den framträder i alla dessa former . . ."

"The people of this country have shown their love and respect for beauty as she is revealed in all her forms. One can especially congratulate the Swedes for having valued and kept their natural heritage. At the United Nations we now listen as Sweden urges the entire world to investigate how all of us can contribute to the maintenance and restoration of the natural condition of our blue planet that looks so inviting, viewed from afar.

"For me it's a great personal pleasure to visit Sweden once more and to see, for the first time, her friendly capital. On behalf of my dear wife, Margaret, my brother, and myself, our heartfelt thanks for all hospitality extended to us. Thank you."

The audiences at these affairs were used to foreigners learning a few token phrases from Berlitz. "Good morning." "Good afternoon." "Please" and "Thank you." But to deliver almost two minutes of an acceptance speech in Swedish was truly a grand gesture. Murray later agonized over having pronounced a word as though it were Danish (a language he was more familiar with). He had practiced the speech over and over with his diplomatic escort, a young man from the Swedish foreign service, and kept making the same error. "No, you mustn't say it that way!" the tutor kept telling him. Murray couldn't believe he had done it now, when it really counted. Otherwise, though, the presentation was very good, and except for one or two prepositions, the grammar was perfect. The choice of words may have seemed a little strange and florid to the reserved Swedes; he clearly wasn't a native speaker. But what a fine tribute to their culture.

Gell-Mann sat down to applause, and soon a long night of drinking and dancing began.

CHAPTER 13

QUANTUM CHROMODYNAMICS

The cosmologist Fred Hoyle once warned Gell-Mann about the perils of fame. If you ever accomplish anything important in life, he said, the world will conspire to keep you from doing anything else again. Everyone wants you to give a speech, write an article, serve on a committee. Before long, Murray was spending more time going to meetings than doing physics—Jason, Nixon's science advisory committee, the Ramsey Commission on the construction of the Fermilab accelerator. And there were other obligations. He was on the council of the Smithsonian Institution and the board of the Council for Advancement of Science Writing. All this in addition to increasingly lucrative government and corporate consulting deals. Even an efficient workaholic would have had trouble keeping up. For a natural procrastinator like Gell-Mann, these mounting burdens made life a constant anxiety dream. In the past he had always taken time to dash off thoughtful recommendations for former students and colleagues. But now he was finding even that too much. "Please feel free to have people call me about you," he wrote to a young physicist at the University of Colorado. "Unfortunately, I have great difficulty writing letters and requests for written recommendations are likely to be ignored."

More daunting still were the endless requests to submit papers. Asked to contribute to the Festschrift volume honoring the Russian physicist Igor Tamm on his seventy-fifth birthday, Gell-Mann was blunt: "Unfortunately when I make promises to contribute articles to such publications, I almost never keep the promises. Therefore, I have learned not to make them in the first place, and I must decline with regret your invitation." Friends of the Italian physicist Edoardo Amaldi, just turning sixty, were given a slightly warmer brush-off: "It would be a great pleasure and privilege . . . but I know from past experience that my promises to write articles nearly always end in a failure to deliver. . . . However, if I do write something in the near future, I shall offer it to you on the chance that you can still accept it." Often he would just scrawl "Sorry" on the bottom of an invitation and leave it to the secretary to send a boilerplate reply.

Even when he had already delivered a lecture, he sometimes couldn't bring himself to write it up. When he spoke at a Solvay conference on group theory and current algebra in 1967, the organizers tape-recorded the lecture and supplied him with a verbatim transcript. All he had to do was edit it. They should have known better. "Manuscript contains many serious distortions," he said in a cable to Brussels. "Do not publish unless and until my corrections arrive."

Not even the honor of a Nobel Prize was enough to overcome Murray's writer's block. One of the proudest obligations of a laureate is to deliver a scientific lecture during Nobel week. Each year, shortly after the ceremonies, the texts are gathered together and published in a celebratory volume, *Le Prix Nobel,* published by the Royal Swedish Academy of Sciences. There could hardly be a more prestigious and select venue in which to appear. Murray, feeling generous amidst his good fortune, had wanted to give a speech that acknowledged all the people who had contributed to SU(3). He even had an assistant working on the details. But with all the other distractions, it finally seemed too much. Feynman told him not to worry, to just talk about the physics. But that too presented difficulties. What was he going to say about quarks? He was as conflicted as ever, worrying over all these problems and philosophical ambiguities. Finally he decided simply to recycle a speech he had given three years earlier at the British Royal Institution and at Caltech's seventy-fifth anniversary (the one with the Updike poem about neutrinos). And so, on December 11, the day after the Nobel banquet, he repeated the talk, retitled "Symmetries and Currents in Particle Physics," at Sweden's Royal Institute of Technology. Then the editors waited for a manuscript. And they waited. What was he going to do? The speech had already been published twice. He had to come up with something new. When the editors inquired in mid-January, Gell-Mann sent a reassuring telegram:

TERRIBLY SORRY STOP TASK OF WRITING HARDER THAN EXPECTED
STOP WILL TRY TO FINISH IN FEBRUARY STOP BEST REGARDS

When March came and there was still no manuscript, the Swedes were getting worried. They dispatched a telegraphic plea:

EDITOR TROUBLED IN VIEW PENDING PUBLICATION PRIX NOBEL STOP KINDLY CABLE IF AND WHEN NEW AIRMAILED MS NOBELLECTURE RE YOUR CABLE JANUARY 29TH

This time Gell-Mann's reply was less than encouraging:

VERY SORRY TO BE EVEN MORE DELINQUENT THAN FEYNMAN STOP HAVING MORE DIFFICULTY THAN EXPECTED STOP NEW TARGET MIDDLE OF APRIL STOP APOLOGIES AND BEST REGARDS

But mid-April came and there was still no manuscript. After he ignored another telegram from the editors, they cabled him again, eliciting his request for yet another extension:

FURTHER ABJECT APOLOGIES STOP WILL TRY TO COMPLETE EVERYTHING DURING MAY

By June, when the volume was supposed to go to press, it was clear that Gell-Mann was simply not going to send them a lecture. The editors gave up:

ASSUME YOUR UNPRECEDENTED DELAY DELIVER NOBELLECTURE MS MUST BE CAUSED CIRCUMSTANCES OUTSIDE YOUR CONTROL STOP IF NOT RECEIVED BEFORE JUNE 10 SINCERELY REGRET LECTURE WILL NOT APPEAR IN PRIX NOBEL 1969

Murray felt terrible, but at least he was off the hook. He finally responded:

VERY MUCH REGRET FAILURE TO MEET DEADLINE AFTER REPEATED EXTENSIONS KINDLY ARRANGED STOP TERRIBLY SORRY FOR DIFFICULTIES I HAVE CAUSED

When the volume was published that fall, the title of Gell-Mann's talk was listed in the table of contents with a matter-of-fact note: The laureate did not present his lecture in time for inclusion in this edition of the *Prix Nobel*.

. . .

Weary from fending off editors, Gell-Mann and his family retreated to Aspen that summer, as they had for the past three years, to hike in the Colorado mountains and picnic with the families of other physicists. In what was becoming a tradition, many of his colleagues spent their vacations on the idyllic grounds of the Aspen Center for Physics. Originally an offshoot of the Aspen Institute for Humanistic Studies, a summer haven that brought together scholars, businessmen, and politicians to discuss the world's problems, the center had been started by two physicists who had studied at Caltech, Michael Cohen and George Stranahan, an heir to the Champion spark plug fortune. The small campus was set in a grove of aspen trees at the edge of town. Rivulets of cold mountain water cut through the grounds on their way to the Roaring Fork River, giving the place the peaceful feel of an alpine glacial meadow. Next door was the Aspen Music Festival, another offshoot of the Aspen Institute, and the sounds of the orchestra practicing for the evening concerts might have made a visitor think he had died and gone to heaven, ready to meet the Great Geometer.

Just a few decades earlier, it would have been hard to imagine that the gritty old mining town would ever recover from the plummet in silver prices and Congress's decision, in 1893, to remove the dollar from the silver standard. But then, after World War II, skiing became a national sensation among the well-to-do. A Chicago industrialist named Walter Paepke led a group of investors who turned the town into a winter playground. But he also had a higher-minded goal. He and his wife Elizabeth, the sister of the investment banker, diplomat, and arms negotiator Paul Nitze, were enthusiasts of the arts and supporters of the University of Chicago's Great Books Program. In 1949 they launched Aspen as an intellectual mecca with, rather oddly, a bicentenary festival celebrating the birth of the German poet and dramatist Johann Wolfgang von Goethe. Heralded worldwide as a premier cultural event, the festival led to the development of the Aspen Institute.

Gell-Mann had first visited the town in 1964 for an arms control meeting and had fallen in love with the place. Where else could one find an island of culture and gentility surrounded by snow-capped peaks? There was even a listing for physicists in the Yellow Pages.

The children loved the Colorado countryside and enjoyed playing with the other physics kids. Lisa, now a young teenager, would ride through the mountains on horseback, her long brown hair flowing behind her. She was quiet, almost painfully shy, but so alive and intense—a magical child, thought Francis Low's daughter Margaret, who idolized her. She seemed to her like a character from *Little Women* or *The Secret Garden*. She was intensely interested in plants and animals—the tiny differences that made each one unique. She immersed herself in the world's details. When the family went to a museum, the others would charge through from beginning to end, hitting the highlights, getting the gist of the story, while Lisa would still be on the first or second exhibit, deliberately reading every little sign. She would be just as meticulous and methodical when skiing on Aspen Mountain, expertly making her way down in what seemed an infinite expansion of precise movements. Margaret gave her a ski cap with a turtle embroidered on it: She was always the last one down. Nick by now was the quintessential adolescent boy, as outspoken and rebellious as Lisa was quiet and accommodating. He was more interested in sports or getting into mischief with his friends than in books and museums. For both Lisa and Nick, Aspen was a time to see more of their father, who the rest of the year always seemed to be traveling to another conference. (It was Margaret who helped Nick build his model airplanes. Murray was too busy.)

For Murray, Aspen was a place where he could find the kind of cross-fertilization of ideas that he had wanted so much to bring to Caltech. The previous summer he had joined an unlikely mix of characters for an Aspen conference on conservation and the population explosion. Among the cast were the environmentalists David Brower (recently deposed as head of the Sierra Club) and Garrett Hardin (who stole the show by calling for a United States birth moratorium to set an example for the rest of the world). They were joined by John Ehrlichman, Nixon's special counsel, who insisted on the need for "politically feasible solutions." This summer the big event was a meeting called "Technology: Social Goals and Cultural Options." Gell-Mann led a faction (which included the writers Mary McCarthy and Paul Goodman) calling for a radical declaration on the need to protect the environment from the mindless march of progress. "The less of us there are," Gell-Mann asserted, "the less we have to tell the poor of the world they must stay that way." But more conservative voices prevailed, and the result was a bland statement noting that, as a reporter from *Science* put it, "the world is indeed beset by many pressing problems, that technology can be employed for both good and evil, and that something should be done both about poverty and threats to the environment."

More and more often, the invitations Gell-Mann accepted had less to do with unifying the forces of nature than with unifying the social sciences with the physical sciences and confronting the world's ills. The year after getting his Nobel, he appeared at a conference on "Science and the Morality of Intellect" at the University of Chicago. He spoke on "The Narrowing Cone: The Use and Misuse of Technology" at a conference at Southern Illinois University. He talked to the Council on Foreign Relations on "Ecology, Science and Their Effects on Foreign Policy."

One clear day in October, Gell-Mann went to the University of California's Santa Barbara campus, where he was asked to speak at the dedication of a new physics building. He had mixed feelings about the way the campus was steadily encroaching on his favorite California bird-watching spot: a stretch of coastline near Isla Vista, some ninety miles northwest of Pasadena. It was a fine vantage point to observe the string of islands—Anacapa, Santa Cruz, Santa Rosa, and San Miguel—that outlined the far edge of the Santa Barbara Channel, its blue water jumping with dolphins, sea lions, and flying fish. Murray liked going there with his field glasses to look for curlews, godwits, phalaropes, and other winged creatures. Just to the east, forming a backdrop to this spectacular aquatic universe, were the Santa Ynez Mountains, home to some of the last of the mighty condors.

Though he always hated to see another patch of wildland paved over, he felt this was a glorious spot for physicists to think about how the precise mathematics of the underlying particles gave birth to a teemingly diverse, unpredictable world. "For me the two things are inseparable, the love of the beauty of nature and the desire to explore further the symmetry and subtlety of nature's laws," he told the Santa Barbara audience. "I have an innocent view of pure science, of inquiry driven by wonder and curiosity, and the thrill of learning enough to make predictions that are astonishingly fulfilled."

But science, he lamented, was under attack. That spring many campuses erupted again in protest after Nixon intensified the bombing of North Vietnam and expanded the war into Cambodia. Four students were killed by National Guardsmen at Kent State University, and in an act that became symbolic of the rejection of science as a tool of the warring elite, protesters blew up the

Army Mathematics Research Center at the University of Wisconsin, killing a young physics postdoc. The year before, protesters at the University of Colorado had disrupted a Jason meeting, shouting "How many people did Jason kill last year?" Gell-Mann didn't specifically mention the tragedies or allude to his own involvement in Jason. But he warned that science and technology were acquiring such a bad reputation that some of the country's best students were abandoning these pursuits, some seeking refuge in astrology, palmistry, and even the pseudoscientific theories of Immanuel Velikovsky.

Some of this backlash arose, he noted, because science was so often badly applied. Pretending that they could analyze and understand the most complex of processes, decision makers had embraced "a kind of narrow rationality" that took into account "only things that are very easy to quantify." Lost in the calculations were factors "like beauty or diversity or the irreversibility of change." The results had been disastrous. "With anything hard to quantify set equal to zero, a highway can be driven straight through a neighborhood or through a rare wilderness because there is no reliable quantitative measure of damage to set against the increased cost of running the road around the outside."

With the pursuit of knowledge fragmented into fiefdoms things could only get worse. He spoke of the photochemical smog that made his own San Gabriel Mountains blurrier every year. For more than twenty years, scientists had known that smog was caused by unburned hydrocarbons and nitrous oxides from car exhaust. Meanwhile health experts warned of another pollutant, carbon monoxide. The government imposed regulations requiring automakers to control both hydrocarbons and carbon monoxide. But to counteract the drag the antipollution devices put on acceleration, Gell-Mann said, the manufacturers simply raised the combustion temperature of the engine, and that just produced more nitrous oxides. Meanwhile, poor urban planning, burgeoning populations, and sprawling cities encouraged more people to buy cars.

No wonder science had such a bad image. "Youngsters tired of the tyranny of badly programmed computers, and of people who act like badly programmed computers, are turning to tarot cards and charlatans," Gell-Mann said, a little melodramatically. There was a danger of being "squeezed to death between bureaucratic automata on the one hand and superstition or raving on the other."

He almost sounded like Theodore Roszak attacking the technocrats. But Gell-Mann was not about to trade his well-tailored sportscoat for a denim workshirt, or mathematics for numerology. The answer, he was sure, was not to reject rationality but to cultivate and expand it. The world needed more scientists and engineers to join with sociologists, economists, legal experts, doctors, and government and business leaders to apply "our awesome technical capabilities" to the task of untangling the interlocking web of problems. One day this multidisciplinary analysis of systems with many countervailing parts would be loosely named complexity studies. For now, Gell-Mann called it "systems analysis with heart." The standards of science education and journalism must be raised, he declared. And most of all, scientists must be given the continued freedom "to follow where curiosity leads and to take the small steps that culminate, once or twice in a generation, in those great universal syntheses like quantum mechanics, relativistic gravitation, the genetic code, the theory of the evolution of the stars."

Gell-Mann's attention was snapped back to the problems of quarks that summer when a determined young theorist named Harald Fritzsch passed through Aspen on his way to do graduate work at the Stanford Linear Accelerator Center. Years earlier Fritzsch and a friend had come to the West after escaping from the East German police, traveling some 200 miles along the Black Sea, from Bulgaria to Turkey, in a kayak equipped with an outboard motor. Recently he had been working on a Ph.D. at the Max Planck Institute for Physics and Astrophysics in Munich, where one of his professors was Werner Heisenberg.

Fritzsch was a fervent believer in quarks, and he was surprised to find that Gell-Mann had only the vaguest awareness of the latest SLAC scattering experiments supporting, more strongly than ever, the notion of pointlike masses inside protons. Of course, it was possible that physicists were misinterpreting the numbingly confusing data. But before long, experimenters at CERN bombarded hadrons with neutrinos, instead of electrons, and found strong corroborating evidence that what Feynman was calling "partons" were indeed fractionally charged, spin-½ particles—quarks. Looking at hadrons from two different perspectives, physicists were seeing signs of the same phenomena. It was as if, as Michael Riordan, one of the experimenters, later wrote, two separate senses had been used to confirm the existence of a single object: "Our eyes alone can easily be deceived by a mirage or hallucination, but if we can also reach out and *touch* this vision, our recognition of its 'objectivity' is greatly enhanced."

Walter Sullivan of the *New York Times* soon broke the story on page one: "Subatomic Tests Suggest a New Layer of Matter." He began with a flourish: "A number of physicists believe that, through a variety of atomic experiments, they have begun opening the door to the innermost sanctum of matter." (In the story Gell-Mann fuzzed things up by insisting that the quarks could simply be manifestations of currents, rather than observable particles.) The *Times* editorial writers followed a few days later with an item called "The Great Quark Hunt," rhapsodizing about quarks as the possible culmination of the great program begun by the original atomist, Democritus: breaking matter into pieces and the pieces into pieces. "It was Dr. Gell-Mann's whimsical notion to name these hypothetical building blocks quarks, a term borrowed from the line in James Joyce's 'Finnegans Wake': 'Three quarks for Master Hark.' " (In addition to botching the quote, the writer gave credit for quarks to Gell-Mann and Ne'eman instead of Gell-Mann and Zweig. "Write about Zweig," Murray scrawled to himself on his copy of the editorial. Then: "No. May 1! Too late!" Apparently he had intended to send a letter correcting the mistake but had waited too long.)

In the face of all this excitement, it was hard for Fritzsch to fathom how Gell-Mann could remain so aloof, talking one moment about quarks as permanently confined particles, the next as abstract symbols, with no explanation for how something "mathematical" and "fictitious" could cause electrons to ricochet. Fritzsch was also struck when Murray told him one day, as though it were a revelation, that a field theory involving quarks and bosons just might explain the strong force after all—a point the young German had long taken for granted. Since his days as a student in East Germany, where he worked on gravity and Yang-Mills models, Fritzsch had believed in field theory as more than a mere calculational device, even for the strong force. He had never been very impressed with the bootstrap approach, which still wasn't yielding precise, testable predictions. For all their efforts, Chew and his disciples still hadn't found the single, self-consistent mathematical edifice—the one true S-matrix—that showed we live in the only possible world. The SLAC results convinced Fritzsch more than ever that the strong force could best be explained as some kind of field. He couldn't understand why Gell-Mann still felt a need to keep his options open. "Hadrons act as if they are made up of quarks but no quarks exist—and, therefore, there is no reason for a distinction between the quark and bootstrap picture," Gell-Mann said sometime later in a lecture. "They can just be two different descriptions of the same system. . . ." He compared the situation to quantum mechanics, in which one could think of the electron as either a particle or a wave.

Fritzsch kept insisting that Murray had to take the SLAC results more seriously. Before long he convinced him that they should try to use a new technique called light-cone algebra (a more generalized form of current algebra) to account for the abstract phenomenon called scaling. Feynman had confronted the problem with his generic partons. Fritzsch was sure he and Gell-Mann could do an even sharper analysis couched in terms of a field theory specifically built from quarks.

By the fall of 1970, he was coming down to Caltech almost every month. Gell-Mann quickly began to appreciate him as an ingenious theorist with a sharp sense of how theory and experiment were intertwined. Sometimes Fritzsch seemed a little careless and rash, Gell-Mann thought, but he always learned from his mistakes. Murray, of course, suffered from being too careful. It was good to have someone like Fritzsch to stir things up. In February, in his final year of graduate school, Fritzsch transferred to Caltech. His arrival was literally earthshaking, coming just hours after the great quake of 1971 rocked the area, knocking the pictures on Murray's office wall askew. He decided to leave them that way in commemoration of his new colleague.

Their work went slowly at first. Gell-Mann always seemed to be traveling to another speaking engagement, breezing into the office every now and then for a few hectic days of blackboard talk. By midspring they had put together two papers. The first was presented in Coral Gables, Florida, at a meeting on "Fundamental Interactions at High Energy" (where Gell-Mann showed up sporting a beard). It was followed by a longer version, which was given at a conference on "Duality and Symmetry in Hadron Physics" at Tel Aviv University in Israel. Starting with the toy field model Murray had introduced years earlier, with a single boson holding the quarks together, the two scientists succeeded in abstracting algebraic relationships that indeed shed some more light on scaling. It was Gell-Mann's old veal-and-pheasant approach: One could use field theory as a scaffolding to build models that, it was hoped, would remain standing even if field theory ultimately was found to be vacuous—"to get the benefits without the embarrassments," as one physicist put it. The paper was less than a landmark. But it helped close the gap between the quark and the parton pictures, showing that they were probably the same. Around this time, Gell-Mann coined one of his more evocative slogans: "Nature reads in the book of free field theory." He still wasn't completely ready to go beyond the veal-and-pheasant approach and embrace field theory wholeheartedly, but he seemed to be leaning in that direction.

There were good reasons to believe, by the early 1970s, that field theory might stage a comeback. In recent years, while most of Gell-Mann's work had focused on the strong force, physicists had come close to unifying electromagnetism and the weak nuclear interaction—and the framework turned out to be none other than an unfashionable Yang-Mills field theory: The new combined "electroweak" force could be thought of as arising from a symmetry and being carried by bosons. Sheldon Glashow had struggled with this on and off since the late fifties, trying to get the group $U(1)$, which described electromagnetism, to mesh with $SU(2)$, which described the weak force. It was a delicate choreography, with the photon whirling around with the three supposed carriers of the weak force (called W^+, W^- and Z° by everyone but Murray, who still called them X's, or uxyls). Abdus Salam had come up with a similar model. But no one had been able to find a solution to the old paradox: The bosons had to be very massive to account for the short range of the force, but they had to be massless, like the photon, in order to work in a Yang-Mills gauge theory. If the bosons were not massless, the theory could not be renormalized, purged of absurd infinities.

The answer turned out to be broken symmetry—that theoretical contrivance physicists use to explain why the world is so much messier than their equations. Salam liked to explain the concept with an analogy involving a circular dinner table, with salad plates placed between each setting. Which is yours—the one on the left or the right? To avoid embarrassment, arbitrary rules of etiquette have emerged. But any guest is free to choose either plate. The arrangement is thus said to be perfectly symmetrical. But once a single diner has made a choice, the effect ripples around the table. The guests sitting in the adjacent seats must now follow suit. They are suddenly constrained in their choices, and the symmetry is broken.

Thus the symmetrical arrangement of the plates gives rise to two possible asymmetrical solutions.

A more physical example is an iron magnet. Here we are confronted with something that is asymmetrical: The spins of the magnet's atoms (each itself like a tiny magnet) are lined up in lockstep so that one pole is north, the other south. But heat up a magnet, and the spins become scrambled, the magnetism goes away. Symmetry is restored, and there is no difference between the magnet's two ends. Let it cool again, and a few atoms will arbitrarily pick one of the possible directions. Like the diner arbitrarily choosing one salad plate over the other, the effect propagates. Nearby atoms are spontaneously compelled to line up in lockstep. In various parts of the magnet, there may be islands that choose to align themselves in different directions. But those domains that happen to grow faster and larger will rapidly predominate, until ultimately the symmetry is broken. The iron bar becomes a magnet again, with a difference between north and south.

Now apply the idea, with a physicist's typical hubris, to the entire universe. At the intense energies that presumably existed moments after the big bang, perhaps the three bosons of the weak force were, like the photon, massless. They were all perfectly symmetrical and indistinguishable. You couldn't tell one from the other. Heating the universe had been like heating a magnet—the differences melted away. At these high energies, the weak nuclear and electromagnetic forces were also indistinguishable, unified. But as the universe cooled, the theory went, the symmetry broke. Some of the massless bosons became massive, and electromagnetism and the weak force split apart and went their separate ways.

Like the Fall from grace in the Garden of Eden, the great breaking of symmetry made a nice creation myth. Pure spirit gives way to material being. The challenge was to explain why the symmetries broke. For the story to qualify as physics, there had to be a mechanism. Coming from two different directions, Steven Weinberg and Abdus Salam found the answer in 1967: a special symmetry-breaking particle called the Higgs boson, named after Peter Higgs, the Scottish physicist who first proposed it. Like the photon, the positively, negatively, and neutrally charged carriers of the weak force, W^+, W^-, and Z°, were massless in their natural state, fulfilling the requirements for gauge bosons. But this hypothetical Higgs particle constantly interacted with them, endowing them with mass and causing the cleavage of the weak force from electromagnetism.

At higher energies, it was proposed, the two forces would flow back into one, like a misshapen chunk of ice melting back into a featureless pool.

For all its cleverness, the model seemed a bit contrived. The three lumbering carriers of the weak force—so far unseen in any experiment—were supposed to be equivalent to the fleet-footed photon; the weak force, which violated parity, was intimately related to the electromagnetic force, which obeyed parity. And darting behind the scenes, creating the illusion that the truly identical were so grossly dissimilar, was this other undiscovered particle, the Higgs boson, which was presumably too massive to be found with the energies of the accelerators. As one theorist put it, the whole contraption seemed to be "such an extraordinarily ad hoc and ugly theory that it was clearly nonsense." (When Gell-Mann, Goldberger, Kroll, and Low had put together their ill-fated theory of the weak force—the one another physicist called "an absolutely disgusting model"—they didn't even cite the Weinberg-Salam symmetry-breaking scheme.)

For that matter, neither Weinberg nor Salam, who still felt a little sheepish about working on field theory, went out of his way to promote the model. It went largely ignored until 1971, when a young Dutch physicist named Gerard 't Hooft astonished almost everyone by showing that the theory could be renormalized. Almost anyone could slap together a semiplausible field theory and offer it as a model of some phenomenon. But those rare few that could be renormalized caused physicists to sit up and take notice. Renormalizability was taken as a sign that the mathematics was sound. And if God was a mathematician, as even the most agnostic physicists implicitly assumed, a renormalizable theory had a chance of saying something true about the world. It would be years before electroweak unification was considered experimentally verified. But it wasn't long before physicists were talking about the "electroweak force" as confidently as they talked about electromagnetism, without feeling a need to distinguish between its two components.

Now that two of the four recognized forces had been joined within the frame of a field theory, it seemed more plausible that the strong force might succumb as well. More than a toy, field theory would be a reflection of an underlying reality. But there were still some missing pieces. If quarks were to be part of the picture, it seemed there would have to be some kind of bosons—carriers of the strong force—binding the quarks together. Here again the scattering experiments suggested an answer. Experimenters measured the total momentum of the quarks inside a proton. Then they compared the number to the momentum of the proton itself. If protons were made of quarks and nothing more, then the two figures should match. But that is not what they found. The momentum of the pieces didn't add up to that of the whole, accounting for only about half the total. This was a violation of the law of conservation of momentum, the kind of seeming lawlessness that had confronted Fermi with beta decay. By now it was second nature for theorists to attempt an escape from this kind of trap by postulating the existence of unseen particles. Neutrinos carried away the missing momentum in beta decay. Maybe there was some other kind of particle, besides quarks, that accounted for this latest discrepancy. Could the extra players be the very bosons holding the quarks together to form protons, neutrons, and the other hadrons?

The possibility had always been there. According to the Eightfold Way, there were eight currents, paths through which the particles in the group could change one into the other, eight ways you could twist the magic eightball. In a real field theory, each of these operations would be associated with a boson. Since the bosons also served to hold the quarks inside a particle together, physicists naturally started calling them "gluons." No one is sure who invented the word. Gell-Mann had used it years earlier for the fictitious boson in his toy field theory, but he vaguely remembered Teller employing the word in an entirely different context. Whoever coined the term, gluons were another step in melding Feynman's and Gell-Mann's theories into one. Partons consisted not just of quarks and antiquarks but also of gluons. (Yukawa's old vision of the strong force, in which pions held nuclei together, was just a cruder approximation. Pions too were made from quarks held together by gluons.)

The next step was to turn this rough idea into a precise model. One of the first to try was Victor Weisskopf, who had once joked about wasting time on a transatlantic call talking about anything so silly as fractionally charged subparticles. Back in Cambridge after his stint as an administrator at CERN, he was eager to return to the cutting edge. He joined with Julius Kuti, a researcher visiting MIT from Hungary, to write a paper developing the idea of the strong force as a field involving quarks and gluons.

Even Feynman was starting to grudgingly come around. He had been as unsure about his partons as Murray was about quarks, fearing they might just be "unnecessary scaffolding . . . used in building our house of cards" or "a useful psychological guide." He would retreat to his beach house in Mexico for days, looking out at the surf and trying to figure out what was happening inside the hadrons. If quarks and gluons were really there, why didn't they pop out when they were struck by electron bullets? In recent experiments, the proton targets had been surrounded by detectors tuned to find fractionally charged particles, but none ever appeared in the debris. And despite a few false leads, experimenters were still not finding free quarks in the world. What could this paradox mean? He felt like a fly banging against a window. Why couldn't he break through?

In the summer of 1971, Fritzsch went back to Europe to finish his Ph.D., at the Technical University in Munich, and Gell-Mann and his family headed back to Aspen. For a while Murray had considered building a summer home on land he had purchased with his Nobel Prize money, eight hundred acres near the town of Montrose. But in recent months he and one of his colleagues, a Caltech chemistry professor, had decided to buy an old Victorian house together in the heart of Aspen's staid old residential neighborhood, the West End, just blocks from the Center for Physics. With the Aspen real estate boom taking off, the price was unbelievably high: $90,000 (it had sold for $9,000 in 1955). But Aspen seemed like the perfect place to spend his summers, surrounded by mountains and interesting, influential people. The house had its problems. It was traditional to paint old Victorians in bright pastels, but this one was a gaudy pink—and not just the exterior but the interior, even some of the furniture. It seemed to Murray that the owner must have gotten a good deal on pink paint and hired a bunch of teenagers to slap it on. He had the house repainted sky-blue and eventually restored it to perfection—so much so that, except for the rusted iron picket fence and the crumbling brick chimney, a casual observer might have trouble telling how old it really was.

In the meantime he had been pulling strings so that he and Fritzsch could spend the next academic year at CERN working on the strong force. He landed a $12,000 grant for himself from the Guggenheim Foundation and helped get a fellowship for Fritzsch. That fall, shortly after arriving in Geneva, they began working on another remaining obstacle: the old problem of quark statistics. No one had yet found a way around the fact that quarks seemed to violate the Pauli exclusion principle, with particles like the delta, for example, apparently made of three identical up quarks, and the omega-minus made of three strange quarks. The problem could be ignored no longer. Gell-Mann's and Fritzsch's interest was rekindled after a physicist named William Bardeen mentioned a confusing anomaly involving the rate at which a neutral pion decayed into two photons. Bardeen had been working at Princeton with a physicist named Stephen Adler, who showed that if one assumed pions were made of fractionally charged quarks, the predicted decay rate was nine times too small. Thus, he concluded, "the quark hypothesis is strongly excluded."

But Adler also raised another possibility: Quarks would give the correct decay rate if they were integrally charged, as in the scheme Han and Nambu had proposed in the mid-sixties, with each flavor of quark also coming in three different varieties of what they called charm. Influenced by the dialectical materialism of the Sakata school, Nambu rejected the notion that quarks were just mathematical symbols. They had to be as real as electrons, and therefore, he thought, they could not have charges measured in thirds.

Gell-Mann and Fritzsch wanted to account for the anomalous decay rate in a way that didn't require integrally charged quarks. Maybe all that was needed was some version of Nambu's charm, a new quantum number that would allow one to distinguish quarks that otherwise appeared to be identical. They had actually begun exploring this possibility in their earlier paper at the Tel Aviv conference. Now, working with Bardeen, they elaborated on the idea. What Nambu called "charm" Murray started calling "color." The three flavors of quarks—up, down, and strange—would also come in three different colors—red, white, and blue. Feeling cosmopolitan there in Geneva, Murray picked the arbitrary labels in honor of the French flag. It was a natural choice of terminology. Gell-Mann and Feynman had used "color" for years at Caltech as a fanciful name for new particle properties—there were red and blue neutrinos, for example. (But it was not an original notion. An Indiana University physicist had written about "color" in a textbook published in 1970. And other physicists had used the term as well.) Like Neapolitan ice cream, every baryon would consist of three differently colored quarks. A delta would be made from a red up, a white up, and a blue up; each would have different quantum numbers, and the Pauli principle would be saved.

The scientists also suggested a solution to another problem that

Nambu had wrestled with. While each individual quark carried the color charge, hadrons (made of three quarks) and mesons (made of a quark and an antiquark) obviously did not. If there were such a thing as color, why was it never seen? Why were there no red protons, consisting of three red quarks, or blue neutrons? If color existed at the level of quarks, then something must suppress it at the level of hadrons. The answer to the problem, Gell-Mann and his colleagues proposed, was a new conservation law: Like electrical charge, color must always be conserved. When the negative and positive charges of the electron and proton came together, they canceled each other out, forming a neutral atom. When a red, a white, and a blue quark came together to form a baryon, the colors also cancelled, vanishing in a puff of mathematical smoke. (In a meson, only two quarks were required to cancel color: A red and an antired, a white and an antiwhite, or a blue and an antiblue would give a net color charge of zero.) The metaphor seemed even more elegant after other physicists replaced Murray's colors with the labels that came to predominate, the primary colors of light: red, green, and blue. On a TV screen, the three mixed to form colorless "white" light.

The theory also accounted for the fact that quarks bunched in twos and threes. Pairs, with a quark and an antiquark, q and $\bar{q}$, formed a meson. Triplets of three quarks, qqq, or three antiquarks, $\bar{q}\bar{q}\bar{q}$, formed baryons. But there seemed to be no case where two quarks, qq, or two antiquarks, $\bar{q}\bar{q}$, came together. Nor were there triplets of two quarks and one antiquark: $qq\bar{q}$. Why, for that matter, were there not particles with four quarks, $qqqq$, or five, $qqqqq$? Physicists called this the saturation problem. In the new scheme, these were forbidden by color conservation. A particle had to consist of each of the three colors, or a color and an anticolor, so that the net amount was always zero.

In his own, older theory Nambu had also proposed a mechanism for suppressing color and for solving the saturation problem. As he saw it, the colorless combinations were the most energetically favorable—and therefore the most likely to appear in nature. Combinations like $qqqq$, in which color would show, could only exist at extremely high energies, Nambu proposed, unreachable by present-day accelerators. In the theory proposed by Gell-Mann, Fritzsch, and Bardeen, combinations like $qqqq$ were illegal not because they were difficult for nature to manufacture but because they violated a conservation law. But Nambu had been on the right track. This early work, published in the mid-1960s, was, in retrospect, amazingly prophetic. He had even anticipated gluons, coming up with a scheme to hold the quarks together using eight different kinds of bosons.

Gell-Mann had long considered Nambu one of the best theorists in the country, placing him in a class just below Feynman and Yang, an elite fraternity that included T. D. Lee, Geoffrey Chew, and Murray's friends Murph Goldberger and Francis Low. But because of Nambu's insistence that quarks were real and integrally charged, Gell-Mann didn't take his theory very seriously. His first contribution, with Han, was dutifully cited in the paper with Fritzsch and Bardeen, but no mention was made of a later, solo paper by Nambu, where he elaborated on his own color suppression scheme. When he was later criticized for not more fully acknowledging Nambu, Gell-Mann said he had never read the second paper. It had been published in a celebratory volume in honor of Weisskopf's retirement as director of CERN. The idea was to collect papers, from physicists who had visited the laboratory, that reflected Viki's intuitive approach to understanding. Lee, Oppenheimer, Heisenberg, Low, Bethe, Pais, and many others sent in papers. Murray, always procrastinating, never got around to it. He felt so ashamed of this lapse that he wouldn't pick up the published volume. If he had, he insisted later, he might have come up with the color scheme even sooner.

With this new ingredient called color, a full-scale Yang-Mills theory of the strong force almost seemed within reach. By the early summer of 1972, as their stay in Geneva was drawing to a close, Gell-Mann and Fritzsch were playing with a model in which quarks were joined into hadrons by eight different gluons. Like the quarks, they formed a group described by the rules of $SU(3)$. They were also leaning toward the idea that the gluons too came in different colors. They called the model QHD, for quantum hadron dynamics. In QED, photons carried the electromagnetic charge. In QHD, gluons carried the color charge.

But Murray, as usual, was plagued by doubts. He still couldn't explain why one never saw individual quarks. Color conservation explained why quarks must always combine in such a way that color canceled out. But it didn't explain why quarks never popped out of the hadrons, even for an instant, to flash a glint of color at the world. After all, if you look closely at a white TV screen, you can make out the individual red, green, and blue dots. Conservation of electrical charge ensured that a negative and positive particle added up to a chargeless particle, but one still could see individually charged particles in the world. If color was going to be part of a theory of the strong force, a mechanism was needed for preventing quarks from ever breaking free.

Murray retreated into more double-talk about "real" and "fictitious" particles. That summer at a seminar at the Majorana school in Sicily, Eugene Wigner pressed him to clear up the ambiguity. "If the quarks are only mathematical entities," Wigner asked, "would it not be necessary to establish a theorem which prevents the dissociation of the hadrons into such quarks, i.e., renders it possible to bind them together permanently?" Gell-Mann replied: "No. I'd consider quarks only as mathematical tools for calculations." His unfortunate terminology was creating more confusion than ever. Nambu then joined in the discussion. "There seems to be some degree of difference between Professor Gell-Mann and me regarding the abstractness of the quarks. I take the quarks (or any similar constituents) more seriously, as real objects. It is partly for this reason that Han and I proposed integrally-charged objects. . . ." Gell-Mann didn't respond to the comment.

His sentiments veered away from field theory again when he began to consider an entirely new way of looking at the strong force that had recently been developed by Gabriele Veneziano, a physicist at CERN who had fallen under the spell of the bootstrap. Like a monk poring over ancient texts, Veneziano searched through old tables of mathematical functions hoping to find one that could be used to realize Chew's dream of crafting an S-matrix that described all the hadrons and their possible interactions. He was surprised to discover that something called the beta function, invented by the eighteenth-century mathematician Leonhard Euler, came close to filling the bill. The result was an arresting new version of the bootstrap called "dual resonance theory"—"duality" for short. Nambu made this highly abstract theory more concrete when he showed that what it was actually describing was not particles—dimensionless points—but "strings": tiny one-dimensional entities oscillating in a superspace of twenty-six dimensions. Each tone in this mathematical music was a different particle. Look more closely at an electron or a proton, and each would turn out to be a different note produced by one of these vibrating strings. It was a tantalizing possibility. If one could get rid of the notion of particles as infinitesimal points, then maybe the infinities that plagued the strong force would finally go away. And where did quarks fit in? A meson could be thought of as a string with a quark on one end and an anti-quark on the other. In a baryon, a string with three ends (remember, this is not happening in ordinary space) would hold the quarks together.

Most older physicists thought string theory was nonsense, a delusion of irresponsible young mavericks. Twenty-six dimensions of space and time? Where were the other twenty-two? Feynman would have nothing to do with strings. But Gell-Mann was intrigued by the idea and its intimate relation to the bootstrap: There were no fundamental particles, just the music of these vibrating strings. And if there were an infinite number of harmonics, there would be an infinite number of particles. But of these, only a tiny fraction were low enough in mass to be created in terrestrial accelerators. Maybe, he was starting to think, the whole field theory of quarks and gluons should be scrapped in favor of this new vision. He liked to place bets on many different squares of the board, refusing to commit himself until the results were in.

In June, with all these thoughts swimming in his head, he drove from Geneva to Paris to give a series of four lectures at the Collège de France, where, a dozen years before, he had spent so many hours struggling with the Eightfold Way. The trip turned out to be a harsh reminder that not everyone saw particle physics as a noble pursuit of pure understanding. When he arrived to give the third talk, he found the room packed with about a hundred self-proclaimed Maoists, the Collectif Intersyndical Universitaire d'Orsay Vietnam-Laos-Cambodge. They were not interested in the symmetries of nature but in the asymmetries of power separating the first and third worlds, and they began berating Gell-Mann for his participation in Jason. He tried to tell them he had come to talk about physics, not the first Indochina war, with France, or the second, with the United States. Noting that the head of Jason, Murph Goldberger, was currently a guest of the Chinese government, Gell-Mann told the protesters they were evidently not Maoists but Lin-Baoists, after the leader of a failed anti-Mao coup who had died mysteriously the year before. He continued the lecture in another room. But when he arrived the next day for the final talk, the protesters were back in force and Gell-Mann was ushered from the auditorium. A Paris newspaper reported that it was the first time a speaker had ever been prevented from lecturing at the school. Murray was especially hurt that some of his French colleagues would not defend him. They seemed as unconvinced as the students that Jason was a moderating influence, working behind the scenes to end the war.

Returning to the United States that September, Gell-Mann presented his and Fritzsch's ideas to a somewhat friendlier crowd at a Rochester meeting at Fermilab. He began with his usual disclaimer: "For more than a decade, we particle theorists have been squeezing predictions out of a mathematical field theory of the hadrons we don't fully believe." He described what he and Fritzsch were calling quantum hadron dynamics, a Yang-Mills–like theory with eight colored gluons and three colored quarks (all, he emphasized, are "fictitious," permanently confined). But he also talked about the veal-and-pheasant version with a single gluon. Finally he raised the possibility that fields should be junked entirely in favor of the newfangled strings. Later he would remember the talk as a more forceful presentation of the "color-octet-gluon scheme," which would turn out to be such an important part of the final theory of strong interactions. But when he and Fritzsch wrote up the talk for publication in the proceedings, he said, he lost his nerve and watered it down. However, others who heard the original version don't remember it as a particularly rousing presentation.

As at Berkeley in 1966, Gell-Mann was given the honor of delivering the final lecture, summarizing what had been learned. (This time he had the advantage of speaking at the end of the conference instead of at the beginning.) He showed the audience a diagram:

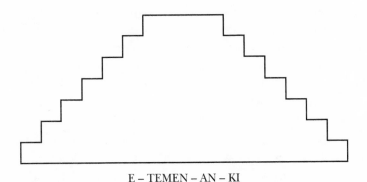

E – TEMEN – AN – KI

It was not a histogram chart, he explained, like those often used to plot the properties of particles, but a picture of the Tower of Babel. He labeled it in Sumerian: E - TEMEN - AN - KI, meaning "House of the Foundation of Heaven and Earth." Quoting the Bible, he recounted how the builders were punished for the audacity of trying to construct a tower to the sky. Their single language was shattered into many, and they were dispersed throughout the world. Gell-Mann warned the physicists not to succumb to the same fate, fracturing into groups speaking separate, mutually incomprehensible languages. "If we allow the same thing to happen to us, we will have to leave off building our temple of the foundations of the universe. If we avoid the fate of the builders of the Tower of Babel, then I see, close at hand, another marvelous dream—a unified theory of the hadrons, their strong interaction, and their currents, incorporating all the respectable ideas now being studied." Among the respectable ideas he mentioned were Zweig's constituent quarks, his own current quarks, the bootstrap, duality (i.e., string) theory, Regge poles, scaling, current algebra—just about everything except Yang-Mills field theory.

The final piece needed to explain the strong force, and why isolated quarks were not found in the world, turned out to be an utterly counterintuitive idea called asymptotic freedom. A force that is asymptotically free paradoxically grows weaker at short distances but stronger at long distances. Gravity and electromagnetism, of course, behave in the opposite manner—the further something is from the source, the weaker the pull. It was natural to assume the same was true with the other forces. But hints were slowly emerging that the strong force was different.

The idea was inspired in part by the paper that Murray, just a few years out of MIT, had written with Francis Low, trying to explain how QED worked at extremely short distances—what would happen if you pushed two electrons closer and closer together, using higher and higher energies. They had found that they could translate the familiar behavior of QED at long distances and low energies to these extreme situations using a device called the "renormalization group." The key idea was that the number describing the strength of the electromagnetic field, the so-called coupling constant, was not constant at all. It was a "running constant" whose value changed at different energy and distance scales.

In recent years a former student of Gell-Mann's, Kenneth Wilson, had been trying to apply the idea to the ill-behaved strong force. Wilson's work led to more refinements by two physicists named Curtis Callan and Kurt Symanzik. From all this theorizing, the notion developed that maybe the coupling constant for the strong force caused it to become weaker as two particles were pushed closer together and stronger as they were pulled apart. When the gap between two quarks became infinitesimal (as they approached each other asymptotically), they would become completely free of each other's influence. This now seemed consistent with what had been happening at SLAC. Recall again that distance and energy are closely related, that it takes higher energies to push things closer together. At lower energies (longer distances), quarks were strictly confined, causing a proton to act like a solid, featureless object. At higher energies (shorter distances), the quarks rattled around more loosely, acting almost like individual particles—targets that could be singled out and struck by an electron bullet. But try to pull one of the quarks away from the others, and the force holding them all together would intensify. There was no way a single quark could ever break free.

What seemed a mere mathematical possibility, almost too good to be true, was confirmed in calculations by two graduate students, David Politzer at Harvard and Frank Wilczek at Princeton, who was working with his adviser David Gross. After a great deal of difficult calculation, they showed that a Yang-Mills theory can indeed be asymptotically free. Gross was shocked by the result. He had actually been trying to kill off field theory once and for all by demonstrating that (1) effects like those found at SLAC required asymptotic freedom, but (2) Yang-Mills theories could not be asymptotically free. "It was like you're sure there's no God and you prove every way that there's no God," he wrote, "and as the last proof, you go up on the mountain—and there He appears in front of you." After several false starts (initially, Gross and Wilczek mistakenly reversed a sign and convinced themselves that a Yang-Mills field could not be asymptotically free), they completed the theory, and in June 1973 their papers and Politzer's appeared back to back in *Physical Review Letters*. It didn't take long for the implications to sink in. With asymptotic freedom, one could finally construct a field theory of the strong force.

Gell-Mann had preprints of the papers in hand that summer when he sat down with Fritzsch and Heinrich Leutwyler, a visiting physicist from Bern, and finally presented a complete description of a Yang-Mills gauge theory with colored quarks and eight colored gluons. But they didn't seem to appreciate the full importance of asymptotic freedom, clinging to the idea that one could use an ordinary field and modify it to keep quarks confined inside hadrons.

Coming so late in the game, the paper they wrote was more a synthesis of what many people had now come to believe. Hadrons were made from quarks held together by a field whose quanta were the gluons. And it was this charge called color that held things together.

After all the exotic avenues physicists had explored in search of a theory of the strong force—dispersion relations, the bootstrap, string theory—the shape of the answer that emerged was essentially old-fashioned QED, with a twist called asymptotic freedom. The result, far more elegant than the lopsided electroweak theory, was a theoretical masterpiece that Gell-Mann would have loved to call his own. Quarks were his and Zweig's, the Eightfold Way was his and Ne'eman's. He could even content himself with knowing that buried within asymptotic freedom was his and Low's old work on the renormalization group. But he was filled with regret that he hadn't seen the whole thing. The new picture of the strong force was the crowning vindication of both quarks and field theory, ideas Murray had been so deeply ambivalent about.

He had been too timid. But ultimately his original instincts proved to be almost right. Quarks were indeed permanently confined, and in that sense, they were more mathematical than real. In any case, his fastidious wording (beginning with his very first paper proposing that quarks might be trapped by some kind of infinite mass) now let him argue, to often dubious listeners, that he had believed in quarks all along. In some ways, the view that finally emerged and is still accepted today owes more to Zweig. Quarks are thought of as little things inside the hadrons. But on the matter of confinement, Murray's earliest inklings, confused as they were, now seem closer to the mark: Quarks are trapped forever. There probably isn't going to be a quarkonics industry.

It was fitting that, in the end, it was Murray who gave the theory its name—not quantum hadron dynamics, which he and Fritzsch had been using, but quantum chromodynamics, QCD. Later, in Aspen, he persuaded the physicist Heinz Pagels to use the name in a major review article. Almost everyone quickly realized it was perfect.

By the autumn of 1973, Feynman too was calling himself a dedicated "quarkerian." He even nominated Gell-Mann and Zweig for a Nobel specifically for discovering quarks. He summed up the new worldview in the first words of an address in Copenhagen: "Protons are not fundamental particles but seem to be made of simpler elements called quarks."

CHAPTER 14

SUPERPHYSICS

As told by historians of science, the rest of the 1970s was the story of how the new field theories of the strong and electroweak forces were pulled, no longer kicking and screaming, into the grand edifice called the Standard Model, a description of all the particles and forces, with only gravity left aside for later consideration. Quarks lay at the foundation. As described in the textbooks, the scheme could hardly have been more straightforward: The hadrons, long thought to be the bedrock of matter, indeed consisted instead of tiny, fractionally charged particles, which came in three flavors (up, down, and strange) and three colors (red, green, and blue). Triplets of these quarks formed the heaviest particles, called baryons, the most familiar being the proton and neutron. The welterweight mesons were made from pairs consisting of a quark and an antiquark. And the color force, which bound the quarks to make the hadrons, was a field carried by eight different bosons—the gluons. A red-blue gluon would change a red quark into a blue quark; a blue-red gluon would change it back again. Shuttling between the shimmering quarks, ferrying the color charge, the gluons held them permanently together. Quantum chromodynamics (QCD) had triumphed. With this neat picture, the strong force, which had eluded explanation for so long, was beautifully tamed.

In addition to the quarks, the Standard Model included two flavors of lightweight leptons: the electron and its partner, the neutrino. They were involved in another process that worked alongside QCD—what Murray liked to call "quantum flavor dynamics." This was nothing more than the weak force, the agent of particle decay. While the gluons could change a quark's color, the carriers of the weak force—W^+, W^-, and Z° (Murray still called them X's)—could change a quark's flavor. Take the combination ddu, which spells neutron, and flip one of its down quarks so it becomes duu, which spells proton. What could be easier? Finally, the photon, the carrier of electromagnetism, was woven into the Standard Model by unifying it with the three bosons of the weak force. Viewed from a higher vantage point, all four were different faces of the same abstract object. (In 1979 the inventors of this electroweak theory, now subsumed as a piece of the Standard Model, were recognized with their own Nobel Prize. Gell-Mann worked behind the scenes to ensure that his old friend Shelly Glashow shared the award with Steven Weinberg and Abdus Salam. Because the W's and Z were so very massive, it wasn't until 1983 that accelerators became powerful enough to find evidence of their existence. And that led to another round of Nobels, for the experimenters Carlo Rubbia and Simon van der Meer.)

While the basic architecture remained intact, physicists soon concluded that three flavors of quarks were not enough to explain all the new, more massive hadrons that crystallized as the particle accelerators reached higher and higher energies. Zweig, in his original aces paper, and Glashow and Bjorken, in a later work, had shown the neat things one could do with a fourth quark, called charm. At first, the appeal was mostly aesthetic. Suppose that the "ordinary" matter that prevails at the frigid energy state the universe currently inhabits is composed entirely of four particles—the up and down quarks (which combine to form the heavy hadrons), the electron, and the neutrino. This quartet would make up a family. Then suppose that the exotic particles that live at higher energies comprise a second family, consisting of combinations of the heavier strange and charmed quarks, along with the "fat electron" called the muon and its closely related muon-neutrino.

FIRST FAMILY	SECOND FAMILY
up quark	strange quark
down quark	charmed quark
electron	muon
neutrino	muon-neutrino

It was hard to resist so nice a picture. Then Glashow and two other theorists, John Iliopoulos and Luciano Maiani, found another good use for charmed quarks. For reasons no one had understood, certain decays that were allowed on the chalkboard (like a neutral kaon breaking up into a positive and negative pion) seemed not to occur in nature. The scientists contrived something called the GIM mechanism, named for their initials, in which the fourth charmed quark would step forth and conveniently suppress the decay.

When the GIM mechanism was unveiled in 1970, most physicists still felt that three quarks were bad enough. Who needed a fourth? As one observer later put it, the cure seemed "worse than the disease." Then in 1974, in what came to be called the November Revolution, teams at SLAC and Brookhaven independently discovered a particle so massive and unwieldy—it was more than three times heavier than a proton—that it should have immediately broken down. Instead it lingered a thousand times longer than the physical laws predicted. As one chronicler later put it: "It was as if anthropologists had stumbled on a tribe of people who lived to the age of 70,000 years." One group insisted on calling the particle the J; the other team called it the psi. In an ugly compromise, it is still known as the J/psi. The situation was reminiscent of the discovery, two decades earlier, of the strange particles, kept from quickly decaying by something called "strangeness." It was now proposed that the long life of the J/psi could be explained by this new quality called charm. In fact, the J/psi could be interpreted as the combination of a charmed and an anticharmed quark, what came to be called "charmonium." (The name was a riff on "positronium," the combination that resulted when an electron and an antimatter positron came together, momentarily forming a weird kind of atom before self-annihilating.) The experiments made a strong case that charmed quarks exist. (There seemed to be no other way to elegantly explain this monster called J/psi.) But more important, it was the final proof many scientists needed that quarks themselves were real. In the J/psi, the quantum number called charm was hidden, canceled out by the anticharmed quark. But before long, particles exhibiting "naked charm" were indirectly observed: a charmed quark combined with an antiup, an antidown, or an antistrange.

And so in addition to strange matter there was now a whole world of charmed matter to contend with. (And what about "Three quarks for Muster Mark"? The line was still perfectly appropriate, Murray would tell people. There were three colors of quarks, if not three flavors, and three quarks came together to form a hadron.)

The tradition Gell-Mann had started with strangeness—bringing order to the subatomic world by inventing new, impishly named quantum numbers—had been taken one step further.

For all the bizarre, Gell-Mannesque lingo and the Rube Goldberg embellishments, QCD was emerging as a compelling way of seeing the subatomic world. It became even more persuasive when experimenters found what could be construed as indirect evidence for gluons. When electrons and positrons came together, they annihilated one another in a spray that could include quarks, antiquarks, and gluons. As these particles sped away from the collision, they left three separate trails of hadrons in their wake. If physicists were doomed never to see the permanently confined quarks and gluons, then these particle jets—which looked like the three-pointed Mercedes-Benz symbol—were about as close as they were going to get.

Early-twentieth-century physicists like J. J. Thomson, Robert Millikan, Ernest Rutherford, and James Chadwick, whose simple, elegant experiments established the existence of the electron, proton, and neutron, might have been astonished at how oblique the "discovery" of particles had become. There were two main ways the game was played. As with the J/psi, experimenters might stumble across an unexpected new find, challenging the theorists to conjure it up from some underlying symmetry—if necessary, a broken one. Or the theorists, playing around with the mathematics of group theory, might hypothesize the existence of a particle—something so logically compelling it just had to exist—sending experimenters to their accelerators to find hints of its presence. Often the particles wouldn't live long enough to leave a track in the detectors. So it would be supposed that, behind the scenes, the unseen particle almost immediately disintegrated into other particles, which might break down into still other particles, until finally you got something that lingered long enough to leave a mark. The meaning of this sign could only be found beneath layers and layers of interpretation.

Through this interplay between theory and experiment, yet another, third family of particles soon appeared. In mid-1976, experimenters discovered the tau, a lepton far heavier than either the electron or its fat cousin, the muon. To complete the new set, the tau was surely accompanied by a tau-neutrino and two more quarks. The fifth quark, called beauty or bottom, emerged the next year in an experiment, led by Leon Lederman, in which particles of "bottomonium" were observed: a bottom quark joined with an anti-bottom quark. Its partner, the top or truth quark, was more elusive. No one could find the scantiest evidence that it existed. But the scheme was so beautiful that the theorists fell back on the familiar old dodge: The top quark must be too massive to create in the feeble accelerator blasts of the day. But in a symmetrical universe it would have to exist. And the universe, the physicists believed, was at heart symmetrical, even if it didn't appear that way, with its four wildly different forces and its hodgepodge of particles whose masses seemed to have been assigned at random. In the cooling of the universe after the primal explosion, the symmetries had broken. More and more, physicists were embracing the almost gnostic belief that behind the flaws of creation was a hidden, crystalline perfection that had shattered to give birth to the real world.

THIRD FAMILY

top quark
bottom quark

tau
tau-neutrino

How hard these ancient patterns were to glimpse. There is no meter that will measure charm or strangeness, no camera that can see a quark or a gluon or even a neutrino. Ultimately all one can detect with any kind of directness are photons and electrons: flashes of light leaving a mark on a photographic plate, or electrical charges moving to create a magnetic pull. The rest of the atomic world is inferred from these observations, and a lot of imagination is required. No one who hadn't been steeped in the details and expectations of QCD could have looked at the outcome of the J/psi experiment and said, "Aha, there are charmed quarks in there." As some philosophers liked to put it, You can't see evidence of quarks until you immerse yourself in the whole sprawling network of theory. The theory, as Einstein once said, allows you to see the facts. As the particle physicists went on to elaborate their abstract, invisible world, it became less and less clear where to draw the line between something that was real and something that was mathematical. For many physicists, like Gell-Mann, this didn't seem to be a meaningful distinction. If one believed deep down that the universe was mathematical, then the equations had to be saying something about reality. The universe was made of crystallized mathematics. That was the cardinal tenet of the faith.

For those who lived with this belief, it was harder than ever to tease apart the distinction between what was invented and what was discovered. Through the eyes of the willfully skeptical, the story of the Standard Model might be caricatured like this: According to the mathematics of group theory, the hadrons are built up from fractionally charged particles called quarks. Since quarks violate the Pauli exclusion principle, it was decreed that they come in three different "colors." Since no one could find particles with fractional charges, it was shown with some fancy mathematics that quarks are trapped forever inside the hadrons—that, going against all intuition, the strong force gets stronger, not weaker, with distance. And since electromagnetism and the weak force don't quite mesh, a new particle, the Higgs boson, was invented to break the symmetry between them. Finally, when new particles popped up that could not be accommodated with these contrivances, the theorists just kept adding quarks—charm, truth, beauty—until everything was accounted for. All this theorizing yielded testable predictions— particles that should show up in the accelerators. But when the experiments required so many layers of interpretation, how could the physicists know when they were reading too much into the lines and squiggles, seeing what their brains were primed to see, like pictures in the clouds? Were these really discoveries, or inventions? It was a question that Gell-Mann, among many others, refused to be distracted by, waving his prescription forbidding him from discussing philosophy.

Whatever one's philosophical inclinations, it was hard not to be in awe of the Standard Model. Discovery or invention, it was a work of art. Whether it was the art of nature or the art of humankind could never be known for sure. But it was pleasing to think that humans, on their tiny planet, with their blinkered senses and animal brains, could weave observation and imagination into such a powerful theory.

Physicists being physicists, this little bit of unification simply whetted the appetite for more. To minds that demanded symmetry, some annoying imperfections remained. In the Standard Model, the electromagnetic force, described by the $U(1)$ group, and the weak force, described by $SU(2)$, were smoothly intertwined. But this electroweak force—SU(1) × SU(2)—and the strong force, SU(3), were still separate, working side by side. The weak carriers would come over from their side of the model and change one particle to another by flipping quarks. But one could not yet talk, in one breath, of a fully integrated electroweakstrong force. For the Standard Model to be truly unified, there would have to be a yet higher symmetry. Just as the photon and the W's and Z had been united, maybe there was an even higher vantage point from which gluons, photons, and the carriers of the weak force could be seen as different faces of the same abstract jewel.

The Standard Model also had other shortcomings. It offered no explanation for the senseless variety of different particle masses. By the time the theory was renormalized, these values were left as arbitrary parameters. They had to be determined from experiment, plugged in by hand. Then there were the strengths of the four forces. Why did they have the particular values they did? And why, for that matter, were there four forces and not three or five or seventeen? Why did there appear to be three families of matter? Nothing at that point ruled out the possibility of a fourth or a fifth. (Some physicists were already entertaining the possibility of two more quarks, perhaps called high and low.)

As it stood, the Standard Model was like a box with all these knobs that had to be manually adjusted. Tweak them a little differently, and the result would be a very different universe. The one true theory, many physicists felt, would have no knobs; it would predict these details from general principles. There would be a grand set of equations with one and only one self-consistent set of solutions: the particles and the forces that we observe. God would indeed have had no choice in making the universe.

This goal, beyond the Standard Model, became known as grand unification. Even before the model was complete, theorists were already trying one unified theory after another—groups called SU(5) and SO(10), a new quantum number called "technicolor." But nothing quite worked. And none of these attempts tried to deal with gravity. This was the biggest stumbling block. After the stunning success of his general theory of relativity, Einstein had assumed that the next step would be to bring electromagnetism into the picture. If gravity was a ripple in four-dimensional spacetime, then maybe the other forces were too. As far back as 1919, a mathematician named Theodor Kaluza thought he had found a way to expand general relativity to include both gravity and electro-

magnetism by adding a fifth dimension. Both forces would be ripples in this larger-than-imagined arena of space-time. But where was
this extra direction? After all, if there were another dimension of
space, then it would be possible to escape from what appeared to be
a closed container—just step out through the fifth dimension.
Another mathematician, Oskar Klein, suggested a solution: The
fifth dimension was curled up in a tiny circle far too small to be
useful to anyone but theorists.

None of these contortions worked. In the following decades, it
became accepted by just about everyone but Einstein that electromagnetism and the nuclear forces were quantum in nature. And
general relativity just wasn't written in the language of quantum
theory. Physicists assumed there must be a quantum field describing gravity, with the force carried by hypothetical particles called
gravitons. But the experimenters could find no way to detect these
extremely weak bosons, and the theorists could find no way to construct a quantum gravity field that was not fatally plagued by infinities, the kind that no amount of shuffling would make disappear. It
sometimes seemed as though the software of the universe had been
created by two teams of programmers using two different languages, quantum mechanics and general relativity, then kludged
together at the end. But surely the Great Plan could not be so ugly.

Any unification theory that left out gravity, Gell-Mann would
grumble to his colleagues, wasn't so grand. He was spending his
time now watching the developments from afar. Except for another
paper with Fritzsch in 1975, he was absent from the physics literature for three years. With no need to prove himself anymore, he satisfied himself with the broad, general lectures expected of a Nobel
laureate. He was still intensely discussing physics (as well as archaeology, politics, linguistics, and much more) with everyone he saw,
but he was not writing anything up. He was pulled back into print
by the dream of forging on beyond the Standard Model and the
flailing attempts at not-so-grand unification—to finally bring all
four forces together.

He still thought that the answer might lie in the bootstrap, and
particularly in the more recent incarnation of this idea, in which
particles emerged from strings vibrating in a superspace of as many
as twenty-six dimensions. If so, then the Standard Model would turn
out to be just a crude approximation, a way station on the road to a
fuller, richer theory. But there were problems with the early version
of strings: The equations contained unphysical never-never-land

quantities like negative probabilities and imaginary masses that, when squared, yielded negative numbers. It was not yet a serious theory, Murray thought. Most physicists dismissed the whole idea of extra dimensions as ad hoc: Add as many as you need to unify the forces, then posit that the extraneous ones are too small to see. And even if one could swallow the idea of extra dimensions, there was another huge shortcoming in the theory: While bosons, the force-carrying particles, were accommodated by the picture, it was not clear where the fermions, the mass particles, fit in.

In 1971 three young physicists, John Schwarz and André Neveu at Princeton and Pierre Ramond at Fermilab, solved some of the problems when they expanded the string model into what would come to be called superstring theory. In the process, they managed to pare the mathematical arena from twenty-six down to ten dimensions. The new model not only accounted for both bosons and fermions, it offered a whole new level of unification: Maybe bosons and fermions were two sides of the same coin, united by a higher symmetry—supersymmetry, it came to be called. It was an impressive mathematical juggling act. Bosons have integral spin, fermions half-integral spin. The result is that a boson rotated 360 degrees will look the same, just as with a classical object. But half-spin particles, weirdly enough, must be rotated through 720 degrees to regain their identity. What the inventors of superstring theory devised was a higher-dimensional space where, at extreme energies, these two kinds of particles, with their very different geometries, would melt together. The solution required that each particle have a supersymmetric partner—what came to be called a sparticle. Quarks would be matched with squarks, leptons with sleptons, bosons with bosinos. The electron's twin was the selectron, the photon's the photino, the gluon's the gluino. Even Murray thought the nomenclature—"slanguage," he called it—was getting out of hand.

And what about these extra dimensions? Ten was better than twenty-six, but this still seemed like six too many. Klein had found a way of sweeping Kaluza's fifth dimension under the rug by shrinking it to a negligibly tiny circle. The advocates of string theory were envisioning a universe in which each point of four-dimensional space-time would have attached to it a tiny six-dimensional ball curled up too small to see. Unlike the four dimensions of space-time we live in, these hidden dimensions had never inflated after the big bang. Gell-Mann liked to explain the idea like this: Imagine that we are flatlanders living in a two-dimensional universe, our movements forever constricted to a plane. If you encountered an object, you could go around it but not up and over. Lines would serve as fences, circles would be prisons. Then a mathematical savant announces that he has good news and bad news: The good news is that all along there has been an extra dimension we didn't know about: height. The bad news is that it is impossible to jump higher than 10^{-33} centimeters. You still couldn't leap outside a circle.

The fanciful new theory remained in the realm of speculation. How would one ever observe a string? The most powerful accelerators were rated in billions of electron-volts. Probing the tiny world of strings would require inconceivable energies of 10^{19} billion volts, an accelerator measured not in miles but in light-years, one the size of a galaxy or even the universe. Adding to the confusion, the equations kept coughing up a massless particle with a spin of 2. The only thing resembling it was the hypothetical graviton. It was hard to see how such a particle—the carrier of gravity—belonged in a theory of the strong force. Then the success of QCD blew string theory out of the water. Who needed all those dimensions and sparticles when quarks and gluons explained so much?

Gell-Mann, never entirely happy with the idea of quarks as fundamental particles, remained enthusiastic about superstrings and supersymmetry. He was perfectly happy to see his own work relegated to a rough approximation of a deeper, more beautiful theory, with quarks flapping around on the ends of strings. He brought John Schwarz to Caltech in 1972 and set him up as a research associate. And he arranged for other string theorists to visit. He wasn't sure superstring theory was right, but he wanted to create a place where the unpopular idea could be explored—a "nature reserve for an endangered species," he called it. The investment paid a small dividend in 1974 when Schwarz and a visiting French physicist, Joël Scherk, decided to take seriously the notion that the massless spin-2 particle that kept popping up in the equations really was the graviton. If this was so, then string theory was not a description of the strong force but something much greater: the way to a fully unified theory that included gravity.

Gell-Mann became more convinced than ever that there might be something to this idea. If strings were too much to swallow, then maybe unification would come from a more conservative version of the idea called supergravity. These theories, which were being promoted by some of Gell-Mann's colleagues at other universities, kept the idea of supersymmetry uniting fermions and bosons but used old-fashioned point particles instead of strings. Supergravity was a subset of superstring theory, with the length of the strings set to zero. The model also had the appeal, in its early incarnations, of requiring only the familiar four dimensions of space and time. (It was soon found, however, that the most elegant version of supergravity required eleven dimensions.) It was basically a super–Yang-Mills field theory. The graviton was paired with eight different gravitinos, and all sorts of particles could be shaken from the equations—twenty-eight bosons, fifty-six fermions, and seventy spin-0 particles that, like the Higgs boson, might be used for symmetry-breaking.

Gell-Mann started exploring supergravity with Ne'eman. Then in 1975 he met Pierre Ramond, one of the pioneers of string theory, at Aspen, and they worked on unification there and at Los Alamos. Murray invited him to Caltech, adding him to the "nature reserve." Before long, Lars Brink, a young physicist from Sweden, came to work with Ramond and Schwarz. Gell-Mann would regularly pop into the office to see what they were up to. Soon they were all collaborating.

Ramond found that even in semiretirement, Gell-Mann lived up to his billing. Working with him, he thought, was like riding in a fast car with the top down. He had never met anyone smarter. If Murray felt deep down that something must be true, he wouldn't let go of the idea until he had either proved his intuition correct or understood why it had been mistaken. He still had an ability to see structures and patterns—unlikely connections—that were invisible to others. At the physics department seminars, Gell-Mann would take a highly abstract problem that one group was working on and connect it to seemingly unrelated experimental work another group was doing—synthesis on a weekly basis. And the man seemed to be engaged in everything. How different he was from Feynman, who seemed interested only in his current obsessions. If you did manage to slip an idea through the fine mesh of Feynman's attention filter, he would take it apart and put it back together so you understood it as never before. And then Murray would show how the idea fit with the rest of the fabric of knowledge. It was an incredible combination of talent.

Ramond was also struck by how little Murray's fame had dulled his competitive drive. Most of the detailed calculations were done by Ramond, Schwarz, and Brink. But when Gell-Mann picked up pencil and paper, he was hard to beat. One day he sat scribbling away in his three-ring binder while Ramond tried to solve the same problem on the blackboard. Ramond was pulling ahead, then he stumbled. "Oh good," Murray said. "For a moment I thought you might be faster than me."

But it was all surprisingly good-natured. His young collaborators knew about Gell-Mann's reputation: the cocksure know-it-all always armed and ready with a devastating remark, always insisting on being in control. But for his students and junior colleagues, with whom he had nothing to prove, he could be a pleasure to work with. Feynman had more flash and the students loved his devil-may-care attitude. But it was difficult to really know him; he was too busy putting on a show. Murray seemed, to some of the young physicists, more like a friend, someone whose defects were worth putting up with. He would come into the office all excited about some new-found bird or an archaeological discovery. There was so much information in his brain that it just fought to get out. He was still, in his forties, the child trying too hard to impress. It didn't seem to occur to him that people could like him for himself and not for what was in his head. But this veneer was just part of the package, the real Murray. You got used to it. Or not. He would walk into the office and look at something Brink and the others had been laboring on. The answer, he would say, must be *this*. After two more days of calculations, they would find that he was right. They had no idea how Gell-Mann did it.

Once the junior physicists had come up with the draft of a paper, Murray would go through it, agonizing over every word, eliminating any redundancies. The web of ideas had to be airtight. Sometimes the result was a little too compact, hard for anyone but the authors to unpack. Murray's greatest contribution to the research was his enthusiasm. With supergravity, he was not so immersed in the details as he had been with $SU(3)$ and quarks. This was new physics for a new generation. But unlike many of his contemporaries, he didn't shut out the radical new possibilities.

The papers with Gell-Mann's name on them comprised but a small part of the literature of supergravity. And Caltech was only one enclave contributing to the effort. By 1980 the excitement over supergravity had reached such heights that Stephen Hawking, on assuming the Lucasian chair at Cambridge (the one Murray had turned down earlier), declared that the end of physics was in sight. He was wrong. In the end, supergravity wasn't enough to unify the forces. No matter how far the algebra was stretched, no one could get all the known bosons and fermions to fit under the same mathematical tent. And the sets of hypothetical particles that emerged in the equations couldn't be made to match up with those in the real world. Sometimes Gell-Mann wondered what it would mean if Hawking were right and physics did someday come to an end. Maybe theorists really would find a unified theory that was compatible with the world as we knew it and also made predictions that could be tested in ever more powerful accelerators. But how could we ever know whether a better theory, with more complete unification, was not just beyond the horizon, residing at still higher energies? "Eventually, there would be a limit to human patience and to the resources that would be expended in trying further to check this successful theory," he said. "Humanity would proclaim it to be the final fundamental physical theory!"

Freeman Dyson once put it like this: "The ground of physics is littered with the corpses of unified theories." The best-known casualty was Einstein's own attempt. Asked by Ne'eman to give the closing address at a centennial celebration for Einstein in 1979 in Jerusalem, Gell-Mann lamented the fact that a commemorative coin had been struck with Einstein's quixotic and incorrect unification equations on the back. One might as well have revered Newton for his forays into alchemy and biblical prophecy. But he reminded the other physicists in the audience—the participants included Glashow, Pais, Nambu, Weinberg, Wheeler, and Yang—of two important lessons Einstein had taught. The first: "While cultivating successful ideas in physical theory we must be careful to prune away any unnecessary intellectual foliage that accompanies them, assumptions that we accept out of laziness or vested interest but that we do not require for success. (It may in fact be the tendency to hang on to such assumptions that makes it harder for us to do good theoretical work as we grow older.)" In his youth, Einstein had been prepared to give up the seemingly obvious notions of simultaneity and the rigidity of space and time. But in his later years, he refused to the end to let go of the idea of classical determinacy. "The second lesson," Gell-Mann noted, "is to take very seriously ideas that work and see if they can be usefully carried much further than the original proponent suggested." It was a fine description of the attitudes that had carried him so far.

The distractions that had begun with the Nobel Prize in 1969 only became greater in the eighties. Gell-Mann continued to serve as a regent of the Smithsonian. And he had recently been named, along with Jonas Salk and Jerome Weisner, the former president of MIT, to the board of the John D. and Catherine T. MacArthur Foundation, which was poised to become one of the most influential philanthropies in the world. The high-powered board soon learned that Gell-Mann was an irresistible force, advancing his environmental agenda by persuading the foundation to fund the World Resources Institute and other projects to preserve the rain forest and the biological and cultural diversity he loved. Sometimes he would find the board stacked heavily against him, but he would hammer away relentlessly until he had shifted the fulcrum. The abruptness with which he could dismiss an idea he didn't agree with, and the tenacity with which he would badger staff members about his pet programs, ensured that his reputation as an intellectual bully spread beyond the physics world. In the early days, board members also sat on the selection committee for the lucrative "genius" awards. Gell-Mann annoyed some foundation members by pushing through grants for scientists pursuing his favorite interests, like string theory and the deciphering of ancient Mayan texts. He also got MacArthur grants for the mathematician Perci Diaconis and the magician James Randi, crusaders dedicated to debunking New Age charlatans.

This had long been another of Gell-Mann's obsessions: the perils of pseudoscience and crank politics. Later he would become an avid member of the Committee for the Scientific Investigation of Claims of the Paranormal, and of the Southern California Skeptics. In 1986 he would personally recruit seventy-two Nobel laureates to sign a brief to the United States Supreme Court urging it to overturn a Louisiana law that required teachers mentioning evolution to give equal time to creation science. (He was angry when Feynman, apparently just to assert his independence, refused to lend his name to the effort.) Gell-Mann assumed that, if he had taught his children anything, it was this distinction between sense and nonsense, a divide that seemed so clear in his own mind. And so, he was completely unprepared to see one of his own children fall into what to Gell-Mann was the depths of irrationality. That it was Lisa made it all the more perplexing.

Nick had always been the stubborn one. Like his father, he never hesitated to speak his mind. If he was tired of being marched off on another backpacking trip or dragged to a new exotic land, he let Murray and Margaret know it. He made it clear early on that he had little interest in pursuing a career in math or science. Lisa was jealous of his bravery. And he envied the praise she always got as the A student, the perfect child—and, it seemed, the one likely to follow in Dad's footsteps. With her love of animals and passion for detail, it seemed natural that she would major in biology at college. To Murray's disappointment, she turned down Stanford. And she refused to apply to Harvard. Both were too snooty, she said. With Stanford no longer an option (he couldn't have cared less about Harvard), Murray had secretly hoped she would go to the University of California at San Diego. She would be nearby and surrounded by nature. Finally she was accepted by his own alma mater, Yale. She hadn't been sure she wanted to go there. Berkeley seemed in many ways more attractive. But she seemed to feel it was expected of her.

Once Lisa was at Yale, Murray was delighted when she expressed an interest in molecular evolution and other far-sighted biological specialties. He was proud of her excellent taste in fields. But then things started to go wrong. First, she wanted to drop biology for another major—she had become convinced that she didn't have the mathematical acumen to be a scientist after all. And the tightly focused, methodical way she explored her interests often led her to turn in papers days late. Procrastination was a Gell-Mann tradition, but Yale seemed less tolerant of tardiness than it had been in Murray's day. He tried to convince her that there were huge areas of biology where one could do important qualitative work: ecology and ethology. He was glad when at last she reluctantly stayed with her major. He had no way of knowing that she was just going through the motions. She had become consumed by other passions.

She had fallen in love with a young man from a wealthy family from the Upper East Side of New York who had rejected his privileged background with a vehemence she found irresistible. Before long, he had persuaded her to dedicate her life to the Revolution, to join the cadres of something called the Central Organization of

United States Marxist-Leninists. Murray himself had dabbled in left-wing politics when he was at Yale. But this group Lisa had joined was so far left on the political spectrum that it threatened to fall off the edge into empty space. Its members considered Mao a dangerous revisionist. They celebrated Joseph Stalin's birthday. And somewhere along the way, the group concluded that Albania, of all places, was the one true people's republic. By 1980 the group, now called the Marxist-Leninist Party of the USA, had issued a manifesto: "Let the ruling class tremble," it read. The members sang songs celebrating Albania and its dictator, Enver Hoxha, for his intransigence toward both the West and the Warsaw Pact.

There was "Albania, Our Beacon":

> *Though encircled by NATO and Warsaw*
> *Trying hard to bring her to her knees,*
> *Still Albania upholds the revolution—*
> *Steers the path to final victory. . . .*

Or "Stand, Brave Albania!":

> *The greed of the Chinese revisionists knows no limit.*
> *No deed's too foul if it serves their wicked ends.*
> *Now they try to silence brave Albania,*
> *And snuff the beacon of the socialist homeland. . . .*

> *Shame on the heads of the Chinese revisionist ruling clique*
> *For attacking Albania, the people's staunchest friend!*
> *World public opinion will condemn you for these crimes.*
> *The heroic Chinese people will hang you in the end!*

They sang "Enver Hoxha, We Wish You a Long Life," "Eternal Glory to J. V. Stalin," and "Down with Ronald Reagan, Chieftain of Capitalist Reaction!" Her parents didn't think much of Reagan (Murray called him "Ray-gun"). But Stalin and Hoxha? They tried to argue with her. This is so unlike you, they would tell her. You've always been someone who is interested in *individual* people, plants, animals. How can you glorify someone like Stalin, who was willing to sacrifice millions of his own people for an abstract cause? Murray had warned himself and so many other physicists not to fall in love with mathematical formalisms, never to forget the real world. Now it seemed to him that Lisa—so smart and studious—was sacrificing herself for an empty ideal, one based on misinformation. Albania a workers' paradise? Murray offered to take her there. "What would you look for?" he asked her. "Would you look for development, improvement in the economic position of the workers?" She said, "No, that's probably what *you* would look for. What I'm interested in is equality."

Hoping the dalliance would pass, Murray helped her get a job, after her graduation in 1978, at the Marine Biological Laboratory at Pacific Grove, California, near Monterey. He hoped she would see that biology could mean doing things outdoors with animals. But she quickly became disillusioned with the lab. She quit and got a job working on the docks.

When her parents asked her to join them and Nick for a year in Geneva, where Murray would be working at CERN, she agreed at first. She had been thinking of using her knowledge of French and Spanish to become a translator, and one of the best schools was in Geneva. But before their departure, she announced that she was staying behind to work for the party. Murray and Margaret were crushed.

They stopped in Paris on the way to Geneva and met the Pineses to celebrate Murray's fiftieth birthday. But it seemed from the beginning that this would be an unhappy year. In addition to the problems of keeping track of a spirited and demanding teenage son, Margaret was not feeling well. When Murray planned an exotic trip to Morocco over Christmas, she begged off and they spent a more relaxing holiday in Provence. They didn't think anything was seriously wrong until that spring when, a few days before they were to return to Pasadena, Margaret began to show symptoms of colon cancer. Murray called his secretary, Helen Tuck, in Pasadena and asked her to make a doctor's appointment. Then they tried to enjoy the next few days, repressing the thought that this was the kind of cancer that had killed her brother. A few days after they arrived in California, their worst suspicions were confirmed. The cancer had already spread to her liver.

Murray approached the illness like a physics problem, learning everything he could about the disease. But quarks and gluons were far better understood than human cells. The symmetry-breaking that caused a cell to metastasize couldn't be tamed by any known mathematics. The only hope was to try and keep her alive on the chance that some researcher somewhere would make a discovery. The colon cancer itself was destroyed with radiation treatments.

The bigger worry was where the cancer had spread. The doctors at the UCLA Medical Center began what seemed an endless series of injections of a tumor-suppressing drug directly into Margaret's liver. It was pure torture. Her body responded well to the treatment, but everyone knew the doctors were only buying time—months at best. Murray was on the phone talking to researchers all over the country. He flew her to San Francisco for a consultation. When Abraham Pais heard that his old friend Margaret might come to New York Hospital for treatment, he wrote Murray a conciliatory letter. Rockefeller University, where Pais was now a professor, was near the hospital. If Murray needed a place to stay, Pais offered, he would ask the university to arrange for an apartment. He never received a reply.

Finally they settled on an experimental treatment at Johns Hopkins in Baltimore. Monoclonal antibodies were used to zero in on the cancerous cells, carrying tiny bits of destructive radioactive material directly to the tumor. The treatment made her sicker than ever. But for one miraculous moment it appeared that it was going to work. The cancer seemed to be in remission. Then the doctors realized they had made a mistake with the imaging. The cancer was as strong as ever. And she had missed the regular treatments she had been responding to.

Some friends thought Murray was only making things worse by encouraging these almost hopeless interventions, keeping her alive a year longer than nature would have normally allowed. They had complained for years that he walked all over her, taking advantage of her good nature. (She once told Lisa, who couldn't understand why her mother had given up her career as an archaeologist, that she wanted to spend her life devoted to a great man.) Others thought his efforts to save her were heroic. Either way, he couldn't help it. He was desperate not to lose his one true friend.

In the end, science failed them. She died on October 9, 1981, in a hospital in Culver City, near Pasadena. Murray tried to get Lisa to move back home and help care for him and Nick, but she had her own life now. She was just beginning what would be years of dedication to the party, organizing workers in Detroit and Oakland. Whole years would pass when the only way Murray could reach her was through a pay phone somewhere. He would leave a message saying he was coming through Detroit. Could she meet him at the airport to talk for a few minutes? Sometimes she would show up, sometimes not. He had lost both his daughter and his wife. For all the symmetry that seemed to hold on the subatomic scale, there was so very little in the macroscopic world of human pain.

The memorial service was held in Aspen on a spectacularly beautiful Indian summer day. Afterward Suzy Pines stayed in the house with Murray for a few days so he wouldn't have to be alone. Then he drove down to Tesuque, a little village between Santa Fe and Los Alamos where he and some friends had recently bought land from his Caltech colleague Max Delbrück. He and Margaret had planned to finally build their wilderness dream house there. He was so distracted and sick from a cold that he almost drove off the highway. When he got to Tesuque, he stayed with the physicist Robert Walker and his wife for a while to recuperate. Then he returned to Caltech.

Murray didn't let his feelings show at work. He continued to give lectures, pushing supergravity and unification theory. He would sit in his postdocs' offices, talking physics. But his heart wasn't in it anymore. Before, he could always be depended on to see right through a vague argument; you couldn't slip anything past him. Now it was easy. A lot of theory that the postdocs knew was pretty flimsy would go by unchallenged. Murray was thinking about other things.

CHAPTER 15

FROM THE SIMPLE TO THE COMPLEX

He felt like an old farmer spinning tales—or, more romantically, like a medieval troubadour traveling from town to town singing of heroic deeds in the faded, forgotten past. It was September 1983 and Gell-Mann was in the old Catalonian fishing village of Sant Feliu de Guíxols, just up from Barcelona on the rugged shoreline of Spain's Costa Brava. He had joined a group of physicists who had come to tell their stories at what was being billed as the First International Meeting on the History of Scientific Ideas. It wasn't really the first, Murray explained to the audience. The previous summer, he had spoken at a gathering in Paris on the history of particle physics. And the meeting in Spain would not be the last. Before long, a series of major conferences on the rivulets of ideas flowing from quantum theory to quarks and the Standard Model would begin in the United States—"an orgy of reminiscence," in Murray's words. For all the mysteries about the universe that remained untouched, particle physics had reached a milestone with the Standard Model. It was time to sort out the flow of ideas, distill raw experience into history.

At the Paris meeting, Gell-Mann had given a fairly upbeat account of his discovery of strangeness. But in Sant Feliu de Guíxols he found himself fixated on all the disappointments of his career. As far as many of his colleagues were concerned, Gell-Mann had dominated particle physics for a decade. Strangeness, V-A, the Eightfold Way, quarks—how many more triumphs could anyone expect? True, each of these discoveries had been made independently by others (and that no doubt added to Murray's melancholy). But only he had discovered every one of them. Who else could claim such a winning streak?

Yet all Murray seemed to see, looking backward, were the other things he could have found if only he had taken his ideas more seriously, thought a little more clearly, worked a little harder. All those lost opportunities and near misses imbued him with a chronic, low-grade mental pain. He tried to make light of it by recalling a jingle from his younger days:

> *As you ramble on through life, Brother,*
> *Whatever be your goal,*
> *Keep your eye upon the doughnut,*
> *And not upon the hole.*

He had not, he ruefully told his listeners, heeded the advice: "I tend to keep my eye on the hole in the toroidal doughnut." In the late 1950s he had seen that there was more than one neutrino—red and blue, he and Feynman had called them. But he let Dick talk him out of writing up the idea. And when he first started dimly fooling with the algebras that led to the Eightfold Way, he should have seen immediately that these were Lie groups. Once he had realized that, why did he waffle so long, terrified of being wrong? He could have published so much sooner. The implication, left unspoken, was that the Eightfold Way could have been his alone.

And how close he had come to inventing color. This turned out to be the piece of the puzzle that had been missing in the early sixties when he and Glashow, between trips to the Beachcomber, were vainly trying to get the strong force to play in the same arena with the weak force and electromagnetism. If only he had seen that the clashing forces just needed a little more mathematical elbowroom—the extra degrees of freedom provided by the quantum number called color. He was sure he would have come to the realization a few years later if he had made himself read Nambu's paper (the one in the Weisskopf Festschrift that he had been embarrassed to pick up because of his missing contribution). Neurosis had defeated him again. And why had he insisted on calling quarks "mathematical" instead of "confined," which is what he really meant? Or so it seemed to him now. (He didn't remind the audience that he had also used the words "fictitious" and "illusory.") He had even hinted at the existence of charm, another of his near-misses, in the report he and Pais had brought to Glasgow that wonderful summer in 1954 when he and Margaret fell in love in the Scottish isles. Though he didn't state it directly, his underlying message was clear: If he had paid more attention to these early inklings, he could have put together the whole QCD scheme years before anyone else.

The problem, he told his audience, was that he had always believed it improper to write about one's uncertainties. You shouldn't publish until you were sure. Back in the fifties, others (he named Yang and Lee as examples) played the game differently. They would think nothing of presenting several contradictory theories, hoping that one would stick. They thought the idea of a physics paper was simply to show your skill in exploring all the consequences of various sets of assumptions. If the assumptions turned out to be right, you would get the glory. If not, no one would blame you. A paper was only wrong, according to this view, if it contained a mathematical mistake. Murray insisted that he had held himself to higher standards.

The agonizing, somewhat subdued in the Costa Brava talk, became a dominant theme in his conversations with colleagues and friends. Some were able to sympathize. They understood that his regrets were deep and genuine. In the tightly knit space of ideas that made up particle physics, everyone was constantly stumbling toward treasures, only to veer off abruptly down a wrong avenue. Murray was less forgiving of himself than most. These missed opportunities were things he would suffer over for the rest of his life.

But for a lot of listeners, Murray's special pleading was getting harder and harder to take. By now many people were prone to think the worst of him, turnabout being fair play. Over and over, he seemed to be saying that he had really discovered just about everything in particle physics, that he really *should* get credit for the scribblings in his notebooks, and even for the electrochemical scribblings in his brain. Where before he had been maddeningly ambiguous about the reality of quarks, now he was obsessed with making sure no one challenged the prominence of his role in their discovery or suggested that he had wavered for a moment in his conviction that they were real. When Zweig was working on a historical account of quarks to deliver at an international conference in Toronto, he made the mistake of showing a draft to Murray. The article included a short account from Robert Serber, who suggested to Zweig that he had first brought the possibility of subparticles to Gell-Mann's attention during the legendary lunch at the Columbia faculty club in 1963. It was a modest and questionable claim. But when Murray read it, he exploded. Zweig had never seen him so livid. *Serber hadn't told him anything he didn't already know!* Zweig, taken aback by Gell-Mann's violent reaction, diplomatically decided to drop the passage and just give an account of his own part in the quark story. (Serber, who died in 1997, never stopped believing he had played a more important role than Gell-Mann would acknowledge.)

For as long as Murray's friends had known him, they had been putting up with his brooding and his hair-trigger temper. As far back as high school, those who could overlook his unfiltered, know-it-all behavior and underdeveloped social skills had found him a charming companion. They were flattered to count themselves among those with whom he chose to associate. For all his abrasiveness, Murray loved companionship and conversation, even if he didn't thoroughly grasp the idea of give-and-take.

But more and more of his colleagues were complaining about how stupid Murray had made them feel in recent encounters, how difficult it was for him to distinguish between established fact and his own opinions and to allow for the possibility of another point of view. The degree to which he let little frustrations drive him crazy could be alarming. He assumed life would go his way, and when it didn't—when a waitress announced that the restaurant was out of his favorite hundred-dollar wine or a reservations agent told him his trip had been delayed—he would explode. People had been on flights with him when they feared he would be evicted from the plane.

No one brought out Murray's bad side as adeptly as Feynman, with his constant needling and carefully practiced eccentricities. Gell-Mann missed collaborating with him but had come to feel that it was impossible, that Dick just wanted a foil for his own ideas. Lately Feynman had been spending more time with artists than physicists. He also liked to visit the Esalen Institute, the enclave of New Age thinking at Big Sur. He dismissed the vibrations emanating from Esalen as pseudoscience, but he loved soaking in the hot tubs and watching young women sunbathe in the nude. Murray bristled every time he heard Feynman recount his latest escapades. The man seemed to live his life to generate anecdotes, Murray complained.

As it turned out, one of Dick's drumming partners, Ralph Leighton, the son of a Caltech physicist (who was one of the editors of *The Feynman Lectures*), had recently started tape-recording Feynman's stories. In 1985 they compiled them in a lighthearted book called *"Surely You're Joking, Mr. Feynman!": Adventures of a Curious*

Character. The title came from a story about Feynman's first day as a graduate student at Princeton, when he was invited to tea at the home of the dean. Asked by the dean's wife whether he wanted cream or lemon, he awkwardly said "Both," eliciting her bemused reply. It had been his first experience with "this tea business," he explained to his readers. He managed to make himself look charmingly gauche and, more important, to make his hostess look petty and ridiculous. And so it went throughout dozens of stories: the precocious, smart-aleck guy from Brooklyn seeing through everyone else's pretenses, never afraid to play the holy fool. The Feynman image had been honed to perfection and packaged for public consumption.

When the book came out, Gell-Mann skimmed through the stories in which he had played a role. Perhaps the biggest slight was how very few of these there were. In a 346-page book, Feynman brought on Gell-Mann for brief cameo appearances on just six pages. Murray quickly fixated on the account of how, in 1957, Feynman had supposedly been the first to figure out that the form of the weak interactions was V-A.

"I was very excited," Feynman had written. "It was the first time, and the only time, in my career that I knew a law of nature that nobody else knew." *That nobody else knew?* Murray was indignant, loudly telling anyone who would listen that he was going to sue. At cooler moments he would sarcastically deride "Dick's joke book." It only made matters worse when it became a best-seller. The Feynman cult had begun. (A year later, Feynman's celebrity would grow during the aftermath of the *Challenger* space shuttle explosion. A member of the investigating team, he took a particular interest in hints that there had been something wrong with the O-ring seals on the rocket boosters. Cutting through the official denials and equivocations, Feynman caused an international sensation when he demonstrated before millions of television viewers that the O-rings lost their resiliency when dipped in a glass of ice water. In the unusually cold weather of the January launch, this weak link had been calamitous.)

In a later edition of *"Surely, You're Joking,"* Feynman added a disclaimer about V-A, acknowledging that Gell-Mann, Sudarshan, and Marshak had also thought of the idea. But that didn't help much. The relationship between Gell-Mann and Feynman, already sour, only got worse with Feynman's increasing fame. The fact that Dick was in the last years of fighting the cancer that would kill him wasn't taken by either man as a reason for a truce.

As the mid-1980s approached, Murray emerged from the long depression, the sense of being emotionally anesthetized, that had followed Margaret's death. It would be hard to imagine someone more profoundly alone, living by himself in the huge, dark Pasadena house, surrounded by his accumulations of artifacts. He was still estranged from Lisa, and he was getting along poorly with Nick. Margaret's old clothes still hung in her closet. He had friends, people who genuinely liked him, but no one who was really a confidant. He had driven so many people away. The most likely place to find him seemed to be on an airplane, traveling to somewhere far away.

He had started dating now and then and was spending most of his time with Lydia Matthews. The ex-wife of a former mayor of Pasadena, she was in charge of protocol at Caltech, arranging for visits by dignitaries, planning celebrations and other events. She had met Margaret when they served on a mental health board together, and had planned her memorial service. She and Murray became great friends. One evening he and Lydia went together to a movie about a child prodigy. Murray emerged from the theater crying. He had never really had a childhood, she thought. In some ways he would always be an adolescent, with all the exasperating, and occasionally charming, behavior one would expect from someone that age. On another occasion, she was touched and surprised by how much he enjoyed, for purely secular reasons, an Episcopal Christmas Eve service she brought him to. He especially liked a line from the Book of Common Prayer praising "the vast expanse of interstellar space, galaxies, suns, the planets in their courses, and this fragile earth, our island home." It would be hard to find a phrase that better encompassed his lifelong interests. Nature fascinated him on every level, from the complexity of the universe to the simplicity of the particles from which it was made. How, he wondered, could one draw a connection between these two extremes?

Thinking back, he realized it was a question that had been gnawing at him for years. He remembered a day in 1956 when he and Margaret had seen a flock of condors. Still newlyweds, they were returning to Caltech from Berkeley, driving through the Tejon Pass just south of Bakersfield in the Hillman Minx with the top down.

He was watching the sky when he thought he saw some big flying thing dart behind a hill. He pulled the car to the side of the road, grabbed his field glasses, and ran in his gray flannel suit through what turned out to be thick, red mud. Margaret, in a skirt and sweater, high heels and stockings, came slogging behind him. At the top of the rise, they looked down on a dramatic scene: eleven condors picking away at a dead calf.

What lingered in his memory was not so much the sheer number of the birds—a large part of the remaining population of this endangered creature—or their huge size and distinctive colors. What impressed him was the surprising ease with which he could tell one bird from the other. Each was an individual. One condor might be missing a few feathers in its left wing, another from its tail. The birds began life more or less the same, but the accidents of history had conspired to make each one unique. You could give every one of them a name. And so it was with all the flora and fauna he had seen in Central Park, Africa, the San Gabriel Mountains, Aspen, and on nature safaris all over the world. Each member of a species was left a little different from the others by its journey through life.

How different this was from the elementary particles. Every electron is identical by definition. Whether it has been pulled from the shell of a sodium atom or has been traveling through an electrical cable between Phoenix and Albuquerque is irrelevant. Electrons don't wear their history; they don't have names. Each is always as good as new. In Oppenheimer's subatomic zoo, the cages only needed to be filled with a single representative of each species. Anything more would be redundant. Yet from these perfectly uniform building blocks—quarks and leptons—a world of great complexity and individuality is made.

Gell-Mann had gained renown for his insights into the world of the simple, for seeing how tiny wisps of energy obeyed the mathematical symmetries of nature. But he was drawn more and more to the manner in which the symmetries inevitably broke. The perfect regularity of the Eightfold Way gave rise to particles all with different masses. And particles formed nuclei and atoms and molecules. These larger constructs were still not individuals, but somewhere, as you climbed the ladder, complex entities emerged, each one utterly unique—birds with distinctive plumage, chunks of granite of disparate shapes, people with different memories and behaviors. How does one get from the symmetries of subatomic particles to the messiness of the world, from unbending equations to the diversity of the rain forest?

He continued to follow the twists and turns of particle physics as it strained to go beyond the Standard Model. By the early 1980s, hopes for supergravity had faded, and superstring theory was making new headway. At Caltech, John Schwarz and a visitor from London named Michael Green helped overcome a huge objection to the theory: that there were a seemingly endless number of different string theories one could construct, with no obvious way to tell which one described the world. Schwarz, Green, and others went on to show that only a handful of them were mathematically self-consistent; the others would come crashing down because of internal flaws.

Gell-Mann could take some satisfaction in the breakthrough. The nature preserve of endangered string theorists he had helped nurture was earning its keep. And he was making small contributions himself, collaborating with a former student, Barton Zwiebach, on three papers about how to accommodate the extra dimensions. But for the most part, he was content to follow the developments from afar, cheering on the younger scientists. He had to keep reminding himself and others of the lesson he had learned from Weisskopf: Don't be overly impressed by beautiful formalisms. As elegant as string theory appeared, the strings themselves were so tiny that it would still take inconceivably high energies to pry them loose, if they were even there. The situation wasn't entirely hopeless. Certain predictions of the theory might be testable, like the existence of sparticles. Or one might be able to tweak string theory here and there and get the Standard Model to emerge as a rough approximation. That would provide some indirect support. In fact, to the optimists, superstring theory could be seen as having already "predicted," in retrospect, the existence of gravity. This was no small feat. Suppose that there are alien scientists living somewhere in a world of weightlessness. (Maybe their laboratory is in a state of freefall, plunging into a black hole.) Relying entirely on their mathematical abilities, they construct a model of the universe in which everything is conjured from vibrating strings. Lo and behold, a massless, spin-2 particle would inevitably come popping from their equations. If they analyzed just what such a boson could do, they might conclude that it was the carrier of the peculiar force we call gravity.

This retrodiction of gravity wasn't enough for the theory to stand on. But perhaps, some hoped, string theory would have cosmological consequences. In the coming years, as their theories outstripped the capacity of their equipment, physicists started using the universe itself as a test bed, taking reality as the outcome of a single ten-billion-year-old accelerator experiment called the big bang. One started with the cosmos as it now appeared and extrapolated backward, speculating on the particles that were created and destroyed in the intense energies present at the moment of creation. If everything was ultimately made from strings, then perhaps today's universe would have certain characteristics that might be otherwise lacking.

This kind of evidence would never be as convincing as that derived from a good run at Fermilab or CERN. After all, the big bang was not a repeatable experiment. Gell-Mann summed up the situation: "There are certainly some indications that our colleagues may have found the 'Holy Grail' of fundamental physics. But only calculations and their comparison with experiment will allow us to tell. In our science, no amount of eloquence will save a wrong theory." No one knew how long it might be before superstring theory was developed enough for even the most indirect testing. Some supporters of the theory suggested that the best hope might lie in meeting up with extraterrestrial scientists, perfectly comfortable with ten dimensions, who had already proved the theory. Then they could tell us the answer.

Less ambitious attempts to fully unify just the strong and electroweak forces were also hitting dead ends, leaving a lot of particle physicists looking for new territories to explore. For Gell-Mann it seemed the perfect time to shift his attention to the opposite end of nature's spectrum and concentrate on the complex phenomena that emerged from combining the simple subatomic pieces. From quarks one ultimately got cells and organisms, which came together to form ecosystems, economies, human cultures. In studying these domains, the strict reductionism of particle physics was at best premature. The human mind could, *in principle,* be described in terms of the electrochemical currents between neurons; and neurons could, *in principle,* be understood by studying quarks. But science was far from being able to do so. These fields weren't ready for the kind of fundamental explanations that had worked so well in particle physics. Pretending that one could precisely quantify mushy things had led to the excesses of reductionism Murray often complained about—calculating the economic need for a highway with no regard for the incalculable value of the pristine lands that would be destroyed. For now, he believed, science needed to study complex phenomena as things in themselves without reducing them to simpler pieces, to take what he called "a crude look at the whole." Since complex systems were found in every conceivable discipline—biology, psychology, linguistics, economics—the best place to explore them would be the kind of institution he had failed to create at Caltech: a university without departmental walls, where people from many different fields would get together and constantly interact.

When he heard that a group at Los Alamos was talking about creating just such a place, he was ready to sign on. George Cowan, former director of research at the Los Alamos lab, had also been thinking about the need for an interdisciplinary center, one that would study the kinds of convoluted systems that resisted the reductionist scalpel. What was needed, it seemed to him, was a scientifically respectable kind of holism, one decidedly different from the New Age version popular down the hill in neighboring Santa Fe. And yet Santa Fe, he believed, would be the perfect place to house such an institute. Set at the foot of the Sangre de Cristo Mountains, it was a cultured town with good restaurants and bookshops, fine museums, and beautiful architecture—a perfect lure for attracting and keeping interesting thinkers. And it was close to Los Alamos, which had been broadening its sights beyond bomb-making. In 1983 Cowan began trying out his idea at the weekly lunchtime meetings of a group of semiretired scientists called the Los Alamos Senior Fellows.

He was heartened by the response. In the years since the Manhattan Project, Los Alamos had become a leader in attempts to use powerful computers to simulate complex phenomena. At first the lab's primary interest was in analyzing shock waves from nuclear explosions. But from this research came studies of other "nonlinear" systems—phenomena, like fluid flow and weather fronts, with so many interdependent parts that the equations describing them could not be solved exactly. The tiniest changes in the numbers plugged into the equations would cause the solutions to career wildly—a phenomenon called chaos. Often the best one could do was to make numerical approximations: Use a computer to inject a steady stream of numbers and observe how the equations behaved—mathematics as an experimental science. If the output of the equations was harnessed to some good graphics software, the result was a simulation, an animated image of the phenomenon you were trying to understand.

One of the favorite theoretical toys for studying complex systems was a grid of squares, usually projected onto a computer screen, called a cellular automaton. Some squares were white, some were black. One would define simple rules for how the patterns evolved. For example, if a square is white and two of its neighbors are black, then it will become black in the next generation. With the right set of rules, the initially simple shapes would blossom into complex, kaleidoscopic structures. These wholes indeed seemed greater than the sum of their parts. But no magic was involved, just computation. The metaphorical possibilities were endless. Each square might be thought of as a biological cell in an organism: It interacts only with its neighbors, according to simple instructions, but from this simplicity life emerges. The squares might be neurons in a brain signaling to one another, or, reaching further, buyers and sellers trading face to face and generating an economy. In all these cases, simple units interacting locally—just following the rules—give rise to rich behavior. Locally simple, globally complex. It was even tempting to think of the squares as subatomic particles and the rules by which they changed as the laws of physics. Could cellular automata provide clues to how the simplicity of quarks and leptons gave birth to a world full of surprise? The physical world itself might be the result of some kind of computation—quarks and gluons as abacus beads.

These kinds of questions were being studied at several universities around the world. But there was no one place where all the aspects of the search came together, none that sought to cut across domains, bringing physicists, biologists, anthropologists, and economists together. Cowan was ready to start such a place in New Mexico. But the lunch-table discussions at Los Alamos had been going in circles for weeks. Some wanted a center specifically devoted to studying the science of computation. Others wanted something broader and grander.

In the midst of this debate, Murray's old friend David Pines visited the group. Since early in his career, Pines had been interested in solid-state or "many body" physics, the study of how the collective interaction of atoms and subatomic particles gives rise to the characteristics of matter. Nothing about a single molecule of carbon could be thought of as hard, clear, or sparkling, but put them together just so and you get diamonds. Arrange the carbon atoms in another way and you get pencil lead. It was almost like running the same pattern on a cellular automaton using two different sets of rules. "More is different," as Pines's colleague Phil Anderson had written in a classic 1972 paper that could have been taken as a manifesto for what the Los Alamites were talking about a decade later: "The ability to reduce everything to simple fundamental laws does not imply the ability to start from those laws and reconstruct the universe." The condors Murray and Margaret had seen together in California were made from matter whose subatomic parts obeyed the laws of quantum mechanics, but you couldn't start with quantum mechanics and predict the existence of condors, or their behavior that day in the Tejon Pass. And people were even more complex. Pines and Anderson were among a group of scientists who wondered whether some of the insights from solid-state physics could be used, metaphorically at least, to understand how, say, individuals flock into societies that have values and customs and behaviors of their own.

Pines was so inspired by the idea of establishing a New Mexico–based institute to study these matters that he called Gell-Mann, and Gell-Mann called Cowan. This, Murray told him, was what he had been looking for all his life. At the MacArthur Foundation, he had worked to set up research networks that cut across disciplines to study things like the psychobiology of depression. He was eager to try this approach on a grander scale. Murray had just finished building a second home on his land in Tesuque. He would be spending a lot of time in northern New Mexico.

Cowan could hardly believe his luck. A famous Nobel laureate with connections to foundation money and the whole world of philanthropy was interested in his institute. He invited Murray to a meeting in December 1983. With his verbal eloquence and instinctive ability to see connections, Gell-Mann laid out a grand vision that brought the factions together. He spoke of the great leaps that had been made in the past when ideas from different worlds collided: Darwinian evolution, molecular biology. Recently particle physics and cosmology had begun merging. The purpose of the new institute should be to identify and nurture these comings-together— "emerging syntheses," he called them. A buzzword was born. In one stroke, computer simulations, nonlinear systems, complexity, and interdisciplinary research all came together under one roof. The Santa Fe Institute would be a place where especially bright graduate students would come and learn how to dig out patterns and make connections. Moreover, Gell-Mann proposed, this should not be a place lost in the clouds of abstraction. Ideally the research would lead the way toward confronting overpopulation, environmental degradation—all the threats to the glorious complex system called earth. Since these concerns went beyond detached scientific inquiry into political advocacy, perhaps questions of policy could be addressed in a separate, parallel institute. However the vision finally came together, this would be like nothing the world had ever seen. In the excitement following Murray's presentation, Cowan was elected president and Gell-Mann chairman of the board.

In the coming months, the vision was scaled down considerably. For all the group's efforts, it was only able to raise a fraction of the seed money needed. What had been envisioned as a hundred-million-dollar institute with a permanent tenured faculty became a million-dollar institute staffed largely by visiting scientists who converged on Santa Fe to briefly interact. Those who didn't already know Gell-Mann quickly learned the price of his involvement. He was impossible as an administrator. There were the usual unanswered letters and unreturned phone calls. In the summer of 1985, at a meeting at Pines's house in Aspen, it was diplomatically suggested that Murray become head of a newly created science board, in charge of the intellectual agenda, and leave the administrative matters to others.

In February 1987 the Santa Fe Institute opened its doors in an old adobe convent on Canyon Road, in the heart of the city's historical district. Despite its relatively small size, the place's allure indeed proved irresistible, attracting a constantly changing cast of scientific iconoclasts. Phil Anderson helped organize a conference in which economists, physicists, biologists, and computer scientists sought parallels between economies, organisms, and the many-body systems of solid-state physics. Another workshop was devoted to artificial life, the attempted simulation of living organisms on computers. In a meeting called "Complexity, Entropy, and the Physics of Information," scientists explored the curious notion that bits of information were as fundamental in the universe as bits of matter and bits of energy. Gell-Mann went to as many of these sessions as he could, still the kid sitting at the front of the class, waving his hand for recognition. When someone misunderstood a point he had tried to make, he would raise his eyebrows and exclaim, "What, did I say that? I don't *think* I said that." Wide-eyed, innocent, the good little boy wrongly accused.

For all their common interests, a deep philosophical difference divided Gell-Mann and many of the Institute's other regulars. Anderson had written his "More Is Different" paper because he was fed up with particle physicists so cocksure that once one understood subatomic particles, "the rest was all chemistry." For all his interest in complexity, Murray was one of the best-known founts of this kind of arrogance. He liked to call solid-state physics "squalid-state physics," just to get a rise out of Pines and Anderson. He was partly kidding. He knew that quantum mechanics alone couldn't predict that, out of all the possibilities, quarks and leptons would combine to form the particular world we inhabit. One had to take into account the accidental twists and turns of history, the wild card of chance. Nothing in the Standard Model could predict how a collection of molecules would aggregate into certain kinds of cells and not others, or how these cells would themselves come together into individual organisms, and these into societies. In that sense, new rules could be said to emerge at each step of the hierarchy. But Gell-Mann wouldn't go as far as those like Anderson who considered such "higher-level" laws irreducible, as fundamental as those of particle physics. To Murray, complexity was simply a matter of symmetry-breaking. Out of an array of equally likely outcomes—all the different ways pieces might come together into wholes—one possibility was randomly chosen. History was an ever-forking path of these chance crystallizations, and the result of the accumulation was a universe of stunning complexity. Understanding this process would be a challenge, he believed, but it wouldn't require a whole new science.

This disagreement led to some unpleasant encounters. At one meeting, Cowan and Gell-Mann were part of a group discussing the Institute's mission. Whenever someone said the main thrust should be to study the sciences of complexity, Murray would interject, with knee-jerk regularity, "and the fundamental principles of which it is composed." The usually reticent Cowan finally blew up and threatened to quit. One of the most fervent of Gell-Mann's opponents was a medical doctor turned theoretical biologist named Stuart Kauffman. Murray became so weary of Kauffman's harangues against reductionism and calls for brand-new laws of thermodynamics describing how order spontaneously arises in the world, that he started derisively calling him "the doctor," mocking his insistence that understanding complex systems required "something else." Kauffman, in retaliation, printed up a new nameplate for his office door: "Stuart Kauff-Mann."

Naturally Murray wanted to be the one to name the new field. No one knew better the power inherent in such an act. He wanted something that captured the notion that complexity arose from simplicity. He suggested "plectics"—from the Indo-European *plek,* he would explain, forerunner to the Latin *plicare,* meaning "to fold." Thus, *simplex* is something once folded and from that we get the English *simple.* But *plek* also comes into Latin as *plexus,* meaning "braided" or "entwined," and *complexus,* "braided together." Simplicity is folded and braided into complexity. The Greek version was *plektos.* In keeping with the tradition of using Greek roots for the names of academic fields—mathematics, ethics, politics, economics—the new science would be "plectics." The reasoning was Murray at his pedantic best, but the idea fell flat. It just didn't have the ring of "quarks" or "quantum chromodynamics." (When Gell-Mann later tried to revive the idea with a paper titled "Let's Call It Plectics," some of his colleagues replied, under their breaths, "Let's not.") People continued to talk loosely about "complexity studies," feeling no need for a Gell-Mannesque neologism.

For all the disputations, Murray loved the intellectual excitement the Institute had brought to Santa Fe. He was especially taken by what he called complex adaptive systems, those clumps of matter that were not only rich and variegated in their makeup but could sense what was going on in their environments. Biological organisms were the obvious example. They perceived the regularities around them and built internal models, maps of the world. "Schema," Murray called them, careful to correct anyone who used the plural "schemas." (It was "schemata," of course, with the accent on the first syllable.) The nucleic acid sequences encoding the blueprint for an organism, the neurons in a brain encoding the grammar of a language—both were schemata. Using this concisely packaged knowledge, creatures competed against one another for survival. Those with the best maps had an advantage. There was nothing fundamentally new in what he was saying. But his terminology helped clarify some of the ideas that had long interested him. Groups of complex adaptive systems could come together to form larger complex adaptive systems: ecosystems, societies, the superorganisms that gathered data, compressing it into values, myths, and traditions to pass down through generations—cultural DNA. (These schemata were not always good ones. Hence the rise of superstition and pseudoscience.) The more he thought about it, this notion of complex adaptive systems seemed like one of the most powerful patterns of all. Here he was, a complex adaptive system, parsing the world into complex adaptive systems. The possibilities were electrifying.

Like everyone else at the Santa Fe Institute, Gell-Mann was feeling his way around an infant science, one that was still far from the state of maturity that particle physics had achieved. How does one even measure this seemingly subjective quality called complexity? Can the concept be defined as rigorously as, say, temperature or entropy? Is complexity really just the result of a sequence of random accidents, as Gell-Mann believed, or were there guiding laws remaining to be found?

Seth Lloyd, a young physicist who had just received his Ph.D. from Rockefeller University, was so impressed with Gell-Mann's pursuit of these questions that he decided to sign on as his postdoc, turning down a job for twice the salary at the IBM Thomas J. Watson Research Center in New York. In the summer of 1988, Lloyd found himself sitting on the hood of his car at the end of the dirt road leading to Gell-Mann's house in the Tesuque foothills. Murray, who was to meet him there, was late and Lloyd passed the time by reading *Damascus Under the Mamluks,* a thin volume about a period of Islamic history that he had borrowed from the library at St. John's College in Santa Fe. Gell-Mann finally arrived in a Subaru station wagon with the QUARKS license plate obscured by the dust. Lloyd was impressed by his dapper appearance and striking curly white hair. "What are you reading?" Murray asked. Lloyd told him. "Oh, that," Murray replied. "An amusing volume, but inaccurate in its particulars."

They got into Gell-Mann's car and drove up to Los Alamos. On the way, he told Lloyd he was mispronouncing his own name and proceeded to give him its etymological derivation. Then, in an office at the lab, they talked about information, complexity, and the foundations of quantum mechanics. Lloyd had written his Ph.D. thesis on these topics and had never met anyone who actually seemed interested in them. He was struck by Murray's good scientific taste, and with the courage it must have taken to enter, late in his career, a whole new domain of science. When Lloyd made an error in the exposition of a formula, Gell-Mann lowered his forehead to the surface of the desk and pounded repeatedly with his fists, shouting, "No! No! No!" Lloyd loved it. It was the beginning of a beautiful relationship, and of several papers on the mathematical definition of complexity.

Around the same time, Gell-Mann and a former student named James Hartle began exploring the ultimate mystery in this relationship between the simple and the complex: How does the indeterminate world of quantum mechanics give rise to the familiar classical world of cause and effect? Gell-Mann shared the increasingly widespread view that the revered Copenhagen interpretation, handed down since the days of Niels Bohr, was inadequate: An electron hovering around a nucleus had no definite position. It was represented by the mathematical wave function describing every possible place it could be—until it was measured, that is. For Bohr it was this very act of observation that somehow broke the symmetry, "collapsed the wave function," as was often said, causing a single state to emerge. Fermi himself had found this unsatisfying. "Since quantum mechanics is correct," he used to ask Murray back in Chicago, "why is the planet Mars not all spread out in its orbit?" Was it really just because people were watching it, collapsing its wave function? What happened when they blinked or turned away? Some physicists and philosophers were even arguing that consciousness was somehow fundamental in conjuring the universe from the haze of possibilities. Murray found these ideas dismaying. "Quantum flapdoodle," he called them.

The Copenhagen explanation completely broke down when one tried to apply it to the whole universe, as was necessary in Hartle's field, quantum cosmology. One could think of the tiny primordial mass that exploded in the big bang as a kind of subatomic particle, subject to the laws of quantum mechanics. Thus one could define a quantum wave that represented all the possible ways the universe could have unfolded from the initial explosion. All these possible histories would hover together in superposition. But how did the actual universe crystallize from this mess? By definition, there could be nothing outside the universe to observe the big bang, no external measurement that could collapse the wave function. How, from the state of mathematical limbo, did a real universe emerge, a place where objects assumed definite positions in space and time?

Gell-Mann and Hartle were among those who thought the key lay in a phenomenon called decoherence. An unmeasured particle was said to be in a state of quantum coherence, with all possible states in superposition. But a macroscopic object like Mars was made of a huge number of particles, constantly bumping together, being struck by photons from the sun, jostled by cosmic rays. All these interactions were like tiny measurements, causing quantum coherence to break down. From the mathematical bundle of possibilities, a solid planet emerged. Maybe it was decoherence—all this microscopic bumping around—that guaranteed that Mars would stay in a single, predictable trajectory, that caused macroscopic objects to appear. Maybe it was decoherence that brought about this stable world where one can measure without disturbing and where complex adaptive systems would naturally arise to exploit the regularities.

He and Hartle liked to think of decoherence in terms of what physicists called coarse-graining. If you look at a flask of gas at a very fine grain, you see billions of molecules darting here and there. It is only when you step back for a coarser-grained look that macroscopic phenomena like temperature and pressure emerge. There are myriad ways in which the molecules can be arranged to give rise to a gas with the same macroscopic characteristics, just as a deck of cards can be shuffled in $52 \times 51 \times 50 \times 49 \ldots$ ways and still appear, from a distance, like the same deck of cards. Maybe something like this is what happens when classicality emerges from quantumness. If you viewed the universe at the finest possible grain, you would see all its possible paths hovering in quantum superposition, particles that were neither here nor there but everywhere at once, impossible to disentangle. However, because of the limitations of our senses, we can follow only a certain amount of detail, ignoring the rest. We are forced to step back and take a coarser look: The pathways begin to blur together; a particular world emerges.

While many physicists felt that decoherence and coarse-graining went a long way toward demystifying quantum mechanics, a bigger question went unanswered. One could explain why Mars had a single orbit and wasn't blurred all over the sky. But why, of all the possibilities the planet could have chosen, did it assume the particular orbit that it did? Why, for that matter, had the universe come out this way and not another? What about the roads not taken, the choices that might have been? Murray was entranced by an idea proposed by the physicist Hugh Everett, who envisioned the universe's unfolding as a labyrinth of possibilities, like Jorge Luis

Borges's "garden of forking paths." There would be a path where the experimenter measured an electron and found it in one position; another where it was in another position. There would be a path where Mars was a little closer to the earth or a little farther away, a path where our sun had acquired thirteen planets instead of nine, a path where the solar system never formed, another where the sun blew up eons ago in a supernova long before life arose. Closer to home, there would be a path where Murray had gone to Harvard instead of MIT, one where his father's language school hadn't collapsed in the Depression, one where he had realized much sooner the importance of color, one where Margaret hadn't died. In the quantum wave function of the universe, all these possibilities would hover together in superposition. Some physicists proposed that all the universe's possible journeys through time be thought of as existing simultaneously—as parallel universes. To Murray that seemed like more quantum flapdoodle. These histories were just different ways things might have been. Only one of the histories was real.

So many questions remained. Couldn't there be more than one way to coarse-grain, to blur one's eyes a little differently and conjure up different universes? "Goblin worlds," Murray had heard them called. Would these be all more or less the same? Or do we select one of many different possible worlds because of the variables we choose to measure? Perhaps we are constrained by our physical nature to measure only certain ones. Even for someone as hard-nosed as Gell-Mann, quantum mechanics led to impenetrable mysteries. Here was another area, like complexity, where he might help push back the frontiers of knowledge, but there were no big break-throughs yet in sight.

Northern New Mexico, with its striking juxtapositions of geography and culture, seemed like the perfect place to contemplate these questions. The setting was at least as beautiful as Aspen's—more dramatic really, with the mountains jutting up from the desert floor. And Santa Fe had a less manicured, grittier, more diverse feel. One day he accompanied Alfonso Ortiz, a professor of anthropology and MacArthur Award winner, to the annual feast day at Tesuque pueblo, down the hill from Murray's mountain hideaway. Ortiz, who had grown up at San Juan pueblo, a few miles north, watched in delight as his friend questioned the Tesuque people about their language, Tewa. He seemed determined to add as many new words as he could to his schemata.

CHAPTER 16

THE QUARK AND
THE JAGUAR

O n February 15, 1988, Richard Feynman died. Abdominal tumors had long before destroyed one of his kidneys. Recently the other one had failed. He had fought hard for years, but more dialysis now seemed ridiculous to him. It was time to let go of the weak grip of medical technology and find out for himself what happens when you die.

Six weeks later Caltech held a grand memorial service, repeating it so that everyone who wished to pay tribute could squeeze into the large auditorium. Organized by Murray's companion, Lydia Matthews, the ceremony was not the funereal kind of affair that Feynman would have avoided. It was more of a multimedia extravaganza, the celebration of a life. When the lights dimmed, bongo music suddenly filled the air, as though Dick himself were on stage slapping away at the skins. Simultaneously images of him flashed on the screen behind the podium.

All his life, Feynman had told stories about himself. Now it was his friends' turn. One by one, physicists, students, fellow artists, and other well-wishers came forward to say good-bye. Ralph Leighton, Feynman's coauthor and drumming partner, told of a Soviet archaeologist visiting the remote hinterlands between Siberia and Mongolia and coming upon a woman sitting outside a yurt reading a large book with a red-and-white cover: a volume of *The Feynman Lectures*. And he tearfully described a recent call of condolence from a blind ham radio operator on Lake Chapala, Mexico, who had read Feynman in braille. So many people had been touched by this difficult man. He had sown his memes everywhere. Interspersed with the talks were film clips of Feynman in action—the waving hands, the rapid-fire talk, the familiar impish smile. Physics, he declared in a line that has become immortal, is like trying to figure out the rules of a chess game between the gods.

At the end of the service, the bongo music returned, more images of Feynman bounced off the wall: Dick at his beach house, on stage in a Caltech amateur theater production, holding up the famous rubber O-ring at the hearings on the space shuttle disaster. Then came a final, abrupt slap of the bongos, a brief interlude of silence, and an amplified burst of his laughter, eerie now because he was gone.

As they emerged from the auditorium, some of Feynman's colleagues wondered about a noticeable gap in the ceremony. Where was Murray Gell-Mann? He had been listed on the program as a speaker, but he didn't show up at either gathering. Mutual friends of Caltech's two great physicists were puzzled. Was Murray too broken up over Dick's death to attend? Or was he still angry at him and jealous of Feynman's fame?

The story turned out to be less complicated, the kind of tale Feynman himself might have enjoyed.

Ever since Murray was a boy, when his cousin Israel had given him some old Roman coins, he had been in love with ancient objects. Year by year, in the course of his travels, his collections grew. Recently, through an art broker in Santa Fe, he had been introduced to a swashbuckling young dealer from Santa Barbara, California, named David Swetnam, who was offering some unusually fine antiquities from South America. Murray quickly became one of his best customers, buying ancient Peruvian ceramics, textiles, and other expensive items.

In California, Swetnam would sometimes hold court at a Peruvian restaurant in Santa Barbara (on a street some called "Smugglers' Cove"), drinking Pisco sours, made from sugar-cane liquor, and showing collectors photographs of the tantalizing items he had for sale. One evening Murray joined a group there as they laughed over Swetnam's latest tale of how he had smuggled a valuable pot into the United States, by covering it with a thin coating of clay and then having it painted to look as though it had been made by contemporary Indians. Once the pot was safely in California, the disguise was easily washed away. Gell-Mann had few qualms about hoodwinking the corrupt Peruvian government, which he believed took poor care of the country's archaeological legacy. Rumors abounded about valuable museum items ending up in the private collections of government officials. The antiquities would be better off, Murray told himself, with serious collectors like himself. It was a rationalization used by wealthy antiquities enthusiasts all over the world. He didn't realize, he would later insist, that knowingly purchasing the smuggled items also put one afoul of United States law.

Swetnam had learned of some priceless treasure—golden necklaces and figurines of warriors, gods, and animals—recently looted by grave robbers from the tomb of a king of Sipan, a ruler of the lost Moche civilization that flourished in northern Peru until the eighth century. He asked some of his better clients, Gell-Mann among them, to front the money needed to get the items into the United States. In May 1987, when Swetnam showed him a picture of a jaguar head decorated with gold and marked with finely etched whiskers and fangs, Murray knew he had to have it. He made three advance payments, totaling $31,000. In the coming weeks, offered additional antiquities, he paid out more.

When he finally saw the jaguar mask, however, Gell-Mann decided he didn't want it after all. He left it with Swetnam, but before he could arrange for a refund or an exchange, federal agents, incongruously dressed in flak jackets and brandishing assault rifles, arrived at his door—just as he was preparing to leave for Feynman's memorial service. An accomplice of Swetnam's had decided to turn informer. In what was the largest customs raid in history, sixty agents converged on the homes of several southern California collectors and dealers, seizing some two thousand artifacts, some from Sipan, worth millions of dollars.

As the agents went through Gell-Mann's private treasure trove, he persuaded them to confiscate only the several dozen Peruvian items, watching over them to make sure the artifacts were carefully packed. Then the legal nightmares began.

In the midst of this drama, Gell-Mann turned the magic age of sixty, when physicists are traditionally subjected to their Festschrifts. He didn't want to be in Pasadena on his birthday, September 15, when the summer heat still lingered, so the celebration was set for January 1989. Val Telegdi joked that it was a celebration of Murray's conception. Like the old television show "This Is Your Life," people from all stages of his career showed up. There were early friends and collaborators like Telegdi, Francis Low, Murph Goldberger,

David Pines, and Yuval Ne'eman. There were later ones like Harald Fritzsch and John Schwarz. There were such renowned colleagues as Abdus Salam, Yoichiro Nambu, and T. D. Lee, and younger ones like Edward Witten, a rising star in superstring theory. And, unlike with most of his colleagues' birthday celebrations, there were people from outside the world of particle physics: from the MacArthur Foundation, the Smithsonian, the World Resources Institute, Wildlife Conservation International. The Santa Fe Institute was represented by George Cowan and J. Doyne Farmer, a young physicist studying artificial life. Murray's Pasadena neighbor and Yale classmate Paul Macready spoke about the nature of creativity. All these people's presence was a reminder of the breadth of Gell-Mann's interests. They were asked to come to Pasadena to talk about where they currently stood in their intellectual efforts and to speculate on what the future might hold. Many of them also took the opportunity to get back at Gell-Mann.

In the introduction to his talk, Ne'eman spoke of a recent visit to the "College of Judea and Samaria"—a dig at Murray's neurotic refusal to say the name "Israel." There were jokes about pronunciation and about Murray's obsession with birds. (The problems with the Peruvian artifacts were politely ignored.) In a talk called "Is Quantum Mechanics for the Birds?" Telegdi made Murray squirm with a mock derivation of his name. Did "Murray" come from the Scottish province "Muraih," or from "Murrey," describing a dark red color, related perhaps to "mulberry" and "maroon"? As for "Gell-Mann," maybe it came from "Hellmann," a corruption of "Heilmann," meaning "healer"—"a profession often practiced by minorities in the empire." And he got in a dig about the hyphen. Murray was not amused, and later made sure a footnote was added to the published volume: "Experts in onomastics assure us that Hellmann is not derived from Heilmann. . . ." But he was delighted with some doctored slides Telegdi showed of Murray directing a symphony, triumphing in a fencing match, preparing an elaborate gourmet meal, balancing like an acrobat ("A remarkable sense of symmetry and equilibrium!"), and taming tigers ("Of course you must know that he never beats these animals; he's the only trainer who can use gentle methods")—a true Renaissance man.

There were actually, after all these years of poking into the universe's nooks and crannies, still many things Gell-Mann could not do. Goldberger (who wished Murray Happy Birthday in Fang, a Bantu language of west equatorial Africa) knew of more than one

Chinese waiter who had been baffled by his friend's attempts at Mandarin. Although his familiarity with so many different languages was impressive and played to maximum effect, he admitted to actually speaking only one perfectly: English. He could speak French, although he made some errors, and to a lesser extent Spanish and Italian. He was pretty good in Danish and from that could convert to Norwegian and Swedish. And though he could read German, Arthur's pedantic attempts at teaching him to speak the language fluently had failed. For almost anyone else, this breadth of linguistic skills would be a matter of pride. But Murray, seeing the hole instead of the doughnut, regretted not having learned more. He dabbled in various English- and French-based creole and pidgin languages and could entertain people at parties by making African clicking sounds. But what got lost in the showmanship was the reason for his linguistic passion: He was fascinated by the underlying patterns, the symmetries he was sure connected all these diverse tongues as surely as the grammer of physics embraced the subatomic particles.

The idea of him conducting a symphony was, of course, ludicrous. Telegdi, like many others, knew that one way to temporarily break down Murray's know-it-all demeanor was to bring up a subject like music, which he didn't have much interest in. But Murray would quickly regroup, managing in a sentence or two to steer the conversation back to a subject he knew about. He always seemed to land on his feet.

By the time the Festschrift ceremony was over, he was beginning to untangle himself from his legal troubles. Some of the collectors contested the seizures in court, arguing that they had bought the artifacts in good faith, and eventually got them back. Swetnam pleaded guilty to customs violations and was sentenced to six months at a federal prison camp. Gell-Mann decided to cooperate with the investigation and forfeit the artifacts to the Peruvian government. He was afraid that the bad publicity from a court fight might hurt some of the MacArthur-funded conservation efforts in Peru, or that he might be barred from the country, never able to go bird-watching there again.

In December 1989, after his lawyer finished negotiating a deal with the prosecutors, Gell-Mann found himself flying to Lima with the artifacts. He was honored by Peru's president, Alan Garcia, at a ceremony at the newly opened Museo de la Nación, praised for setting a good example, and given a key to the city. In his talk Gell-

Mann spoke admiringly of the ancestral Indians who had produced such fine art. And he called for new laws that would allow collectors to own some of the work while ensuring that scientists and historians kept track of its whereabouts. Creating a legal, regulated market might actually discourage looting, he argued, and stem the loss of archaeological knowledge.

Murray had survived the ordeal, just barely. In a matter of weeks he had gone, he would jokingly say, from felon to national hero. But his reputation had clearly been tarnished. In some ways, the pottery incident was perfectly consistent with the self-absorbed manner in which he so often led his life, blind to any but his own perspective, assuming the rules didn't apply to someone as enlightened as himself. It was also true that his motives had been pure. He had no intention of profiting from the artifacts. He loved them and wanted them to be part of his life. Before he had been driven down the path of physics, he had longed to be an archaeologist. Margaret had pursued archaeology before she married him. He was in love with these ancient cultures, fascinated that so long ago human minds and hands had transformed shapeless matter into things of beauty, teasing out their symmetries. The obsession cost him dearly. Sadly, he resigned from the board of the Smithsonian Institution, on which he had served for fourteen years. Neither the jaguar mask nor the money he had paid for it was ever recovered.

It was a very different jaguar that was on Gell-Mann's mind when he decided to write a book, a broad-ranging work that would stand as a kind of grand unified picture of his inner world. The cat he was thinking of had leapt from a poem by Arthur Sze, a young Chinese-American writer Murray had met in New Mexico. He had sold Margaret's loom to Sze's wife, a Hopi weaver; and Sze, who had once studied physics at MIT, had given Gell-Mann a copy, just published, of his new anthology called *River River*. Murray was struck by a verse in the first poem of the collection, "The Leaves of a Dream Are the Leaves of an Onion," which spoke of the interconnectedness of everything in the universe:

> *The world of the quark has everything to do*
> *with a jaguar circling in the night.*

These two lines seemed to capture so well the mystery of how simplicity leads to complexity, with the precise laws of physics giving rise to creatures with minds of their own. Gell-Mann would call his book The Quark and the Jaguar. He was disappointed that he had never actually seen one of these distinguished animals in the wild. But he had come close. A couple of years earlier in an Ecuadorian rain forest, after enjoying a fresh piranha dinner, he and some companions were walking on a jungle trail when they heard the crackling sound of a large animal stealing through the underbrush and smelled the gamy odor of a jungle cat. Murray just caught a glimpse of the tip of its tail before the jaguar disappeared into the forest. A few years later in Belize, he saw a smaller cat called a jaguarundi. Walking among the Mayan ruins, he had been lost in thought, contemplating the oddities of quantum mechanics and the accidents of nature. Why does a parrot randomly pick a certain piece of fruit to eat? Why does a tree growing through a stone wall crack it in a certain manner? And how did it come to be, in the unfolding of the universe, that a hundred yards in front of him a jaguarundi was now standing, staring at him with a curiosity that seemed to match his own?

He assumed his book would be a best-seller. After all, Feynman had written one. Stephen Hawking's *Brief History of Time* was a publishing phenomenon that sold a million copies in hardcover. And the fact was that Murray needed the money. The debacle with the Peruvian artifacts had cost him a small fortune. Around this time in Aspen, he saw his colleague Jeremy Bernstein, who was successfully pursuing his parallel career writing books and *New Yorker* articles about physics. How, Gell-Mann asked him, does one go about making money on books?

Anyone who knew about Murray's intractable writer's block and infinite capacity for procrastination would have to be dubious of such a venture. But for John Brockman, a New York literary agent who specialized in popular books by scientists, Gell-Mann's decision to commit authorship was news he had been awaiting for years. Armed with an M.B.A. from Columbia, Brockman had gotten his start in the 1960s promoting what used to be called "happenings" at places like Andy Warhol's Factory in downtown Manhattan. As a literary agent he had made a killing representing writers like Fritjov Capra, best known for *The Tao of Physics,* and Gary Zukav, author of *The Dancing Wu-Li Masters,* books Murray disdained because they seemed to take seriously his whimsical reference to Buddhism in the Eightfold Way. Didn't they know he was joking?

Brockman's own attempts at authorship—he fancied himself an

intellectual for whom agenting was just a fantastically lucrative day job—had not been as successful. His book *Einstein, Gertrude Stein, Wittgenstein & Frankenstein: Re-Inventing the Universe,* billed on the jacket flap as "a dramatic and dynamic portrait of the universe modern science is creating," was recalled by the publisher in 1986 for "incomplete permissions clearance" after it was discovered that certain passages bore a striking resemblance to a previously published magazine article by James Gleick. Robert Wright and other science writers soon also found similarities with their own writings. Brockman blamed the problem on sloppy work by a research assistant, inspiring one of the writers whose work had been emulated to comment wryly that the author's excuse was that he hadn't actually written his own book.

If Brockman was having difficulty being accepted as a serious intellectual, at least he could act as a kind of facilitator for others, making a lot of money along the way. Over the years he had accumulated an all-star cast of scientist-authors. In Manhattan and later at his estate in Connecticut, he would gather them for meetings of an intellectual salon he called the Reality Club. Later he uploaded the operation to the Web. He touted these thinkers as the new intellectuals, conversant in complexity theory, superstrings, and artificial life—an exciting new expanse of discourse that replaced musty old ideas like Marxism and Freudianism. C. P. Snow had famously complained that the intellectual world was divided between two cultures, humanists and scientists. Brockman promoted his clients as members of a Third Culture, bringing new ideas directly to the public with their books.

If a scientist's name appeared on the cover of the *New York Times*'s weekly *Science Times* section or on page one of the *Wall Street Journal,* he or she might expect a phone call from Brockman Inc. with a terse promise: "I can get you $300,000 for a book." Or more. Publishers would complain about the prices he asked for what were sometimes brief, slapdash proposals sent simultaneously to a dozen publishers in hopes of igniting a bidding war. But they were afraid of missing the next big thing. Brockman was among the agents and publishers who had been courting Gell-Mann for years. He was sure he could get him a million dollars, of which he would keep about 15 to 20 percent.

Murray found this proposition most intriguing. Brockman was brash, fast-talking, manipulative. This was the kind of bulldog he wanted representing him. But there was no way he was going to be able to deliver the kind of proposal that would command a seven-figure advance. Someone else would have to write it. Brockman called Roger Lewin, a respected science writer who had held top editing jobs at the journals *New Scientist* and *Science.* Lewin had written the acclaimed *Bones of Contention,* about the search for the origins of man, and collaborated on books with the anthropologist Richard Leakey. Brockman agreed to pay him $1,500 to dash off a book proposal.

Lewin had met Gell-Mann before and had no illusions about how difficult it would be to work with him, even briefly. Murray sent him some recent articles and lectures on complexity, and they talked on the phone. Every time Murray launched into a disquisition on some interesting piece of arcana, Lewin would adroitly steer him back on course. A thorough professional, he knew just what he needed in order to write the sales pitch.

He was a little miffed when Murray objected that the fee, a fraction of what a writer of Lewin's stature would expect for a magazine article, was too high. But Lewin quickly hammered out a very engaging proposal. It began with a bold overture that Murray himself, with all his hedging and hair-splitting, never could have written on his own: "I am embarked upon a personal voyage of scientific discovery, an adventure that just a few years ago was impossible. . . . My hope is as simple as it is ambitious: that through the science of complexity we will reach an understanding of the way the universe works, from the structure of galaxies to the moment of a creative thought in the human mind, from prebiotic evolution to the rise and fall of prehistoric societies." The book, it was promised, would go beyond Hawking's and Feynman's best-sellers. "I don't believe I am being excessively immodest if I say that my name is at least as prominent in our field as theirs. And I wouldn't be Murray Gell-Mann if I didn't believe that my insights into the nature of the universe will make just as compelling a book as theirs." He made a coy promise to include anecdotes that Feynman had left out of his own books—"the ones that didn't leave him one up on everybody else!"

Never mind that this hardly sounded like Murray. (Later in the document, Lewin wove in some of Gell-Mann's own previously written words.) The proposal was sent out in the fall of 1990 as though it had come hot off the physicist's own word processor. One of the many desks it landed on was that of a young editor at Bantam

Books, which, after much heavy editing, had unleashed Hawking's best-seller. Brockman called and barked into the phone, "We want a million dollars," then hung up. They finally settled on $550,000 for the rights to publish the book in the United States and Canada. Gell-Mann would get 25 percent of that immediately, from which Brockman would take his cut. Delivery of the manuscript was scheduled for June 1992.

With that auspicious send-off, Brockman headed to the annual international book fair in Frankfurt, Germany, where *The Quark and the Jaguar* became the hottest item. Everyone was looking for the next Hawking. Brockman sold the book in about a dozen foreign markets, securing promises of around a million additional dollars in advances. This, it was widely reported, was the most ever paid for a popular science book. Lewin happened to be in Santa Fe when the offers came rolling in. Murray was contrite about having balked at the $1,500 fee he had paid for such a successful proposal.

Now he had to write the book. And so, with the stakes higher than ever, he was forced to undergo the familiar agony of facing the blank page. He was cursed with knowing too much. What a nightmare it was trying to pin words to paper when each one shimmered with connotations, long etymological histories—semantic auras that interfered with each other constructively and destructively like quantum waves. The sentences could never quite say what he meant. Doing theoretical physics was easy. You would get up late in the morning, and wander into the office for a few hours to talk with colleagues about some crazy new idea that had popped into your brain while you were shaving. You could work while hiking through the mountains or dozing on a plane. Only every once in a while did one have to suffer the torture of writing up a few pages for publication (or editing what the postdocs had drafted). But this book-writing business—he needed to produce hundreds of pages!—seemed inhumane. He couldn't believe that one was actually expected to sit at a desk for hours at a time and write.

For years journalists and science writers who called Gell-Mann for an interview or encountered him at a conference had been subjected to diatribes about what he considered the inadequacies of their craft. In 1983, when Robert Crease and Charles Mann were working on *Second Creation,* their history of particle physics, the only time Gell-Mann would agree to talk to them was on a cross-country flight from New York to California. They booked two tickets and rushed to the airport.

We found him on the aircraft already buckled in and tapping his fingers with impatience; the only trace of the exhausting series of lectures and seminars that he had just delivered in half a dozen cities was a slight darkening under his eyes. His California tan had not faded from the gray spring skies of New York City. "It's a long flight," he said. "I hope you're interesting." We asked him a question. He groaned, looked out the window. "People talk a great deal of nonsense about science," he said. "And you have obviously reaped the consequences of that." The cabin doors clicked shut, and we flew from New York to Los Angeles.

When their manuscript was done, Gell-Mann would call Crease late at night excoriating him over the draft pages he had been sent to comment on. Now Murray himself was having to confront the difficulties of taking the precise language of mathematics and translating it into the inevitable ambiguities of English. When he thought of physics, it was not so much as pictures and metaphors, the tools of the trade of science writers, but as abstract relationships—algebras of currents that could only be described mathematically. He warned against physicists being overly formal because he knew it was such a danger in himself. At the same time, he eschewed overly heuristic approaches—Feynman and his parton-studded flapjacks. All this made it even more difficult for him to convey his work to the mass audience he and his publisher envisioned. And, as usual, he was terrified of being wrong. He remembered a story he had been told about an apocryphal Norwegian lighthouse keeper who whiled away the endless winter nights reading books and marking all the errors. He imagined some of his colleagues turning the tables and gleefully pointing out his mistakes. And there was still the shadow his father cast onto the pages, scrutinizing every word.

He found some consolation in sharing his literary troubles with a beautiful poet he had met a couple of years earlier in Aspen, a town frequented by writers as well as physicists. Marcia Southwick, who was about twenty years younger than Murray, had grown up in New Haven, where her father was the head of orthopedic surgery at Yale. She was a graduate of the prestigious Writers Workshop at the University of Iowa, and had later taught there. When Murray met her, Marcia's marriage to the poet Larry Levis had ended, and she was teaching at the University of Nebraska in Lincoln. She loved the town—her family, which owned a newspaper business, originally hailed from there—but couldn't stand the hot, humid summers. She started going to Aspen to cool off, teaching sometimes at the Aspen Writers' Conference. A fan of the public lectures held by the Center for Physics, she had gotten to know some of Murray's friends. After the talks, the physicists often flocked to a downtown bar called the Ute City Banque. The writers also gathered there.

One evening Gell-Mann was at the bar with Lydia and some friends when he and Marcia started talking. Murray made a bad first impression. He was so opinionated and outspoken, overflowing with information. Marcia found him a little intimidating. The intensity of his eyes zeroing in on her was almost frightening. And he wasn't exactly smooth, approaching women in the awkward manner of someone who had been single for too long. She fended off a couple of invitations. Murray didn't push too hard. He had heard she was dating the owner of the bar, a friend of his. But he called her a few times in Nebraska during the following year. The next summer she agreed to go out with him. They started with a walk along a trail near the top of Independence Pass, a twelve-thousand-foot divide southeast of Aspen. From there they went to a concert, then to dinner and a movie, *Dead Poets Society,* talking nonstop all the while. She was exhausted by the time the date was over but impressed with how sweet Murray could be and surprised by how much they had in common. Having grown up in New Haven, she knew all about the Yale experience. And she too had a son named Nicholas. They shared a taste for luxurious things—and for true-crime books. Moreover they were both interested in the creative process, how a poem or a theory evolves, the different versions mutating and competing in a Darwinian competition until the finished product emerges: a compromise between randomness and regularity. From his MacArthur Foundation connections, Murray knew more than she expected about the current poetry scene. And she seemed very interested in his ideas about complexity and the patterns woven deeply throughout the world.

After a year of carrying on a long-distance romance, he proposed in the summer of 1991. They were at his house in Aspen. "Come in the living room and close your eyes," he said. "I have something for you." When she looked, he was holding one of those devices used by jewelers to measure fingers. Then he pulled out a ring with an enormous emerald-cut diamond. It was too big, but she wore it with a Band-Aid wrapped around the back.

Around that time, with the deadline approaching and so much

money already on the table, Bantam was getting worried about its investment. Gell-Mann's editor was sent to Pasadena on a scouting mission and returned to New York with the impression that the book was barely off the ground. It seemed clear to her that Murray needed help. Brockman agreed and called Roger Lewin, who said no thank you. Working on the proposal had been bad enough, and he was writing a complexity book of his own. Finally an editor named Ed Dobb at *The Sciences,* an award-winning magazine published by the New York Academy of Sciences, agreed to act as a kind of editor/ghostwriter for a fee of $50,000. He soon came up against the insurmountable wall faced by anyone trying to work with Murray to produce something on time. Gell-Mann would tell Dobb a story, like the one about his father trying to talk him out of studying archaeology at Yale. Dobb would turn it into a narrative, which Murray would brusquely reject as inaccurate, purple prose. Dobb finally quit in frustration. The result was an angry impasse, with Gell-Mann talking about suing for return of the money and Dobb apparently feeling that $50,000 was small compensation for the verbal abuse he had endured.

By now it was early 1992. Murray had until June to produce a manuscript. Bantam was in no mood to grant an extension. Two competing books on complexity were coming out that fall, Lewin's and another by a writer for *Science* named Mitchell Waldrop. In addition, Brockman had signed up two of Murray's Santa Fe Institute colleagues, including Stuart Kauffman, to write books of their own. The $550,000 was beginning to seem to the editors like a bad investment. If they didn't have a manuscript they liked by deadline, Gell-Mann would have to repay the money that had been advanced to him so far. Not even counting Brockman's commission, more than $50,000 was already gone, most of it for Dobb's fee.

In mid-February, Brockman called a young writer named Rick Lipkin and described his dilemma. If Bantam pulled out, many of the foreign publishers might join the rush for the door. The whole house of cards could collapse. At first Lipkin was reluctant to take on more work. He was involved with another Brockman project, the cognitive scientist Marvin Minsky's memoirs. But collaborating with the great Gell-Mann sounded interesting. When Murray himself called, Lipkin asked how much of the book was done. Gell-Mann hemmed and hawed, then said he would have his assistant send him the work in progress. When Lipkin saw the package, a few inches thick and weighing several pounds, he was encouraged. There must be a lot of pages in there. Then he opened it. The "manuscript" consisted largely of a thick stack of Murray's articles and speeches clamped together in a three-ring binder. "Is this all you have after almost two years?" he asked. Murray, embarrassed, admitted that it was.

Lipkin realized it would be impossible to write an entire book in the few remaining months. The only realistic goal, he proposed, was to produce a hundred or so unpolished pages good enough to persuade Bantam to grant a short extension. Then they could finish the book over the summer for publication the following spring. He agreed on a fee of $50,000, with $10,000 paid up front. (Feeling burned by their experience with Dobb, this time Gell-Mann and Brockman were taking no chances.) Lipkin got on a plane and flew to Santa Fe. Murray picked him up at the airport, and the young writer moved into his house in Tesuque. For the next three months, they were together almost constantly.

They got along well at first, occupying bedrooms at opposite ends of the house. Each morning, with the sun coming up over the Sangre de Cristos and lighting the Rio Grande Valley, they would rise and eat breakfast together and try to work. Afterward they would go for long walks in the piñon-studded foothills. They even shopped for groceries together. But it quickly became clear that Murray's mind just wasn't on the book. Lipkin began to get really nervous when he realized that the wedding was planned for June, the same month the book was due. And Murray was in the midst of plans to retire from Caltech, sell the Pasadena house, and move to Santa Fe to live full-time with Marcia.

After several weeks in Tesuque, they took a short break and resumed work in Pasadena. Murray seemed even less focused than before. The plan had been for him to talk into Lipkin's tape recorder. Lipkin would then draft something for Murray to read and revise. But it was almost impossible to get him to sit still. It was like trying to make a hyperactive teenager do his homework. Murray would launch into disquisitions on this and that, padding around the house in his slippers. He seemed more interested in planning an upcoming bird-watching trip. One day as Lipkin was editing some pages, he listened in exasperation as Murray, dressed in full bird-watching regalia, stood in the next room playing a recording of birdcalls. He didn't seem to realize that he really had to get this book done. As Lipkin tried to rein him in, Gell-Mann became increasingly short-tempered. He would read something

Lipkin had written and tear it up. Deep down, Murray seemed to resent the position he had been put in, or put himself in. He didn't want someone else writing his book.

Lipkin decided to try a different approach. He put a pencil and writing pad in Murray's hands and said, "Write a paragraph." Murray groaned and complained, finally squeezing out a few sentences. Lipkin edited the page and handed it to Murray's assistant, who typed it up. This went on for weeks and weeks until they finally had about a hundred pages for Bantam. By this time Gell-Mann had grown even more resentful. He sent Lipkin home at the end of May, saying he would keep him on call.

Just over a decade after Margaret's death, Murray married Marcia Southwick on the lawn of his home in Aspen. She came around the corner in a horse-drawn carriage, alighting like a dream come true. Her son, Nick Levis, and her father gave her away. At the rehearsal dinner the night before, Ben had teased Murray for always telling people how his big brother had been his inspiration, generously taking him on trips to the park or the New York City museums, teaching him about science. The way he remembered it, Ben said, was that their mother would say, "Get that kid out of the house." And he would dutifully comply. Murray, a little miffed, insisted that wasn't true. But not much could bother him on such a happy occasion. He was delighted that both Lisa and his own son Nick were at the wedding. The family was finally starting to pull back together.

Bantam, mollified by the draft Murray had submitted, gave him until November to finish the book. Though Lipkin was consulted occasionally, the plan now was for Marcia to take over the role of writing coach and trainer. She would read his pages and mark places where there were not enough examples, or where an idea hadn't been developed or conveyed very well. He would counter, *Well, how should I say it?*, and she would be at a loss. She was a writer, not a scientist. One day when he was trying to work on the book in Aspen, she told him he needed to get away from the distractions, to sit for five straight hours and write. But the phones kept ringing. Marcia packed them all up in her car and drove away. When she returned, she found he had made a little progress.

When November came, Gell-Mann still had only a rough, incomplete manuscript. Bantam rejected it, and he had to return the money. (Brockman kept his commission, Murray agreeing that the debacle was his fault. He hadn't believed the deadline was serious.) Some of the foreign publishers were also threatening to pull out. Around Thanksgiving, the pressure building, he had a mild heart attack in Tesuque. Marcia, terrified, had to drive him down the winding dirt road to the hospital in Santa Fe. He quickly bounced back from the experience, but he knew that from then on he would have to be more careful about all the rich, gourmet dinners he enjoyed so much and the stress he constantly inflicted on himself.

Brockman tried to get other publishers interested in the project, but what had been the big book at the Frankfurt fair in 1990 now had a reputation as an impossible project involving an impossible author. People in the book trade started calling it "The Jerk and the Quagmire." Finally W. H. Freeman, a small scientific publisher, agreed to pay $50,000.

The editor, Jeremiah Lyons, was confronted with something that he didn't consider even an approximation of a manuscript. Armed with advice from Lewin about how to handle Gell-Mann, he was determined to push the book through to completion. And so he signed on for the final roller-coaster ride, the thrill of hearing physics firsthand from a master, the grinding frustration of working with someone who never met a deadline. But Lyons was hooked. In the course of the editing, he took a job at another publisher but kept working with Murray pro bono. Gell-Mann would call him at 10 p.m. to go over a section word by word. He was impressed that Murray seemed to be able to hold the entire manuscript in his mind's eye like a program loaded into RAM. He wanted an explanation for every suggestion Lyons made. "It just sounds better that way" was never good enough. Lyons would hang up after midnight and realize he had been on the phone for two hours and gone through one and a half pages.

Others also tried to help. Though disappointed that he was never paid the balance of his fee (Gell-Mann and Brockman felt he hadn't completed the job he had been hired for), Lipkin still cared enough about the book to read and comment on a draft. The novelist Cormac McCarthy, whom Murray had met through the MacArthur Foundation, line-edited the entire manuscript, but Gell-Mann was too rushed and disorganized to take advantage of the suggestions. He complained bitterly about the copyeditor and the production editor, driving them crazy by refusing to accept that there was a point of no return where one couldn't keep making changes, and changes to the changes.

Finally Freeman had to take the book away from him and put it into print. Though they had skimped on the advance, they spared no expense on the presentation. The cover, printed with a large color picture of a jaguar, was encased in a transparent plastic jacket emblazoned with the title. On the back was a quote from Carl Sagan: "A stimulating, provocative, and uncommon cut across compartments of human knowledge that are usually hermetically sealed. It is always a pleasure to see a first-class mind grappling with the greatest mysteries. . . ."

On April 26, 1994, the same week *The Quark and the Jaguar* was published, physicists at Fermilab announced strong evidence that the top quark had finally been found, some two decades after it was proposed. The last piece of the Standard Model was in place. Gell-Mann found his name on page one of his hometown paper, the *New York Times*.

On the evening of the discovery, he was in his old neighborhood, the Upper West Side of Manhattan, not far from the public library branch he had used as a child, giving a reading at Shakespeare & Company booksellers. Marcia had made him trade in his nerdy black glasses for more elegant frames. They went well with his thick, snow-white hair. He looked almost le nine as he read about the jaguarundi and the time he saw the condors. He was warm, charming, funny. The audience loved him, and when the reading was over they formed a long line to have him autograph their copies. Many were flattered when he commented on the linguistic subtleties of their names.

He didn't know that his own name would be in the *Times* again that Sunday in a review that could be described most generously as mixed. Even more hurtful, the book had been "briefed," given but a single paragraph in a column called "In Short." His publicist, who had an advance copy, was afraid to show it to him. She hid it amid a stack of reviews, many of them more favorable, which she thrust into his hands as he boarded a plane for the rest of the book tour.

He was more fortunate with his other reviewers. The *Washington Post* assigned the book to his old friend Sheldon Glashow, who owed his Nobel to Murray's intervention. *Science* gave it to one of Gell-

Mann's former students, the Harvard physicist Sidney Coleman. They both thought the book was fine. Other, more independent reviewers also found things about it to admire. In the United States it was a small success, selling more than thirty thousand copies in hardcover, and it was a best-seller in Germany. But for someone hoping to match Hawking's and—more important—Feynman's royalty statements, it was a huge disappointment. Murray was also dissatisfied with the text. If only he had been given another year, he would complain, the result could have been so much better. But despite the flaws, he felt that he had produced a good book. And he was proud that, in the end, he had done it himself. He had never worked so hard on anything in his life.

VALENTINE'S DAY, 1997

The Gell-Manns' sprawling new house on Santa Fe's east side was decorated for a party. Shiny Mylar Valentine balloons were tethered everywhere, even on the tall wooden West African bird-god that stood guard in the grand entranceway. Murray and Marcia's parties had become familiar social events. On the list of guests, one would find an eclectic mix that might include such friends as William Zeckendorf, the Manhattan real estate developer; Roger Kennedy, former head of the National Park Service; Sir James Murray, whose past diplomatic posts included that of British ambassador to the United Nations; Tom Udall, the environmentally minded congressman from northern New Mexico; and Michael Nesmith, the musician, producer, and former guitarist for the Monkees. (A seeker after cutting-edge ideas, especially those with a New Age sheen, Nesmith often came down from the nearby village of Nambe to sit in on lectures at the Institute.) Interspersed with the celebrities would be a sprinkling of scientists, artists, and writers. One never knew whether the next person encountered would be known for work in complexity, physics, poetry, film, or wildlife conservation, or was just famous for being famous or being rich.

This evening was no different. The trees around the glowing indoor swimming pool sparkled with white lights as the caterers rushed in and out of the kitchen. A six-man mariachi band with a woman singer—reputed to be the best in Santa Fe—was entertaining at the water's edge, while in the library, lined with rows and rows of pottery, Murray tried to demonstrate his new giant-screen satellite-dish TV. He couldn't quite get it to work. He and Marcia were in fine spirits (at one point she pulled him onto her lap). During a lull in the party, she wandered through the rows of Murray's library, recalling aloud how much she had missed him when he was away on a recent trip. Alone in the big house, she felt his books radiating his presence.

Late in the evening, he walked over to where I was sitting with my wife and some friends on the far side of the pool. "It's too bad you've been over here all evening," he said. "It would have been nice to introduce you to people as my biographer." He laughed at his joke. We had come a long way in the years since I started the project. But I knew he was apprehensive about the results, and I wondered how likely the good feelings were to survive beyond publication. Would he soon be calling me "that Johnson person"?

Holding a drink in his hand, he complained that he had been lecturing too much, spreading himself thin, but (he said it again) he had to make enough money to support his "expensive wife." Lately he had been noticing that his powers of recall were slowing. Names from the past that once popped into his mind instantly sometimes took hours to retrieve. But he tried to take it in stride. When he was in his twenties, he said, he thought that turning thirty would be horrible, that he would rather be dead. He felt the same about forty and fifty. But lately he had concluded that each decade had been better, especially his sixties now that Marcia had come into his life. Even when you are in your twenties, he said, there are days when you feel like you are ninety. In your thirties, there are more of these days, and still more in your forties, and so forth. But he was far from the point where the proportion was approaching 100 percent.

As the trumpets blared in the background, he talked a little about the promise of superstring theory. For years after the conceptual breakthroughs of the mid-1980s, physicists had been stuck with five different ten-dimensional string theories, four too many, all competing to describe this single world. In recent years, to the astonishment of many theorists, these had been shown to be nothing more than different perspectives on a single underlying theory—an eleven-dimensional version of supergravity. All the blind men had been seeing the same abstract elephant. Maybe this would be the true theory after all.

But what had emerged was far more complex than anyone had bargained for. Along the way, theorists found that the superstrings were accompanied in the mathematical spaces by even weirder objects called branes, short for "membranes." One might, for example, have a nine-dimensional surface, a nine-brane, flapping in a space of ten dimensions. There was a special kind of brane called a

d-brane (after the nineteenth-century mathematician Peter Diri-
chlet) that acted something like an infinitesimal black hole, chomp-
ing off the ends of strings, defining their end points. Taken together,
this new arena was sometimes called M theory, the M standing for
"Magic," "Mystery," or "Mother of All Theories." Gell-Mann was as
optimistic as ever that this extension of superstrings might be the
approach that would finally lead to unifying gravity with the three
forces in the Standard Model. It seemed like the only viable possibility. (It still wouldn't be a "theory of everything," he reminded me,
for one had to take into account the branching Borgesian labyrinth
of quantum accidents that led to this particular world.) But for all
his enthusiasm, he admitted that much of the mathematics was now
beyond him. He could follow the rough outlines of the arguments,
but he was no longer in a position to help fill in the blanks.

He retained a lingering annoyance at Glashow, who for more
than a decade had been mockingly comparing the ideas about
superstrings to medieval theology. But even the theory's strongest
supporters reluctantly admitted how speculative it still was. String
theory must be right, they believed, because it was so beautiful, and
surely the same was true, deep down, of the universe.

Gell-Mann had been lucky to live in a time when society paid people handsomely to engage in such pursuits, in what was essentially
(though he would probably groan at the term) applied philosophy:
the search for hidden platonic symmetries. Pursuing these abstractions had made him wealthy, able to spend money as though it
were as abundant as neutrinos. But there were signs that the era
was coming to an end. Just before *The Quark and the Jaguar* was
published, Congress scrapped the plans for the Superconducting
Supercollider, which would have been the largest accelerator ever,
measuring some fifty-three miles in circumference. For half a century, the physicists had lived off the glory they earned by delivering,
on schedule, the nuclear bombs that ended World War II. Recently,
it seemed, their political capital had been spent. The public was
no longer willing to accept without question their demands for
another atom smasher. With the end of the cold war, it was harder
to argue that the enormous costs of this purely theoretical work
would be repaid with technological spinoffs. Supporters of the
multibillion-dollar Supercollider had tried to convince Congress by
touting by-products like extremely powerful magnets and tunnel-
digging techniques. But in the end, the true goal—exploring the
higher energies that might lead to unification—proved impossible

to sell. When the project was abandoned, several miles of tunnels had already been excavated beneath the Texas plains. One company inquired about using the ruins for a mushroom farm.

Physicists hold out hopes that the less-powerful Large Hadron Collider under construction at CERN might be energetic enough to find the Higgs particle and maybe even some of the sparticles of superstring theory. And they dream of a day when the political climate changes so that bigger machines can be built. It is even conceivable that a new way will be found to cheaply re-create intense cosmological energies. But it is hard to see beyond the present day when talk of strings and branes and the magical, mystical theory of almost everything is tethered by only the finest of threads to anything resembling the real world. The Standard Model, for all its flaws, might be as good an explanation as we are going to get.

Gell-Mann's own life (he turned seventy in September 1999) also was heading toward a new denouement. In the years after his second marriage, he and his daughter Lisa slowly began to reconcile some of their differences. In 1998 they set off on a backpacking trip together, almost like old times. (The Marxist-Leninist party, to which she had devoted so much of her adulthood, had lasted until 1993, when it disbanded at its Fifth Congress; the final edition of its paper, the *Workers' Advocate*, sadly declared that "We no longer are what we once were" and "no amount of tinkering, adjustments or reorganization can patch things back together again.")

He and Nick were also becoming closer. Murray tried to help him break into filmmaking, and Nick began to develop a deeper appreciation of his father's good points. He told me during a telephone conversation that he had come to realize what an extraordinary opportunity he had been afforded as a child, traveling (even if not always voluntarily) to so many different places. He had absorbed so much just from basking in his father's curiosity. And if Murray wasn't always the most attentive parent, Nick now understood that it was because his father hadn't had much of a boyhood himself, that he didn't really know what it was like to be a child.

Gell-Mann was returning to his roots in other ways. He and Marcia bought a pied-à-terre in a white-brick building in lower Manhattan, on a gentrified street a couple of blocks from where he had been born. Murray had wanted a place on the Upper East Side, Marcia preferred downtown. And she prevailed. At first he hated the idea of going home again. But soon he was enthusiastically showing her around the old neighborhood. He urged her to rejoin the National Arts Club on Gramercy Park so she could get a key to the gates that had kept him out as a boy. Marcia was not someone, like Margaret, who would subdue her own will. And Murray tried to make sure that her career was not completely eclipsed by his own. Late in 1998 they collaborated on a symposium in Santa Fe called "Simplicity and Complexity in the Arts and in the Creative Process," a continuation, in public, of the conversation that had captivated them both on their very first date. How, they asked, was creating a theory like creating a poem? (The symposium was held along with an exhibition of works by a robot-maker and a video artist. Murray and Marcia held a reception for them on Halloween, the day after her birthday. He was dressed as the lord chancellor of England, with a wig of white locks cascading down to his waist—the enforcer of order. She was in her old wedding dress and a black wig: the bride of Frankenstein.) Murray was convinced that the essence of both science and art was discerning regularities in the world. But why were some people drawn to writing, seeing metaphors and similes, while others were drawn to physics, seeing algebraic symmetries? He was so taken by the question that he was even talking about writing another book, enduring the pain again.

He was traveling more than ever: In a little more than a year, he visited Spain, Italy, England, Chile, Switzerland, Venezuela, Sweden, Cuba, and Egypt. The boy who had started bird-watching in the Bronx and Central Park had now seen almost four thousand of the nine thousand species in the world. He still hoped he could pass the halfway mark in the years ahead. He would need to go to Australia again, New Zealand, North Africa. At the very least he hoped to glimpse a member of each family and subfamily, to embrace the whole pattern.

After decades of slights, both real and perceived, Gell-Mann's relationships with many of his former colleagues were more strained than ever. Fred Zachariasen rarely had a good word to say about his old friend. And one New Mexico neighbor unlikely to be found at the Gell-Manns' parties was George Zweig. Though Zweig lived in the dry, rolling countryside just north of Santa Fe, commuting to his job as a theorist at Los Alamos National Laboratory, he and Gell-Mann rarely spoke to each other. This was particularly noticeable since Zweig's wife, Erica Jen, was now vice president for academic affairs at the Santa Fe Institute. When *The Quark and the Jaguar* was published, Zweig glanced at the index, confirming what he suspected: that Murray hadn't mentioned him or aces. Zweig had never been convinced for a moment that Murray believed in quarks until it became expedient for him to do so. Insisting (perhaps a little too strongly) that he wasn't disappointed in the way fame had eluded him, Zweig would tell people that he even felt a little sorry for Murray: With all his equivocations, Gell-Mann had denied himself the intellectual thrill of suddenly stumbling across something so new and profound.

Early in this project, Gell-Mann described our relationship as biographer and subject as a complex adaptive system. We were forming ever-changing schemata of each other, in an uneasy give-and-take. As I was putting the finishing touches on my grand unified theory of Murray Gell-Mann, I was driven back to the old question of nature versus nurture, cast in a slightly different way. How much of who we are is innate, genetic, like local laws of physics, and how much is the result of the accidents of history?

There is a photograph in his library: Little Murray, dressed in a suit and tie, is leaning over a desk and intently drawing with a pencil and straightedge as though he were trying to square the circle. He looks angelic and self-possessed, and maybe a little worried. I've rarely seen a picture that captures someone's essence so perfectly— the hunger for pattern, for connections, and the anxiety over making mistakes, over seeing connections that aren't really there.

Did his drive for wealth begin with that memory of gazing through the locked gates of Gramercy Park? Was his possessiveness about ideas ignited by the falling-out with Abraham Pais? How much of his fear of writing came from Arthur's stern criticisms? And, most important, did his passion for order really arise from his confused childhood, those disjointed scenes (sitting on a pony, on Santa Claus's lap) that didn't quite seem to hang together? Jerry Lyons, the editor who helped him finish *The Quark and the Jaguar,* told me that after a few hours of listening to Murray's seamless disquisitions on every subject under the sun, he almost expected the world to come with annotations attached. Life seemed emptier when Murray left and the narrative suddenly ended. Was it that feeling of emptiness, of being born into an unannotated world, that compelled Murray to constantly fill it with words? Or is that just an interpretation, too neat, imposed in retrospect, what he would call a false schema? The same goes for the memories themselves. In our interviews, when he would recall something from his childhood, he would worry about whether it was a reconstruction, as all memories are to some extent. How do we tell what we remember from what we think we remember?

He captured the dilemma of both science and biography, of life itself, in a line that forms one of the epigraphs to this book:

> In our work we are always between Scylla and Charybdis; we may fail to abstract enough, and miss important physics, or we may abstract too much and end up with fictitious objects in our models turning into real monsters that devour us.

Knowing this full well, we do our best to carry on.

Terms in italics are defined elsewhere in this glossary.

Accelerator. A machine, usually of gargantuan size, used to accelerate particles to high speeds and crash them into targets or into one another. The debris is then studied for clues about nature.

Antimatter. The oppositely charged counterpart of ordinary matter. Every particle is associated with an antiparticle, and vice versa.

Associated production. A process causing *strange particles* to be created only in pairs, allowing them to resist immediate disintegration by the *strong force*.

Asymptotic freedom. A paradoxical phenomenon whereby the *strong force* grows weaker at shorter distances and stronger at longer distances.

Baryon. A heavy particle like the *proton* or *neutron* that feels the *strong force*. A baryon is made of three *quarks*. See *hadrons*. Contrast with *mesons*.

Beta decay. A form of radioactivity in which a *neutron* disintegrates into a *proton*, an *electron*, and an antineutrino, the *antimatter* equivalent of the *neutrino*.

Bootstrap. An outgrowth of *S-matrix* theory developed as an alternative to *quantum field theory* for describing the *strong force*. According to the bootstrap, there are no fundamental particles; instead the *hadrons* are each made from one another like nodes in a fishing net.

Bosons. Particles whose *spin* is measured in whole numbers. They include the *gauge bosons*, i.e., particles acting as the carriers of forces: the *photon* (for the *electromagnetic force*), the gluon (for the *strong force*) and the W^+, W^-, and Z^0 particles (for the *weak force*).

Bottomonium. The combination of a bottom *quark* and an antibottom *quark*.

Broken symmetry. A *symmetry* of physics that becomes flawed, or broken, when manifested in the real world. For example, it is believed that in the beginning there existed a single, perfectly symmetrical superforce, which subsequently split into the four very different forces of nature that appear today.

Bubble chamber. A device used to detect particles created by *accelerators* by means of the trail of bubbles they create in a near-boiling liquid.

Charm. A characteristic, like *strangeness*, used to describe certain *quarks*. See *flavor*.

Charmonium. The combination of a charmed and an anticharmed *quark*.

Cloud chamber. A device used to detect particles by means of the condensation trail they leave in a fog-filled container.

Color. A quantity, somewhat analogous to electric charge, through which the *strong force* wields its influence. It comes in three varieties, usually called red, green, and blue.

Conservation law. A *symmetry* in which a quantity like energy, *spin,* or *color* remains unchanged.

Cosmic rays. Particles that shower the atmosphere from outer space.

CP conservation. A *symmetry* in which particles in a reaction are exchanged with their *antimatter* partners and then reflected in the *parity* mirror. If the outcome is identical, then CP symmetry is conserved.

CPT conservation. A *symmetry* in which particles in a reaction are exchanged with their *antimatter* partners, reflected in the *parity* mirror, and reversed in time. If the outcome is identical, then CPT symmetry is conserved.

Cross section. The probability that a particle is produced in a certain reaction.

Current. The path through which one kind of particle is converted into another. Equivalent to a rotation in *group theory.*

Current algebra. A mathematical system, equivalent to a *group,* describing the relationships among a family of particles.

Decuplet. A version of $SU(3)$ with ten members. Sometimes called a decimet.

Deep inelastic scattering. Experiments in which *electrons* were fired at high energies at *protons* ricocheting in a way that indicated *quarks* were inside.

Dispersion theory. An outgrowth of *S-matrix* theory developed as an alternative or supplement to *quantum field theory* for describing the *strong force.* Unlike *perturbation theory,* it has the goal of giving precise, not approximate, answers.

Eightfold Way. A classification scheme for organizing the plethora of *baryons* and *mesons* into closely related families called *multiplets.*

Electromagnetic force. The long-range force—light, radio waves, x-rays, and *gamma rays* are examples—that arises from the interaction of electricity and magnetism.

Electrons. Negatively charged lightweight particles belonging to the families of *leptons* and *fermions.*

Electroweak force. A unified force combining the *electromagnetic* and the *weak force.*

Fermions. Particles whose *spin* is measured in half units. They are generally thought of as the building blocks of matter: *electrons, protons, neutrons, quarks,* etc. Contrast with *bosons.*

Feynman diagram. A vivid pictorial method for describing and calculating the various ways that particles can interact.

Flavor. A variety of *quark:* up, down, strange, charmed, bottom, or top.

Gamma rays. High-frequency electromagnetic radiation that appears when a particle and its antiparticle collide. See *antimatter.*

Gauge boson. See *boson.*

Gauge field theory. Same as *Yang-Mills theory.* A special kind of *quantum field theory* in which forces are carried by particles called *gauge bosons* and serve to enforce a phenomenon called *gauge symmetry.*

Gauge symmetry. A *symmetry* in which a quantity remains constant no matter what scale or yardstick is used to measure it.

General relativity. Einstein's theory for describing the behavior of systems moving past one another at accelerating velocities. It forms the framework of the modern theory of gravity as curved space-time. See *special relativity.*

Ghosts. Hypothetical particles, predicted by equations, that have nonsensical properties like negative energies or even negative probabilities of existing. See *never-never land.*

Gluons. The carriers (called *gauge bosons*) of the *strong force.* They hold *quarks* together to form *hadrons* and bind *nucleons* to form the nuclei of atoms.

Grand unified theory. An attempt to go beyond the *Standard Model* and show that the *strong force* and the *electroweak force* are not just related but can be unified into a single phenomenon. See *unification.*

Gravitons. The hypothetical carriers of gravity.

Group theory. See *Lie group.*

Hadrons. Collective term for all particles (both *baryons* and *mesons*) that feel the *strong force.* Called "stronglies" by Murray Gell-Mann and Richard Feynman. *Hadrons* are made of *quarks.* Contrast with *leptons.*

Higgs boson. Hypothetical particle that interacts with other particles to give them mass.

Hyperons. *Strange particles* consisting of heavy *fermions.*

Isospin. A *quantum number* describing, for example, *protons* and *neutrons* as opposite manifestations of a single particle called a *nucleon.* In the scheme, a *nucleon* with a "clockwise" isospin is a *neutron;* a "counterclockwise" isospin yields a *proton.*

Kaons. *Strange particles,* called K's for short, consisting of middleweight *bosons.*

Leptons. Light particles that do not feel the *strong force:* the *electron,* the *neutrino,* and their cousins the *muon* and mu-neutrino and the *tau* and tau-neutrino. Leptons are not made from *quarks.* Contrast with *hadrons.*

Lie group. A mathematical system like the one used to describe the *Eightfold Way,* in which members of a family of particles can be rotated in mathematical space to convert one into the other.

Mesons. Middleweight particles that feel the *strong force.* Mesons are made from a *quark* and an antiquark. See *hadron.* Contrast with *baryon.*

Multiplet. A family of particles related by some *quantum number.* The *neutron* and *proton,* for example, form a two-member family, or doublet, related by their opposing *isospins.* A three-member family is a triplet, a single-member "family" a singlet.

Muons. Short for mu-mesons. A heavy version of the *electron,* mistakenly thought for years to be the *pion.* Just as the *electron* is associated with the *neutrino,* the muon is associated with the mu-neutrino.

Neutrinos. Chargeless particles with little or no mass that belong to the families of *leptons* and *fermions.*

Neutrons. Chargeless heavy particles belonging to the families of *baryons* and *fermions.*

Never-never land. A mathematical realm where unphysical quantities like negative mass and negative probabilities show up in the equations. Also called "physics off the mass shell." See *ghosts.*

Nucleons. Collective term for *protons* and *neutrons,* the components of the atomic nucleus.

Omega minus. A particle predicted by the *Eightfold Way.* Its discovery convinced many physicists that the theory was correct.

Parastatistics. A scheme proposed for explaining how *quarks* could be brought into compliance with the *Pauli exclusion principle.*

Parity. A *quantum number* intimately related to reflection symmetry, in which a reaction and its "mirror image" are identical.

Partons. Tiny particles proposed by Feynman as the building blocks of *hadrons.* Partons turned out to be another name for *quarks* and *gluons.*

Pauli exclusion principle. The dictum that no two *fermions* can be in the same state, with the same quantum numbers.

Perturbation expansion. An equation in which each term represents a *Feynman diagram* of the increasingly arcane ways that the particles in question can conceivably interact.

Perturbation theory. A scheme for making approximate calculations using a *perturbation expansion.*

Photons. The carriers (called *gauge bosons*) of the *electromagnetic force.*

Pions. Short for pi-mesons, particles believed for many years to be the carrier of the *strong force.* See *gluons.*

Positronium. The combination of a *positron* and an *electron*.

Positrons. Positively charged *antimatter* twins of the *electron*.

Protons. Positively charged heavy particles belonging to the families of *baryons* and *fermions*.

Quanta. Plural for "quantum." Often used to refer to the packets of energy that are the currency of *quantum mechanics*. *Photons*, for example, are the quanta of the *electromagnetic force*. See *bosons*.

Quantum chromodynamics (QCD). The *quantum field theory* describing the *strong force* as being carried by *quanta* (specifically *gauge bosons*) called *gluons*.

Quantum electrodynamics (QED). The *quantum field theory* describing the *electromagnetic force* as being carried by *quanta* (specifically *gauge bosons*) called *photons*.

Quantum field theory. A mathematical framework in which forces are carried by *virtual particles*, called *quanta*. See *bosons*.

Quantum mechanics or **quantum theory.** The theory, underlying all of modern particle physics, which holds that quantities like energy and *spin* are not continuous but come in discrete packets called *quanta*.

Quantum number. A characteristic, like charge, mass, or *spin*, used to describe particles. Subsequently, more abstract quantum numbers like *isospin, strangeness*, and *charm* were added to the list. Quantum numbers are conserved in conservation laws.

Quarks. Particles that come in six *flavors* and three *colors*, and combine to form all of the *hadrons*.

Regge theory. A scheme for classifying particles. If *hadrons* are plotted on a graph with energy on one axis and *spin* on the other, they align themselves into patterns called Regge trajectories. The points on the trajectories where particles appear are called Regge poles.

Renormalization. A mathematical technique for eliminating infinite quantities that appear in the equations of *quantum field theory*, thereby producing sensible results out of otherwise meaningless calculations.

Renormalization group. Mathematical device used to describe how the strength of a force varies with distance.

Representation. One of the different forms or manifestations that a *Lie group* can assume. For example, *SU(3)*, which describes the *Eightfold Way*, can be represented as a group with three, eight, ten, or twenty-seven members.

Resonances. Particles whose lifetimes are so fleeting that they don't leave tracks in particle detectors. Their existence must be inferred from their decay products.

S-matrix. Short for "scattering matrix," a theoretical framework offered as an alternative to *quantum field theory* for describing the *strong force*. See *bootstrap*.

Self-energy. A phenomenon through which a charged particle interacts with itself, leading to an endless regress that would theoretically result in charge and mass becoming infinite. These infinities are removed by *renormalization*.

Special relativity. Einstein's theory for describing the behavior of systems moving past one another at a constant velocity. See *general relativity*.

Spin. A quantum-mechanical analogue of rotation. A particle that spins clockwise is said to have down spin; counterclockwise, up spin.

Standard Model. The *quantum field theory* that combines the *electroweak force* and the *strong force* into a single framework without completely unifying them.

Strange particles. Particles first found in *cosmic rays* that have unexpectedly long lifetimes.

Strangeness. A *quantum number* independently invented by both Gell-Mann and Kazuhiko Nishijima to account for the unexpected longevity of *strange particles*.

String theory. See *superstring theory*.

Strong force. The short-range force that binds *quarks* together to form *hadrons* and holds the *nucleons* together to form the nucleus. See *gluons*.

Sum over histories. The method developed by Feynman for calculating how a particle will behave by taking into account all the possible trajectories it could follow (its "histories"), even those in which it would move backward in time.

SU(2) and SU(3). "Special unitary" groups in two and three dimensions. SU(2) describes the *isospin* symmetry: the ways a nucleon, for example, can be rotated to get a *neutron* or *proton*. SU(3) describes the eight ways different *hadrons* in the *Eightfold Way* can be rotated to convert one into the other.

Supergravity. A rival to *superstring theory* as a method for fully unifying all four forces of nature into one. See *unification.*

Superstring theory. The theory that all four forces can be unified by assuming that particles are made from two-dimensional entities called superstrings vibrating in a multidimensional superspace. See *unification.*

Supersymmetry. A theoretical device underlying *superstring theory* that unifies *bosons* and *fermions.* In the scheme, every particle has a supersymmetric partner called a "sparticle": *quarks* are paired with squarks, *electrons* with selectrons, *photons* with photinos, *gluons* with gluinos, and so forth.

SVAPT (scalar, vector, axial-vector, pseudoscalar, and tensor). Mathematical terms used to describe the different kinds of interactions that convert one particle into another.

Symmetry. Something that remains constant while something else changes, for example the circular shadow of a rotating ball.

Tau. A *lepton* heavier than either the *electron* or *muon,* and associated with the tau-neutrino. (Not to be confused with the tau in the *tau-theta puzzle.*)

Tau-theta puzzle. A mystery in which two otherwise identical *kaons,* the tau and theta, seemed to differ in *parity.* It was concluded that they were actually the same particle, and that *parity* was not conserved in interactions involving the *weak force.*

U(1). This stands for "unitary (one-dimensional group)," the name of the *Lie group* describing electromagnetism.

Unification. The act of showing mathematically that two or more seemingly unrelated phenomena are just different manifestations of the same thing. The *electromagnetic* and the *weak force,* for example, have been shown to be the two faces of a unified *electroweak force.*

Universal Fermi Interaction. A unified theory accounting for all the various types of *weak force* interactions. Called "UFI" or "oofi" for short.

Uxyls. See *W particles.*

V-A. The theory describing the *weak force* as the intertwining of two *currents* called vector and axial-vector. See *SVAPT.*

Veal and pheasant. A technique used by Gell-Mann for abstracting truths from *quantum field theory* while remaining agnostic about whether it was a valid framework for describing the *strong force.* He compared it to cooking pheasant between slices of veal and then discarding the veal.

Virtual particles. Particles that flit in and out of existence just long enough to act as carriers of forces. See *quantum field theory.*

W particles. Positively and negatively charged particles (called *quanta* or *gauge bosons*) that, along with the neutral *Z particles,* carry the weak force. Gell-Mann and Feynman called them *X particles* or *uxyls.*

Weak force. The short-range force responsible for various kinds of radioactivity.

X particles. See *W particles.*

Yang-Mills theory. See *gauge field theory.*

Z particles. Neutral particles (called *quanta* or *gauge bosons*) that, along with the *W particles,* carry the *weak force.*

SOURCES AND ACKNOWLEDGMENTS

In trying to place Murray Gell-Mann in the history of twentieth-century physics, I found several sources so vital that they deserve special mention. I constantly referred to well-worn copies of Robert Crease and Charles Mann's *Second Creation,* James Gleick's *Genius,* Abraham Pais's *Inward Bound,* Jeremy Bernstein's *The Tenth Dimension,* Michael Riordan's *The Hunting of the Quark,* and Andrew Pickering's *Constructing Quarks.* John Polkinghorne's *Rochester Roundabout* provided fine summaries of the Rochester conferences, important milestones in the development of particle physics. I can barely imagine having gone these five years without the three volumes of physicists' reminiscences that emerged from the historical symposiums organized by Laurie Brown, Lillian Hoddeson, Max Dresden, and Michael Riordan: *The Birth of Particle Physics, Pions to Quarks,* and *The Rise of the Standard Model.*

Gell-Mann disputes many of these published accounts. It has been a challenge to triangulate among the various tales and recreate what I believe really happened. (Sometimes, to be safe, I have second-guessed myself in the Notes.) Sorting out conflicting claims sent me beyond the historical distillations to the raw material of physics: the papers published in the *Physical Review* and other journals, and the proceedings of various meetings, particularly the Rochester conferences and the physics summer schools held at the Ettore Majorana Centre in Erice, Sicily.

I have fleshed out these published documents by interviewing, and in a few cases corresponding (usually by E-mail), with more than a hundred people who have known Gell-Mann, from the time he was born on 14th Street in Manhattan until his present life in Santa Fe—relatives, childhood acquaintances, colleagues, journalists, friends, and an occasional enemy. (These sources include a number of people at the Santa Fe Institute who have conversed with me about Gell-Mann since I moved to New Mexico in 1992.) I would like to thank Bruce Abell, Philip Anderson, Brian Arthur, David Benninghoff, Jeremy Bernstein, Hans Bethe, Richard Brandt, Lars Brink, John Brockman, Kevin Brook, Laurie Brown, Keith Brueckner, John Casti, Geoffrey Chew, Michael Cohen, Sidney Coleman, George Cowan, Jim Crutchfield, Alfred Davis, Lincoln Day, Charles Desoer, Joe Devaney, Freeman Dyson, Doyne Farmer, Glennys Farrar, Florence Freint, Harald Fritzsch, Jim Furman, Ben Gelman, Nick Gell-Mann, Sheldon Glashow, Ellen Goldberg, Michael Goldhaber, Marvin Goldberger, Gene Goldfarb, Robert Gomer, George Gumerman, James Hartle, Roger Hildebrand, Lillian Hoddeson, Justin Israel, David Jackson, Stuart Kauffman, Diane Lams, Christopher Langton, Leon Lederman, Gloria Marie Campbell Lent, Roger Lewin, Rick Lipkin, Seth Lloyd, Francis Low, Celia Lowenstein, Jeremiah Lyons, Judy Macready, Paul Macready, Dan Martin, Lydia Matthews, Jay Meili, Leslie Meredith, Sydney Meshkov, John Miller, Melanie Mitchell, Harold Morowitz, John Morris, Yoichiro Nambu, Yuval Ne'eman, Richard Norton, Abraham Pais, David Pines, John Polkinghorne, Pierre Ramond, George Rathjens, Michael Riordan, Bud Rosenbaum, Robert Rosenbaum, Arthur Rosenfeld, Jack Ruina, Edwin Salpeter, John Schwarz, Abner Shimony, Mike Simmons, Margaret Low Smith, Marcia Southwick, Norman Spencer, Arthur Sze, Valentine Telegdi, Joseph Traub, Helen Tuck, Ann Walker, Harry Walker, Israel Walker, Robert Walker, Steven Weinberg, Robert Weinman, Victor Weisskopf, Geoffrey West, Frank Wilczek, John Noble Wilford, Kenneth Wilson, Frank Yang, Joel Yancey, Fred Zachariasen, Wojciech Zurek, George Zweig, and Barton Zwiebach.

Gell-Mann himself generously agreed to what eventually amounted to more than twelve hours of taped interviews. He also allowed me unrestricted access to his personal archives, a garageful of letters and documents that go back to the beginning of his career. I hope he will not be too disappointed that the story I have finally told does not always accord with his memories.

Several of my fellow journalists generously provided tapes or transcripts of interviews with Gell-Mann: Robert Crease and Charles Mann (conducted in the 1980s for *Second Creation*), Ronald M. Schultz (for an *Omni* interview published in May 1985), John Horgan (for a *Scientific American* profile published in March 1992), and David Berreby (for a *New York Times Magazine* profile published

May 8, 1994). Alun Anderson and Liz Else of *New Scientist* gave me a transcript of an interview done in 1994 for a story that didn't materialize. These records provided important views of how Gell-Mann appeared to various people at different times of his life, and gave me further glimpses of his always prickly relationships with science writers. Toward the end of the project, I benefited from oral history interviews with Gell-Mann by Joel Gardner for the MacArthur Foundation and Geoffrey West for the Science Archive.

The American Institute of Physics oral histories with Richard Feynman, Robert Marshak, and other physicists were also valuable, and the AIP's Emilio Segré Visual Archives were a godsend. Spencer Weart, Michele Blakeslee, Jack Scott and the rest of the staff were extremely helpful. Judith Goodstein and Shelley Erwin at the Caltech archives guided me through a number of important resources, including Richard Feynman's papers and films of Gell-Mann lecturing in the late 1960s. A trip to the archives of Columbia Grammar School in Manhattan gave me a vivid glimpse of Gell-Mann as a schoolboy; thanks to Patricia Levin and Sara Ziff for their assistance there. The Los Alamos National Laboratory Research Library, with its open stacks of journals, was indispensable.

I was fortunate to be invited to Stockholm in 1997 to speak to the Association of Swedish Science Journalists and attend the Nobel Prize ceremonies and banquet, getting a glimpse of what Gell-Mann had experienced twenty-eight years earlier. Many thanks to Joanna Rose for making this possible. In Stockholm, Marie Granmar helped me track down important documents about Gell-Mann's Nobel Prize, as did Christina Tillfors of the Nobel Foundation. Thanks also go to Susan Ballati and other members of the staff at the Santa Fe Institute, especially Gell-Mann's assistant, Marla Karmesin, for their help and encouragement during this long project.

I would also like to thank Steve Bello for making available the excellent photos his father, Francis Bello, former science editor of *Fortune,* took of Murray Gell-Mann, Julian Schwinger, Freeman Dyson, and Geoffrey Chew. Eight of Bello's photos of physicists are in the permanent collection of the National Portrait Gallery in Washington.

Several people generously read the manuscript while it was in preparation, helping me to fine-tune it. To help me get the physics right, James Amundson, Jeremy Bernstein, James A. Carr, Seth Lloyd, Michael Riordan, and Geoffrey West spent many hours with the book. And to help ensure that the explanations would be clear to a general audience, James R. Carr, J. E. Johnson, Douglas Maret, and Nancy Maret read and commented on various versions. The book wouldn't be the same without them. At Knopf, I would like to thank my editor, Jonathan Segal, his assistant, Ida Giragossian, the production editor Melvin Rosenthal, and the designer Anthea Lingeman.

Finally I would like to thank my readers. I enjoy hearing from them and can be reached through my pages on the World Wide Web (talaya.net), where I will post any necessary updates or corrections; by E-mail (johnson@nytimes.com or johnson@santafe.edu); or in care of my American publisher, Alfred A. Knopf.

Complete citations to books can be found in the Bibliography. Otherwise sources are identified fully the first time they occur in the notes to each chapter. The Bibliography does not, however, list each separate volume of the proceedings for the international meetings of particle physicists informally known as the Rochester conferences. There were no proceedings for the first conference and only mimeographed proceedings for the second (*Meson Physics,* edited by A. M. L. Messiah and H. P. Noyes). Proceedings for Rochester 3 through 7 each appeared under the title *High Energy Nuclear Physics,* with various editors and publishers. After that the proceedings are called just *High Energy Physics.*

Unless otherwise stated, all interviews were conducted by the author. "MGM" stands for Murray Gell-Mann.

page

vii **first epigraph:** from MGM and E. P. Rosenbaum, "Elementary Particles," *Scientific American,* July 1957.
second epigraph: Gell-Mann, "Quarks," *Acta Physica Austriaca,* Supplement 9 (1972): 760.

PROLOGUE *On the Trail to La Vega*

5 **MGM's conflicts over the reality of quarks:** Documentation for this and other observations made in the Prologue can be found later in the book where the issues are explored in detail.

10 **His legendary linguistic skills:** Almost everyone who has encountered Gell-Mann has a favorite story. Crease and Mann recount a Chinese restaurant incident in *Second Creation,* p. 307; Low related the Chinatown incident (interview 3/8/94), Zachariasen the one about Kathmandu (4/1/94).
Gell-Mann cancels his talk at *Nature* **centenary:** telegram from MGM to John Maddox dated 10/28/69; **insists that the timing was a coincidence and the illness was real:** letter from MGM to Maddox dated 11/6/69.
MGM fails to submit his Nobel lecture: The saga is documented in ten telegrams sent back and forth between MGM and the Nobel Foundation, with dates ranging from 1/19/70 to 6/5/70. They are cited more fully in the notes to chapter 13.

10 **with abject apologies:** This is documented in the correspondence mentioned above.

11 **MGM gives advice:** letter to a boy in Medford, Oregon, dated 9/14/60; letter to a woman in Klamath Falls, Oregon, dated 7/17/56. (The two names are withheld here out of respect for the recipients' privacy.)

recommendations for Sudarshan: 8/13/63 to Brandeis University; 1/25/61 to the University of Rochester. **MGM's recommendations for Zweig:** 1/7/63 to the National Academy of Sciences; 11/18/65 to the Sloan Foundation.

"seminal work" on quark model, etc.: telegram from MGM to A. Zichichi, Instituo di Fisica dell' Universita, Bologna, Italy, dated 1/9/70.

MGM meant to write about the mistake in the *Times*: A photostat of the editorial, 5/1/71, is in his files; on it is a note in his handwriting ("Write about Zweig"), then another, apparently added later: "No. May 1! Too late!" No letter from him was found in a search of the *Times* morgue.

"I'm getting to be as bad": letter from MGM to S. M. Berman, CERN, dated 12/15/60 (answering a letter of 10/10/60).

12 **Gell-Mann blows up at a secretary:** anonymous interview with her, 7/21/98. The conference where I met Gell-Mann is documented in Cowan et al., eds., *Complexity: Metaphors, Models, and Reality.*

11–12 Wilford's *New York Times* article on the fifth force was published 1/8/86. He gave more details about the encounter in an E-mail, 11/4/98.

15 **"I don't know any *funny* stories":** interview with Zachariasen, 4/1/94.

CHAPTER 1 *A Hyphenated American*

19 **MGM called quarks "red," "white," and "blue":** for example, in a lecture he gave in Schladming, Austria, in 1972: "Quarks," *Acta Physica Austriaca*, Supplement 9 (1972): 736.

The term "color" was commonly (and facetiously) used at Caltech by both Feynman and MGM in many different contexts. In the late 1950s, for example, MGM referred to red and blue neutrinos. In *Second Creation*, p. 156, Crease and Mann say the tradition began at Caltech in the 1930s, when experimenters Carl Anderson and Seth Neddermeyer speculated that cosmic rays consisted of "red and green" electrons, impinging on the earth with different energies.

The term "flavor" is often attributed to Yoichiro Nambu; "charm" was coined by Sheldon Glashow.

MGM describes parents as "immigrants from Austria": in, for example, *Current Biography* (1966), p. 124; a *New York Times* profile, 31 October, 1969, p. 20; and his autobiographical "Personal Statement," p. 2, distributed by his office in the early 1990s.

"the son of a forester, living in the beech woods": MGM, *Quark and Jaguar,* p. 340.

20 **MGM and young woman at party:** interview with Glashow, 3/8/94.

Lederman pretends to hyphenate his own name: Lederman, *The God Particle,* p. 299.

21 Details of Isidore Gellmann's life in Galicia and Bukovina were compiled from an interview with MGM, 5/8/95, with follow-up conversations on 7/19/95, 3/6/96, and 4/8/96, and from birth certificates, letters, and other personal documents supplied by his office. A few very sketchy details are in MGM, *Quark and Jaguar,* p. 340. For descriptions of Bukovina, I also relied on Chalfen, *Paul Celan: A Biography of His Youth*, pp. 3–19; Goldsmith, *Architects of Yiddishism*, p. 185; and Stenberg, *Journey to Oblivion*, pp. 42–67, as well as sev-

eral recent articles about Czernowitz: Peter Pulzer, "Return to Little Vienna," *Times Literary Supplement,* 23 September, 1994; Dan Petreanu, "A Pale Memory," the *Jerusalem Report,* 3 June 1993; Henry Kamm, "Lvov Journal: The Survivors/Grand Buildings and the Ghosts," *New York Times,* 7 May 1987; Henry Kamm, "Jews' Fate Unmarked in Soviet-Ruled Region," *New York Times,* 11 May 1987; and Henry Kamm, "Chernovtsy [Czernowitz] Journal; A Garden of Yiddish, with a Soil That Bred Poets," *New York Times,* 25 February 1992. Thanks to Iosif Vaisman for pointing me to many of these sources.

22 **Origin of Gellmann and Gelman:** Patrick Hanks and Flavia Hodges, *A Dictionary of Surnames* (New York: Oxford University Press, 1989).

"I seldom went into a Jewish home": The quotes from *Der Pojaz* are in Stenberg, *Journey to Oblivion.*

23 **"The bourgeoisie spoke German and were driven to synagogue":** *New York Times* article by Kamm, 25 February 1992.

24 **Isidore's denial that the family fled persecution:** MGM interview, 5/8/95.

25 Descriptions of the Viennese intellectual milieu were inspired by Janik and Toulmin, *Wittgenstein's Vienna,* and Beller, *Vienna and the Jews.* Details of Isidore Gellmann's life in Vienna are from an interview with MGM, 5/8/95.

25–6 **MGM is irritated by his father's talk of monads:** from an unpublished interview with MGM by Crease and Mann for *Second Creation.*

26 Isidore's letter home: from MGM's personal files. It was written in German, with blue ink, in a meticulous script, with every line perfectly level and spaced. It is dated 1911. The return address is "Isidor Gellmann, Vienna V, [street address illegible]." It is addressed to "M. Gellmann, 310 E. 4th, New York, Amerika."

Moses and Celia's problems in the U.S.: MGM interview, 5/8/95.

Isidore Gellmann's immigration had to be after 1911 since that is the date of his letter to New York from Vienna. I searched the *Index to Passenger Lists of Vessels Arriving at New York, N.Y.,* July 1902–1943, but found no Arthur Gell-Mann (or any of the Soundex variants).

Isidore adopts "Arthur" as his first name: MGM speculated in an interview (5/8/95) that his father might have started using the new name at Ellis Island.

the job at the Hebrew Orphan Asylum: The details are from a reference letter for Arthur Gell-Mann from Dr. A. Stern, Superintendent, Hebrew Orphan Asylum, Philadelphia (Green Lane, Germantown), dated 8/12/20. MGM believed the orphanage was Arthur's first job (interview 5/8/95), but the letter of recommendation implies that he started there in 1916 or 1917.

Details of the Reichstein and Fellner family history are from MGM (interview 5/8/95); "Genealogical Information on the Descendants of Moses Fellner," an unpublished manuscript by his great-great-grandson Kevin Brook (2/17/96); and an interview (7/18/95) with another Fellner descendant, Norman Spencer.

Details about the job at Raabe, Glissmann are from a letter of reference dated 5/3/22 for Arthur Gell-Mann from A. Raabe and M. Frieden, Raabe, Glissmann & Company, Inc., 20 Broad Street, New York, N.Y.

Arthur and Pauline's secret marriage: MGM interview, 5/8/95.

Arthur starts using the hyphen: According to Gell-Mann family lore, the hyphen was carried over from Europe. However, there is no written evidence of this, and his father's 1911 letter uses the spelling "Gellmann" (see earlier reference). There is, however, some precedent for Bukovinian Jews to hyphenate their names. In *Paul Celan: A Biography of His Youth,* pp. 26–27, Israel Chalfen says that Celan's father, born in 1890 in a village north of Czernowitz, used the surname Antschel-Teitler, a combination of his parents' family names. MGM thought he remembered that the wedding announcement used the hyphen (5/8/95 interview), but he was unable to find a copy.

28 **Old friends and relatives speak of the "Gelmans":** interviews with the Gell-Mann's cousins, Israel Walker (1/23/95 and 1/25/95), his daughter Ann Walker (1/25/95), and his brother Harry Walker (1/25/95), and with MGM's high school classmate, Justin Israel (3/21/94).

The brokerage company goes bankrupt: mentioned in the Raabe, Glissman letter of reference.

Arthur's job at a toy company: MGM interview, 5/8/95.

the Gell-Manns' various addresses: from the New York City telephone directories and interviews with Ben Gelman (3/6/95) and MGM (5/8/95 and 11/3/98). It has been difficult to piece together the precise chronology, so I have kept it a bit vague. Ben's birth certificate shows he was born in Brooklyn, and he vaguely remembers hearing that the family lived on 27th Street when the language school first began, on 15th Street before living on St. Mark's Place, and then on 14th Street and 15th Street again. The first telephone listing I found for Arthur Gell-Mann was November 1920 at 233 East 14th Street. But from autumn 1924 to winter 1927–28, he is listed at 225 East 14th Street. The classified listings for the late twenties show the Arthur Gell-Mann School at 233 East 14th Street. And in the mid-twenties, there is an unidentified S. Gellmann on the same block, at 216 East 15th Street and then 227 East 14th Street. Adding to the confusion, 216 East 15th Street may have been the address where the family moved in 1931 after Murray was born.

Pauline's restlessness: interview with MGM, 3/6/96.

29 **"When a vowel is followed by 'rr' ":** from Arthur Gell-Mann's unpublished syllabus, in MGM's files. Though I am assuming he wrote this himself, it is possible that he copied it from a published book, as he had his pharmacology text.

30 **description of Stuyvesant Square:** from personal visits and interviews with Ben Gelman (3/6/95) and MGM (5/8/95).

MGM's first memories: interviews with MGM, 5/8/95 and 11/2/98.

"The lights of Babylon": Ben Gelman told this story in an article in the Southern Illinois University *Alumnus* magazine: "My Brother Murray," Fall 1993, p. 20.

31 **story of the eclipse:** MGM interview, 5/8/95.

feelings of confusion: MGM interview (11/2/98); unpublished MGM interview by Alun Anderson and Liz Else for *New Scientist,* summer 1994.

collapse of the language school: MGM interview (5/8/95); MGM, "Personal Statement," p. 2; MGM, *Quark and Jaguar,* p. 15.

life in the Bronx: MGM interview, 5/8/95.

32 **Arthur's mathematical diversions:** MGM interview, 5/8/95. The notebooks are in MGM's files.

33 **complaints about Arthur as a German teacher:** unpublished interviews with MGM by David Berreby for "The Man Who Knows Everything," *New York Times Magazine,* 8 May 1994.

Pauline's mental state: Berreby interviews.

"I seem to have inherited a mixture of both": MGM, "Personal Statement," p. 2.

exploring New York with Ben: MGM, "Personal Statement," p. 2; MGM, *Quark and Jaguar,* p. 12.

34 **Details about the Walker family and other friends:** interview with MGM, 5/8/95, and with the Walkers, cited above.

playing chess with Arthur: interview with Israel Walker, 1/23/95.

35 **Roman coin story:** interviews with Israel Walker (1/23/95) and MGM (5/8/95).

MGM's mathematical skills: Ben Gelman, "My Brother Murray," pp. 20–21.

the spelling bee: Gelman article and letters from Florence Freint to Diane Lams, cited below.

"I knew that!": Recollections of Murray at P.S. 44 are from his classmate, Gloria Marie Campbell Lent (personal communication 9/3/98).

35–6 **Freint's friendship with MGM:** interview (1/18/95) and letters from her to MGM (May or June 1994) and his assistant, Diane Lams (July and August 1994). Freint also sent a photograph of herself as a teacher at the settlement house.

36 **Arthur no help in getting MGM into private school:** MGM interview (5/8/95); Berreby interviews.

36–7 **getting MGM into private school:** Walker interviews; Freint interview and letters.

37 **"School days, school days":** Freint letter to Lams, 1994; interview with Freint. (She couldn't remember what a "uni-project pal" was.)

skipping grades: MGM interview (5/8/95); Berreby interviews.

CHAPTER 2   *The Walking Encyclopedia*

38 Details of life at Columbia Grammar come from the *Columbiana* yearbooks, the *Columbia News* (billed as "Oldest School Newspaper in America"), *The Columbia Grammar Bulletin of the Parents and Teachers Association,* and newsletters and brochures from the school archives. I also interviewed some of MGM's classmates: Jeremy Bernstein (2/21/94), Alfred Davis (3/2/94), Gene Goldfarb (3/23/94), Justin Israel (3/21/94), and John Morris (3/22/94). MGM also recalled the school in interviews (5/8/95 and 4/8/96).

39 The Columbus Avenue El was discontinued in 1940.

"dangerous riffraff": Bernstein, *The Life It Brings,* p. 19.

The limericks are reprinted from *Columbia News Magazine,* 1943, p. 11.

39–40 Beene anecdotes are from *Columbia News,* 16 October 1942, and 20 November 1942.

40 **Murray correcting teachers:** Justin Israel interview.

"pet genius": Israel interview.

"C.G.S.'s own small-scale model of Einstein": *Columbia News,* 2 October 1942, p. 2.

"What, in your opinion, is the hardest course": *Columbia News Magazine,* 1943, pp. 12–13.

MGM came home after being roughed up: Ben Gelman, "My Brother Murray" (cited in notes to chapter 1), p. 21.

40–1 **Columbia Grammar at war:** described in various issues of *Columbia News* and *Columbia News Magazine.*

41 **MGM would scurry home:** Israel interview.

Arthur's strictness: MGM interview (5/8/95); Berreby interviews (cited in notes to chapter 1).

"Events Around Town," "waxinations off the griddle," "Platter Chatter": *Columbia News,* 22 December 1942.

Arthur didn't believe they were at the right house: MGM interview, 11/2/98.

42 **MGM's uneasy relationships with wealthier friends:** interviews with Israel and other classmates, cited above.

Ben graduated early: MGM, "Personal Statement," p. 3; MGM interview (4/8/96).

visiting the museums with Ben: MGM, "Personal Statement" (cited in notes to chapter 1), p. 2; MGM, *Quark and Jaguar,* p. 12.

42 **Ben buys** *Finnegans Wake:* Jeremy Bernstein, "The Quark," *New Yorker,* 18 July 1977, pp. 22–23.

The Walkers were worried about Ben: Israel Walker interviews 1/23/95 and 1/25/95.

44 **"In a shower of votes":** school yearbook, 1944, p. 28; class prophecy, pp. 29–30.

MGM's utopian ideas: MGM, "Personal Statement," p. 3; MGM, *Quark and Jaguar,* p. 14; MGM interview (4/8/96).

the daring valedictory address: Israel interview. No copy was found in the Columbia Grammar archives. The *New Yorker* cartoon appeared 6/24/44 on p. 25.

"radical, left-wing stuff about world government": MGM interview, 5/8/95.

Normandy invasion memory: MGM interview, 4/8/96. Freint said she didn't remember the incident (interview 1/18/95).

45 **argument with Arthur over college major:** MGM, *Quark and Jaguar,* pp. 15–16; MGM, "Personal Statement," p. 3; Berreby interviews. MGM told this story as early as 1967 in a television profile, "Particles Are a Family Affair," for the NET series *Spectrum.* A tape of the interview, with David Prowitt, is in the Caltech archives.

tediousness of high school physics: Merle Rubin, "A Physicist of Many Parts: Interview with Murray Gell-Mann," *Christian Science Monitor,* 28 February 1989.

46 **working for Roosevelt:** MGM, "Personal Statement," p. 3; MGM interview (4/8/96).

46–7 The description of MGM at Yale during World War II comes from interviews with his classmates Richard Brandt (4/11/95), Justin Israel (3/21/94), Paul Macready (5/19/95), Harold Morowitz (4/24/95 and 5/23/95), George Rathjens (3/17/95), Bud Rosenbaum (3/28/95), Robert Rosenbaum (3/28/95), and Abner Shimony (3/31/95); and letters from David Benninghoff (12/95) and Lincoln Day (8/18/95). I also relied on two interviews with MGM (4/8/96 and 5/8/95), and his "Personal Statement," pp. 3–4.

47 **MGM meets John Knowles at breakfast:** Berreby interviews.

The young musical prodigy, Kenny Wolf, described the insult matches in a letter to MGM dated 6/22/56.

Incident with Dean Buck and anti-Semitism at Yale: Israel and Shimony interviews. Also see Oren, *Joining the Club.*

48 **MGM holding forth on the election returns:** Morowitz interviews.

49 **memories of Kovarik:** Morowitz interviews.

MGM cutting Margenau's class: Shimony interview. MGM said (4/8/96) that he didn't recall the incident.

MGM's respect for Margenau: MGM, "Personal Statement," p. 3.

50 **Suddenly this young kid shows up:** Brandt interview.

MGM never seemed to study: Bud Rosenbaum interview.

Breit anecdote on MGM demanding a grade: unpublished Crease and Mann interviews for *Second Creation.*

50–1 **Leigh Page seminar:** Morowitz and Shimony interviews.

51 **"lazamerchi" story:** Morowitz interviews.

52 **climbing East Rock:** Morowitz interviews; MGM interview (5/8/95).

adventures with Haggerty: Morowitz and Shimony interviews.

"a small, militant group": Buckley describes the Political Union on pp. 107–10 of *God and Man at Yale.*

53 **memories of Carteret Lawrence:** MGM interview (5/8/95); MGM, "Personal

Statement," p. 4; Berreby interviews; Crease and Mann interviews for *Second Creation.*

MGM's ingrained pessimism came from Arthur: Berreby interviews.

MGM rarely talked about parents: Morowitz and Shimony interviews.

53–4 **Ann Walker's memory of riding on the bike:** Ann Walker interview, 1/25/95.

54 **MGM's trouble getting accepted into graduate schools:** MGM, "Personal Statement," p. 4; MGM interview (4/17/97).

55 **"How could I go to that grubby place?":** MGM, "Personal Statement," p. 5. He also tells the suicide story in "Remarks Given at the Celebration of Victor Weisskopf's 80th Birthday," CERN, Geneva, 20 September 1988, typescript in MGM's files, p. 1; and in "The World as Quarks, Leptons and Bosons," a talk given at MIT in 1974, later published in Kerson Huang, ed., *Physics and Our World: A Symposium in Honor of Victor F. Weisskopf* (New York: American Institute of Physics, 1976), p. 83.

The road trip through the West: MGM interview, 3/20/98.

MGM arrives in Cambridge looking Ivy League: interview with David Jackson, 3/21/95.

CHAPTER 3   *A Feeling for the Mechanism*

56 **"the network of supply canals":** Hapgood, *Up the Infinite Corridor*, p. 56.

57–8 The story of Weisskopf's childhood is based on his autobiography, *The Joy of Insight*, pp. 1–29, and an interview (3/7/94).

59 **the question to Planck about relativity:** Physicists now distinguish between invariant mass (zero for the photon), which does not increase with velocity, and relativistic mass, which does.

60 Weisskopf's training under the great physicists of Europe is described in the early chapters of *Joy of Insight*.

"When it comes to atoms": Heisenberg, *Physics and Beyond*, p. 41.

61 **"What Professor Einstein has just said":** This commonly repeated story is recounted in Crease and Mann, *Second Creation*, p. 95, and Bernstein, *Tenth Dimension*, p. 10.

For a nice, brief history of quantum field theory, see Steven Weinberg, "The Search for Unity: Notes for a History of Quantum Field Theory," *Daedalus*, Fall 1977, pp. 17–35. For a full treatment, see Schweber, *QED and the Men Who Made It.*

63 **MGM thought theoretical physics was in disgrace:** "From Renormalizibility to Calculability?" in Roman Jackiw, Nicola N. Khuri, Steven Weinberg, and Edward Witten, eds., *Shelter Island I, Proceedings of the 1983 Shelter Island Conference on Quantum Field Theory and Fundamental Problems in Physics* (Cambridge, Mass.: MIT Press, 1985), p. 3.

Weisskopf's work on purging infinities: Weisskopf, *Joy of Insight*, pp. 80–81, 95–97; Crease and Mann, *Second Creation*, pp. 100–1, 106–9, 132–33.

64 **accounts of Shelter Island:** Weisskopf, *Joy of Insight*, pp. 166–72; Crease and Mann, *Second Creation*, pp. 125–28; Gleick, *Genius*, pp. 232–34.

66 **"Sensible mathematics":** Dirac, quoted in Kaku and Trainer, *Beyond Einstein*, p. 68.

"I prefer to know nothing about everything": Crease and Mann, *Second Creation*, p. 109.

67 **"Weisskopf units":** Not to be confused with the real Weisskopf unit, used to specify gamma decay rates in nuclear physics.

Weisskopf's loose calculational style and other traits: Goldberger interview,

4/11/96; MGM, "The World as Quarks, Leptons and Bosons," pp. 83–84 (cited in notes to chapter 2).

"Physics is simple but subtle": quoted in Weisskopf, *Joy of Insight*, p. 39.

68 **MGM's relationship with Weisskopf:** Goldberger interview (4/11/96); Weisskopf interview; MGM interview (4/8/96); and the two reminiscences by MGM fully cited toward the end of the notes for chapter 2: "Remarks Given at the Celebration of Victor Weisskopf's 80th Birthday" and "The World as Quarks, Leptons and Bosons."

Weisskopf thought MGM was a show-off: Weisskopf interview.

Weisskopf watching Nazi thugs: Weisskopf, *Joy of Insight*, p. 51.

68–9 **Weisskopf's good effects on MGM:** Weisskopf interview; MGM, "Personal Statement," p. 4 (cited in notes to chapter 1).

69 **"for his ideas about the substructure of the proton":** Weisskopf, *Joy of Insight*, p. 178; **Weisskopf found "quark" ugly:** ibid., p. 237.

What MGM was like on campus: interviews with classmates Charles Desoer (4/9/96), Joe Devaney (6/19/95), David Jackson (3/21/95), Jay Meili (4/9/96), and Robert Walker (4/8/96); letter from Joel Yancey (7/20/95).

MGM and Wiener: MGM, *Quark and Jaguar*, p. 72.

"There, there, little boy": Goldberger interview.

MGM and Goldberger at the blackboard: MGM, "The World as Quarks, Leptons and Bosons," p. 83; Goldberger interview.

70 **visiting Schwinger's class:** Goldberger interview.

the story about the boron isotope: MGM, "Personal Statement," pp. 4–5.

71 **ethanol parties at MIT:** Meili interview; unpublished MGM interview by Alun Anderson and Liz Else for *New Scientist* (cited in notes to chapter 1).

72 **MGM takes his majors:** MGM interview, 4/8/96.

road trip to Mexico: interviews with Meili and with MGM, 4/8/96.

73–4 **troubles with thesis and security problems:** MGM interview (4/8/96); Berreby interviews (cited in notes to chapter 1).

74 **appointment to Institute for Advanced Study:** MGM interviews, 4/8/96 and 10/17/96.

75 **New Year's Day hangover:** *New Scientist* interview by Anderson and Else.

CHAPTER 4 *Village of the Demigods*

76 **"Concentration Camp, Princeton":** quoted in Regis, *Who Got Einstein's Office?*, p. 34.

"quaint, ceremonious village": letter from Einstein to the queen of Belgium, quoted in Pais, *Subtle Is the Lord*, p. 453.

"evil quanta": ibid., p. 465.

"I have locked myself into quite hopeless scientific problems": quoted in ibid., p. 453.

"a haven where scholars and scientists": Regis, *Who Got Einstein's Office?*, p. 33.

77 **Einstein's problems with Flexner:** ibid., pp. 33–34.

78 Gell-Mann describes watching Einstein and Gödel in MGM, *Quark and Jaguar*, p. 39.

The account of MGM's days at the Institute is based on my interviews with him (10/17/96), Low (3/8/94), Dyson (1/15/98), and Pines (11/13/96).

Gell-Mann didn't want to seem like a sycophant: interview with MGM, 10/17/96.

79 The impression Oppenheimer made at the Institute for Advanced Study is described in a painfully frank account in Pais, *A Tale of Two Continents*, pp. 239–43, and in Regis, *Who Got Einstein's Office?*, pp. 139–40. Nice portraits

of Oppenheimer can also be found in Rhodes, *Making of the Atomic Bomb,* pp. 443–444, and Crease and Mann, *Second Creation,* pp. 96–98.

The name for the Trinity Site came from a poem by Donne: Rhodes, *Making of the Atomic Bomb,* pp. 571–72.

"I am become Death," etc.: quoted in ibid., p. 676, and many other places.

doubts that Oppenheimer was fluent in Sanskrit: Crease and Mann, *Second Creation,* p. 96, footnote 13. They cite J. Mehra and H. Rechenberg, *The Historical Development of Quantum Theory* (New York: Springer-Verlag, 1982), 1A, preface, p. xxv.

80 Dyson describes his insight about QED in *Disturbing the Universe,* pp. 54–7, 67–8.

Low's first impressions of Gell-Mann: interview with Francis Low, 3/8/94.

"They put this child in my office": The colleague was Jeremy Bernstein; interview with Bernstein (2/21/94).

82 **Oppenheimer berates a visitor:** interview with Low, 8/6/96. Pais also describes Oppenheimer's ruthless questioning in *Tale of Two Continents,* p. 239.

82–3 **Salpeter's visit to the Institute:** Salpeter interview, 8/21/96. He also told the story in an unpublished oral history interview by Spencer Weart, 3/30/78, pp. 32–33, American Institute of Physics, Center for History of Physics.

83 **Gell-Mann's style of doing physics:** Low interview.

"Why don't we do this": ibid.

Gell-Mann's first paper (with Francis Low): "Bound States in Quantum Field Theory," *Physical Review B* 84 (1951): 350–54.

83–4 **MGM's impressions of Oppenheimer and Dyson:** interview with MGM, 10/17/96.

84 **Some thought Dyson deserved the Nobel Prize:** Silvan Schweber, for example. See his *QED and the Men Who Made It,* p. 575.

Gell-Mann's letter to his parents from Princeton: undated, in MGM's personal files.

CHAPTER 5   *The Magic Memory*

86 The story of Enrico Fermi's life is presented in Rhodes, *Making of the Atomic Bomb,* pp. 204–9, and Crease and Mann, *Second Creation,* pp. 198–201. For full-scale biographies, see Laura Fermi, *Atoms in the Family,* and Emilio Segrè, *Enrico Fermi, Physicist.*

86–7 **"about as catastrophic as if the moon struck the earth":** quoted in Rhodes, *Making of the Atomic bomb,* p. 209.

87 **beta decay:** Strictly speaking, the neutron decays into a proton, an electron, and an antineutrino.

87–9 The impression Fermi left on some of the young Turks of Chicago was described in interviews with Goldberger (4/11/96), Telegdi (3/31/94), Hildebrand (11/4/96), and MGM (10/17/96). I also referred to C. N. Yang, "Introductory Notes to the Article 'Are Mesons Elementary Particles?' " in Yang, *Selected Papers,* pp. 305–6.

89 **Pauli called Fermi a "quantum engineer":** Rhodes, *Making of the Atomic Bomb,* p. 206.

"It was a time when you could be proud": Telegdi interview.

89–92 My story of MGM in Chicago is based on the interviews with the "young Turks" mentioned above and one with Robert Gomer (10/31/96). MGM talked about the suicide of his roommate in interviews on 10/17/96 and 3/20/98.

90 **"The only thing that is wrong with this young man":** Telegdi interview.
91 **impressions of MGM's physics style:** ibid.
MGM's dark view of Hyde Park: MGM interview, 10/17/96.
92ff. The early history of pion physics and isospin is described in Crease and Mann, *Second Creation*, pp. 157–74; Pais, *Inward Bound*, pp. 423–36; Polkinghorne, *Rochester Roundabout*, pp. 8–9; and a reminiscence by Nicholas Kemmer, "Isospin," in *Proceedings of an International Conference on the History of Particle Physics*, Paris, July 1982; published in *Journal de Physique* 43, colloque C-8, supplement to no. 12 (1982): 359–94. Pion experiments at the University of Chicago are described in the same issue: H. L. Anderson, "Early History of Physics with Accelerators," pp. 101–64.
96 **"ten-year joke":** Oppenheimer, *Physics Today*, November 1966, p. 58. Quoted in Crease and Mann, *Second Creation*, p. 164.
98 **Gell-Mann and Telegdi paper:** "Consequences of Charge Independence for Nuclear Reactions Involving Photons," *Physical Review* 91 (1953): 169–74.
98ff. The history of dispersion relations and the S-matrix is described in Pickering, *Constructing Quarks*, pp. 73–78; Pais, *Inward Bound*, pp. 492–505; Schweber, *QED and the Men Who Made It*, pp. 154–55, 527–52; and in articles by Helmut Rechenberg, Andrew Pickering, and Geoffrey Chew in Brown et al., eds., *Pions to Quarks*, pp. 551–607. Also see the following two presentations in *High Energy Nuclear Physics*, Proceedings of the Sixth Rochester Conference (1956): Goldberger, "Introductory Survey; Dispersion Relations," Section I, pp. 1–20, and MGM, "Dispersion Relations in Pion-Nucleon and Photon-Nucleon Scattering," Section III, pp. 30–36. Gell-Mann gave a reminiscence of these times in a historical talk at a September 1983 conference in Sant Feliu de Guíxols, Catalonia, Spain: "Particle Theory from S-Matrix to Quarks," later published in M. G. Doncel, A. Hermann, L. Michel, and A. Pais, eds., *Symmetries in Physics (1600–1980)*, Proceedings of the First International Meeting on the History of Scientific Ideas (Barcelona: Belaterra, 1987), pp. 474–97.
103 **"Of course, it may be that someone will come up," etc.:** Fermi is quoted in Pais, *Inward Bound*, p. 494.
104 **"little more than a vessel containing well-arranged experimental data":** ibid., p. 498.
"hardly anyone would challenge": Rochester 6 Proceedings, Section III, pp. 31.
104–5 **Goldberger's impressions of MGM doing physics:** interview with Goldberger, 4/11/96.
105 MGM's papers with Goldberger on dispersion theory include "The Formal Theory of Scattering," *Physical Review* 91 (1953): 398–408; "The Scattering of Low Energy Photons by Particles of Spin 1/2," *Physical Review* 96 (1954): 1433–38; and "On the Scattering of Gamma Rays by Protons," in E. H. Bellamy and R. G. Moorhouse, eds., *Proceedings of the 1954 Glasgow Conference on Nuclear and Meson Physics*. (London: Pergamon Press, 1955), pp. 267–71. MGM and Goldberger also wrote a paper with W. E. Thirring: "Use of Causality Conditions in Quantum Theory," *Physical Review* 95 (1954): 1612–27.
"This is an unusual and gratifying position": Rochester 6 Proceedings, Section I, p. 10.

CHAPTER 6   *"No Excellent Beauty"*

107 **"If you run out of Greek letters":** Von Neumann made the comment to Pais; it is quoted in Pais, *Tale of Two Continents*, p. 333.
The title of the original paper on the discovery of strange particles: George

Rochester and Clifford Butler, "Evidence for the Existence of New Unstable Particles," *Nature* 160, no. 4077 (1947): 855–57.

108 Oppenheimer coined the term "subnuclear zoo" in a public lecture at the Rochester 6 Conference. See Section VIII, page 1, of the proceedings (cited in notes to chapter 5).

MGM's *Scientific American* article: MGM and E. P. Rosenbaum, "Elementary Particles," July 1957, *Scientific American*, pp. 72–88.

The narrative of MGM's strangeness discovery is based on his own account, "Strangeness," in *Proceedings of an International Conference on the History of Particle Physics*, Paris, July 1982; published in *Journal de Physique*, 43, colloque C-8, supplement to no. 12 (1982), pp. 395–408; Crease and Mann, *Second Creation*, pp. 172–78; Pais, *Inward Bound*, pp. 519–23; and interviews with MGM (10/17/96) and Pais (2/28/94). I also relied on MGM's July 1957 *Scientific American* article and Bernstein's account of the science in *Tenth Dimension*, pp. 43–57.

109 **"some kind of dirt effect":** Richard Dalitz used these words in the discussion after MGM's "Strangeness" presentation; see p. 406 of the proceedings.

110 **Spin-1 bosons can have as many as three degrees of freedom:** But not necessarily. The fact that the spin-1 photon is massless limits its options to two. The pion, with spin-0, has no degrees of freedom.

113 **"By the way, a few minutes ago":** MGM, "Strangeness," p. 396. The account of MGM's slip of the tongue is taken from this talk and from the profile of MGM in Berland's *Scientific Life*, p. 56. In "Strangeness," MGM remembered Low, Dyson, and Lee being present during the incident. Neither Dyson (1/15/98 interview) nor Low (8/6/96 interview) remembered the talk. Lee did not respond to several requests to be interviewed.

Oppenheimer coined the word "megalomorphs": A. M. L. Messiah and H. P. Noyes, eds., *Meson Physics*, Proceedings of the Second Rochester Conference (1952), p. 50. Copies of this mimeographed publication are few and far between. I found one among Marshak's papers at Virginia Polytechnic University. Polkinghorne also reports the "megalomorph" episode in *Rochester Roundabout*, p. 28.

Pais's talk, "An Ordering Principle for a Megalomorphian Zoology," is in the Rochester 2 Proceedings, pp. 87–93.

114 Pais's even-odd scheme is described in his Rochester 2 talk and in Pais, *Inward Bound*, p. 518; it is summarized in Polkinghorne, *Rochester Roundabout*, pp. 29–32.

Gell-Mann's impressions of Pais's relationships with Einstein and Oppenheimer: MGM interview, 10/17/96. Pais tells what he really thought of Oppenheimer in *Tale of Two Continents*, pp. 239–43.

115 MGM describes his first visit to Europe in "Strangeness," p. 398.

"splendid isolation," etc.: Leprince-Ringuet, quoted in Crease and Mann, *Second Creation*, p. 170; translated by the authors from Leprince-Ringuet, *Les rayons cosmiques: les mesotrons* (Paris: Editions Albin Michel, 1945), pp. 137–38.

116 **MGM in Urbana:** interview with Pines, 11/13/96.

117 **his paper with Low on renormalization group:** "Quantum Electrodynamics at Small Distances," *Physical Review* 95 (1954): 1300–12.

"one of the most important ever published": Steven Weinberg, "Why the Renormalization Group Is a Good Thing," in Alan H. Guth, Kerson Huang, and Robert L. Jaffe, eds. *Asymptotic Realms of Physics: Essays in Honor of Francis E. Low* (Cambridge, Mass.: MIT Press, 1985), pp. 1–19.

118 **While in Urbana, MGM also collaborated with David Pines and M. Ferentz:** "The Giant Nuclear Resonance," *Physical Review* 92 (1953): 836–37.

Pais's Netherlands talk: "Isotopic Spin and Mass Quantization," *Physica* 19

(1953): 869–87. He gives his own account of the talk in *Tale of Two Continents*, pp. 303–5.

"in which one talks of families": Pais, *Tale of Two Continents*, p. 293.

119 Feynman tells the tale of Pais in Kyoto in *"Surely You're Joking, Mr. Feynman!,"* pp. 242–43. Pais gives his version in *Tale of Two Continents*, pp. 311–18.

"sufficiently pompous": MGM, "Strangeness," p. 400.

"Isotopic Spin and New Unstable Particles" was published in *Physical Review* 92 (1953): 833–34. The seeds for the strangeness idea were also sown in the unpublished preprint, "On the Classification of Particles," Department of Physics and Institute of Nuclear Studies, University of Chicago, which I obtained from MGM's office files. In "Strangeness" (p. 399), he also mentions the third unpublished paper, and says it was never widely circulated. He told me later (10/17/96 interview) that it was eventually lost.

To make Gell-Mann's abstractions clearer, I am describing the strangeness scheme as it was expanded shortly afterward in the paper he delivered at the 1955 Pisa conference: "The Interpretation of the New Particles as Displaced Charge Multiplets," *Nuovo Cimento Supplement* 4 (1956): 848–66.

122 **Experiments establishing associated production:** W. B. Fowler, R. P. Shutt, A. M. Thorndike, and W. L. Whitemore, "Production of Heavy Unstable Particles by Negative Pions," *Physical Review* 93 (1954): 861–67.

"What possible difference does it make?": quoted in MGM, "Strangeness," p. 400.

124 **Pais thought strangeness was a new quantum number:** Pais tells this story in *Tale of Two Continents*, pp. 334–37.

"quite enthusiastic about": letter from MGM to Pais dated 4/30/54, quoted in ibid., p. 335.

125 **MGM's courtship of Margaret Dow:** interviews with MGM (10/17/96 and 1/22/97), Goldberger (4/11/96), Pais (2/28/94), and Pines (11/13/96).

125–6 **Trip to Scotland:** MGM interview (1/22/97); Hildebrand interview (11/4/96).

126 **Glasgow paper with Pais:** "Theoretical Views on the New Particles," in *Proceedings of the 1954 Glasgow Conference on Nuclear and Meson Physics* (cited in the notes to chapter 5), p. 342.

127 **The naming of the sigma and xi and the argument about quantum numbers:** Pais, *Tale of Two Continents*, p. 336. Pais actually refers to the song as "The Darling of Sigma Xi."

128 **MGM at Columbia:** interview with Leon Lederman, 1/19/98.

Gell-Mann and Pais had thought of the K-zero scheme back at Princeton: Pais, *Tale of Two Continents*, pp. 335, 337.

Gell-Mann and Pais's K-zero scheme: "Behavior of Neutral Particles under Charge Conjugation," *Physical Review* 97 (1955): 1387–89. Though the idea was left out of the Glasgow report, a footnote referring to the forthcoming paper appeared in the proceedings. The theory is also described in Pais, *Inward Bound*, pp. 521–23; Pais, *Tale of Two Continents*, pp. 337–39; Polkinghorne, *Rochester Roundabout*, pp. 68–69; Bernstein, *Tenth Dimension*, pp. 56–57, and Crease, *Making Physics*, pp. 228–33.

Lederman was part of a team of experimenters that confirmed the particle-mixing phenomenon at Brookhaven: K. Lande, E. T. Booth, J. Impeduglia, and L. M. Lederman, *Physical Review* 103 (1956): 1901–4; and M. Bardon, K. Lande, L. M. Lederman, and W. Chinowsky, *Annals of Physics* 5 (1958): 156–58.

128–9 Gell-Mann's falling-out with Pais was described in Pais, *Tale of Two Continents*, p. 339, and in interviews with Pais, Goldberger, Telegdi, and Pines. Gell-

Mann was barely able to talk about the matter, referring to him only as "this Pais person" (interviews 10/17/96 and 2/19/97).

129 **MGM said he wasn't influenced by Pais's work on associated production:** "Strangeness," p. 397.

"It's terrible": interview with Telegdi, 3/31/94.

Pais pleaded with his colleagues not to use "strange particles": *High Energy Nuclear Physics,* Proceedings of the Fifth Rochester Conference (1955), p. 135. Pais describes his psychoanalysis and creative block in *Tale of Two Continents,* pp. 245–48.

130 **visit to Fermi's deathbed:** MGM interview, 10/17/96; Yang, "Introductory Notes" (cited in the notes to chapter 5), p. 306.

Gell-Mann compares Fermi with a meson: Berland, *Scientific Life,* p. 64.

131 MGM's exploratory visit to Pasadena is recounted in Gleick, *Genius,* pp. 310–11; MGM's reminiscence, "Dick Feynman—The Guy in the Office Down the Hall," *Physics Today,* February 1989, pp. 50–54; and MGM, *Quark and Jaguar,* p. 208.

undated letter from MGM in Pasadena to Margaret: MGM, personal files.

Harvard trying to hire MGM: documented in various correspondence in MGM's personal files for the year 1955; for example, in a letter from Goldberger dated 2/6/55.

132 **Telegdi's and Goldberger's efforts to keep MGM in Chicago:** letter to MGM from Goldberger dated 2/6/55.

"adjust your appointment": letter to MGM dated 2/24/55 from the University of Chicago.

Caltech's employment offer: letter to MGM from Lee DuBridge, president of Caltech, dated 3/8/55.

"one of the most brilliant": letter from Oppenheimer to the Selective Service, Local Board No. 13, New York City, dated 3/20/55.

MGM's wedding: interviews with MGM (10/17/96 and 1/22/97); interview with Pines (11/13/96). MGM also describes the wedding in a tribute to Pines: "Reminiscences," *Philosophical Magazine B* 74, no. 5 (1996): 431–34.

salary figures for Caltech and Chicago: documented in various letters in MGM's files.

The Gell-Manns move to Pasadena: MGM interview, 1/22/97.

133 **Telegdi's impressions of MGM and wife in Copenhagen:** Telegdi interview, 3/31/94.

"temporary kind of solution": Rochester 6 Proceedings, Section VIII, p. 1.

Experimenters found particles predicted by strangeness: Two were the xi-zero and the sigma-zero. See MGM, "Strangeness," p. 402.

CHAPTER 7 *A Lopsided Universe*

134 **The neighborhood with the mountain lion:** MGM interview, 1/22/97. Descriptions of the Pasadena lifestyle were inspired by Starr, *Inventing the Dream,* pp. 99–127, and my own visits there.

135 **Description of Athenaeum:** personal visit; Christianson, *Edwin Hubble,* pp. 195–96.

135 **Feynman liked to show up in shirtsleeves:** MGM, "Dick Feynman—The Guy in the Office Down the Hall," *Physics Today,* February 1989, pp. 50–54.

The Gell-Manns give Feynman a new set of bongo drums: The story was told to me by Pines (interview 11/13/96), who was visiting Caltech at the time. The bidding war is documented in letters to Gell-Mann from Harvard (3/23/56), the University of Chicago (4/5/56), and Caltech (5/8/56).

MGM would miss working with Murph: He says so in an undated letter to Goldberger.

MGM's relationship with RAND and LASL is documented in contracts and correspondence in his personal files for 1956 and thereafter.

MGM's security investigation: MGM interview, 10/17/96. Rosenfeld confirmed the story about the car in an interview (3/31/97). Nina Byers did not respond to a request for an interview. In a letter dated 1/14/56 to Gell-Mann, Carson Mark of LASL writes: "at long last your AEC 'Q' clearance has been granted." MGM declined my request for a copy of his FBI files.

136 MGM's consulting deals with the Convair and General Atomic divisions of the aerospace company General Dynamics are documented in his correspondence with the corporation.

a third of Gell-Mann's earnings were from consulting: letter from MGM to Keith Brueckner, dated 11/7/61.

"a malfunctioning computing machine" or "an atom that is bounced around": ibid.

Reading Latin poetry with Margaret: see profiles of her in the *Los Angeles Times,* 9 November 1969, and *Pasadena Star-News,* 4 November 1969; also in unpublished transcript of Crease and Mann interviews for *Second Creation.*

MGM was seeing an analyst: Berland, *Scientific Life,* p. 64.

How MGM felt about working with Feynman: MGM interview, 2/19/97.

"twisting the tail of the cosmos": MGM, "Dick Feynman—The Guy in the Office Down the Hall," p. 51.

"No! You can't do that," etc.: Gell-Mann described a typical Feynman argument this way in "Strangeness Minus Three," a documentary film about the Eightfold Way produced for the BBC *Horizon* series, 1964.

137 Though the mirror metaphor works only partially (because mirrors don't reverse up and down), almost everyone, including Feynman (*Feynman Lectures,* vol. 1, chapter 52), has used the analogy, because it vividly captures the flavor of the concept.

Feynman's ideas on symmetry and alarm clocks: ibid., vol. 1, chapter 52, pp. 4–5. The saga of the downfall of parity must be one of the most thoroughly mined fields in the history of physics. A somewhat technical but exciting account, including reminiscences by Telegdi, Wu, and Garwin, as well as excerpts from Yang and Lee's Nobel lectures, can be found in vol. gamma of the journal *Adventures in Experimental Physics* (Princeton, N.J.: World Science Communications, 1972). Historical accounts by Dalitz, Fitch, Sudarshan, and Telegdi can be found in Section VII of Brown et al., *Pions to Quarks,* pp. 434–94. I also relied on the clear explanations in Gardner, *The New Ambidextrous Universe,* pp. 165–230, and the dramatic recounting in Crease and Mann, *Second Creation,* pp. 203–14.

Weyl on symmetry: paraphrased in *Feynman Lectures,* vol. 1, chapter 11, p. 1. Feynman's Martian story is adapted from chapter 52 of the same source. It is now widely believed that Mars does not have a significant magnetic field.

139 Noether and the connection between symmetries and conservation laws is nicely described in Crease and Mann, *Second Creation,* pp. 184–89.

141 MGM described his parity doublet scheme in off-the-cuff remarks at the 1956 Rochester conference: *High Energy Nuclear Physics,* Proceedings of the Sixth Rochester Conference, Section VIII, pp. 23–27, and in an interview (2/19/97).

MGM's timidity about publishing something wrong is described in "Strangeness" (cited in the notes for chapter 6). The subject repeatedly came up in interviews and conversations with him (10/17/96 and others), and in an interview with Goldberger (8/6/96).

The dispute between MGM and Lee and Yang: interviews with MGM (2/19/97), Yang (3/1/94), and Goldberger (4/11/96 and 8/6/96). It is also documented by correspondence: letter from Lee and Yang (2/8/56); apology from MGM to Lee and Yang (2/14/56); acceptance from Lee and Yang to MGM (3/20/56).

142 **"I do not believe either of you capable of stealing":** letter from MGM to Lee and Yang dated 2/14/56.

Lee and Yang's parity doublet paper: "Mass Degeneracy of the Heavy Mesons," *Physical Review* 102, no. 1 (1956): 290–91.

Goldberger tries to play Dutch uncle: Goldberger interviews, 4/11/96 and 8/6/96.

The discussions on parity at Rochester 6 are reported in Section VIII of the proceedings and in Polkinghorne, *Rochester Roundabout*, pp. 56–57. (This was also the conference where MGM and Goldberger gave important talks on dispersion relations. See notes for chapter 5.)

The story about Feynman and Block is told in Gleick, *Genius*, p. 333; Crease and Mann, *Second Creation*, p. 205; Mehra, *Beat of a Different Drum*, pp. 462–64; and Feynman, *"Surely You're Joking,"* pp. 247–48.

143 **"Why don't you ask the experts?" etc.:** Feynman, *"Surely You're Joking,"* p. 248.

Feynman runs the broken parity idea by MGM: MGM interview, 2/19/97.

MGM wrestling with parity in graduate school: He tells this story in a historical essay, "Progress in Elementary Particle Theory, 1950–1964," in Brown et al., *Pions to Quarks*, p. 697.

"O.K., that's my opinion too": MGM interview, 2/19/97.

144 **"The tau meson will have either domestic":** Rochester 6 Proceedings, Section VIII, p. 22.

MGM's description of parity doublets: ibid., pp. 23–27.

The other physicist who came up with the tetrahedron was N. Dallaporta. See MGM, "Progress in Elementary Particle Theory," p. 706.

144–5 **"some strange space-time transformation properties," etc.:** Rochester 6 Proceedings, Section VIII, p. 28.

145 **Oppenheimer says it's time "to close our minds," etc.:** ibid., pp. 28–29.

Feynman and MGM return to Caltech excited by parity nonconservation: Polkinghorne, *Rochester Roundabout*, p. 57.

Pais and Yang bet Wheeler a dollar: Pais, *Inward Bound*, p. 525; Polkinghorne, *Rochester Roundabout*, p. 57.

MGM described his trip to the Soviet Union in "Particle Theory from S-Matrix to Quarks" (cited in the notes for chapter 5), p. 479; a "Report on the Moscow Conference on High Energy Physics," submitted to the National Science Foundation; and an interview (2/19/97). Feynman alludes to MGM's Russian talk in *"Surely You're Joking,"* p. 248. I also relied on the colorful descriptions in Pais, *Tale of Two Continents*, pp. 352–54, and on coverage by the *New York Times:* "Physicists Free to Visit Moscow," 24 April 1956, which gives the roster of participants (mistakenly calling Pais "A. Tais"), and "Americans in Soviet Report Science for Peace Stressed," 22 May 1956, p. 1 (dateline Moscow, 21 May 1956). The visit is also documented in correspondence in MGM's files for 1956.

145 **Many of the American scientists were impressed by Soviet peacetime nuclear efforts:** Marshak, Pais, and Wilson expressed these opinions in "Physicists Free to Visit Moscow."

146 Yang gave his account of the discovery of parity violation in his *Selected Papers*, pp. 26–31. Lee told his side of the story in "Broken Parity," in G. Feinberg, ed., *T. D. Lee: Selected Papers*, vol. 3, pp. 498–509. Good popular accounts

are in Regis, *Who Got Einstein's Office?*, pp. 140–74, and Bernstein's "A Question of Parity," in *A Comprehensible World*, pp. 35–73.

Lee found the give-and-take exhilarating: Feinberg, *T. D. Lee: Selected Papers*, p. 497.

147 Wu's account of her role in parity violation is in *Adventures in Experimental Physics* (cited earlier in the notes for this chapter). Her meeting with Lee is reported there (pp. 101–2) and in Feinberg, *T. D. Lee: Selected Papers*, p. 498.

The tome Wu gave Lee: K. Siegbahn, ed., *Beta- and Gamma-Ray Spectroscopy* (Amsterdam: North Holland, 1955).

"Is Parity Conserved in Weak Interactions?" appeared in *Physical Review* 104, no. 1 (1956), pp. 254–58.

The editor didn't like titles with question marks: Yang, *Selected Papers*, p. 30.

Lee and Yang expected parity conservation would not be violated: Yang interview, 3/1/94.

Wu experiment: *Adventures in Experimental Physics*, pp. 109–18. The results were published as C. S. Wu, F. Amber, R. W. Hayward, D. D. Hoppes, and R. P. Hudson, "Experimental Test of Parity Conservation in Beta Decay," *Physical Review* 105, no. 4 (1957): 1413–15.

148 Telegdi recounts his parity experiment in *Adventures in Experimental Physics*, pp. 131–5, and in Brown et al., *Pions to Quarks*, pp. 464–84. The results were reported in J. I. Friedman and V. L. Telegdi, "Nuclear Emulsion Evidence for Parity Nonconservation," *Physical Review* 105, no. 5 (1957): 1681–82.

148–9 The Columbia experiment is described by Garwin in *Adventures in Experimental Physics*, pp. 124–27, and in Lederman, *The God Particle*, pp. 265–73. It was published as R. L. Garwin, L. M. Lederman, and M. Weinrich, *Physical Review* 105, no. 4 (1957): 1415–17.

149 **As Feynman would later put it:** *Feynman Lectures*, vol. 1, chapter 52, p. 9.

Telegdi's problems with *Physical Review* are described in his article in *Adventures in Experimental Physics*, pp. 134–35. The *Physical Review*'s editor, Samuel Goudsmit, gives his side of the story on p. 137 of the same issue.

"I am only the president": Wigner quoted in *ibid.*, p. 134.

150 Zee tells his story about parity violation in *Fearful Symmetry*, p. 34.

"I do not believe that the Lord is a weak left-hander": Yang, *Selected Papers*, p. 30. He cites a letter to Victor Weisskopf dated 1/17/57.

"Walked through door": Crease and Mann, *Second Creation*, p. 210.

"Gentlemen, we must bow to nature": ibid., p. 217.

The birth of Elizabeth (Lisa) Gell-Mann: described in a letter from Margaret Gell-Mann to Arthur and Pauline Gell-Mann dated 10/30/56.

151ff. The V-A story is described in Crease and Mann, *Second Creation*, pp. 210–14; Gleick, *Genius*, pp. 336–39; and Mehra, *Beat of a Different Drum*, pp. 465–68. MGM gave his version in an interview (2/19/97). Feynman's version is in *"Surely You're Joking,"* pp. 247–55, and on pp. 171–81 of the transcript of an American Institute of Physics oral history interview with Feynman conducted 6/28/66 by Charles Weiner. Sudarshan's story is in "Midcentury Adventures in Particle Physics," Brown et al., *Pions to Quarks*, pp. 485–94. Marshak's is given in another unpublished AIP interview conducted on 9/19/70 and 10/4/70 by Weiner, pp. 49–68.

152 **Lee's lament at Rochester 7:** Crease and Mann, *Second Creation*, p. 211; *High Energy Nuclear Physics*, Proceedings of the Seventh Rochester Conference, Section VII, pp. 1, 7.

153 **Sudarshan and Marshak's trip to Los Angeles:** Described in Sudarshan, "Midcentury Adventures," Brown et al., *Pions to Quarks*, p. 488, and in Marshak's AIP oral history.

MGM had been toying with axial currents since Pisa: MGM interview,

2/19/97; Crease and Mann transcripts for *Second Creation.* Marshak seems to corroborate this in his AIP oral history, p. 48. In a letter dated 5/17/55 to Marcello Conversi, the organizer of the conference, MGM wrote that he was currently working on "a new theory of the weak interactions of elementary particles, including the decay mechanisms of K-particles and hyperons."

154 **MGM was fascinated by Sudarshan's name:** MGM interview, 2/19/97.
Sudarshan was pleased that MGM seemed so cordial: Sudarshan, "Midcentury Adventures," Brown et al., *Pions to Quarks,* p. 488.
MGM says he probably won't write a paper: MGM interview, 2/19/97. The review article he wrote with Rosenfeld was "Heavy Mesons and Hyperons," *Annual Review of Nuclear Science* 7 (1957): 407–79; the reference to V-A is on p. 433.
The Gell-Manns' vacation: MGM interview, 2/19/97.
Lee tells Feynman to flip a coin: Feinberg, *T. D. Lee: Collected Papers,* p. 483.
Feynman thinks he's the first to know a new law: Feynman, *"Surely You're Joking,"* p. 250; Gleick, *Genius,* pp. 337–38.

155 **MGM couldn't believe Feynman's bombast:** MGM interview, 2/19/97.
MGM's regrets about the promise to Marshak and Sudarshan: ibid.
MGM objected to Feynman's formalism: ibid.
Feynman and Gell-Mann's V-A paper: "Theory of the Fermi Interaction," *Physical Review* 109, no. 1 (1958): 193–98.
"One of the authors has always had a predilection": ibid., p. 194.
"One of us (M.G.M.) would like to thank": ibid., p. 198.
MGM felt a little guilty: MGM interview, 2/19/97.
"exceedingly nice," etc.: letter from Goldberger dated 9/24/57.
"I have flipped my coin": in Feinberg, *T. D. Lee: Selected Papers,* p. 483.

156 **Marshak's and Sudarshan's disappointment:** Sudarshan, "Midcentury Adventures," Brown et al., *Pions to Quarks;* Marshak, AIP oral history.
"originality, imagination, and physical insight": letter of recommendation from MGM to the University of Rochester dated 1/25/61.
"This F-G theory of beta-decay is no F-G": letter from Telegdi to MGM dated 11/24/57. Also quoted in Feynman, *"Surely You're Joking,"* pp. 253–54.
Telegdi asks about "janitorships" at Caltech: letter to MGM dated 10/10/56.
MGM's opinion of Telegdi: recommendation to Keith Brueckner, University of California at La Jolla (3/18/60); memo to Robert Bacher, Caltech physics department (2/21/61).

157 **"Genius at Cal Tech":** *Newsweek,* 18 June 1956, p. 114.
MGM's congressional testimony on antimatter: John W. Finney, "Inquiry Members Get Lost in Space: Physicists Take Panel on a Theoretical, but Vexing, Tour of Universe," *New York Times,* 12 February 1958.
MGM's job offers: letter from the University of Pennsylvania (12/23/58); letter from the University of New Mexico (12/7/56); letter from Breit (2/20/58); undated letter from Goldberger (apparently from 1958).
"I would guess that it is pretty much hopeless," etc.: The letter is dated 7/17/56.

158 **MGM deconstructs a Bible passage:** documented in correspondence with the Reverend Harry Valentine of Springfield, Pa. (2/12/58 and 2/21/58).
"all quantum, electromagnetic," etc.: letter dated 10/14/58.
cover story on Teller: *Time,* 18 November 1957.
MGM still hated Teller: interview, 10/17/96.
"heir apparent to the mantle of Einstein," etc.: *Time,* 18 November 1957, pp. 24–25

159 **MGM objects to the quote in *Time*:** A manuscript of the letter to the editor is in MGM's personal files.

CHAPTER 8   *Field of Dreams*

164 The comparison between local gauge symmetry and the invisible hand of economics is made by Crease and Mann in *Second Creation,* pp. 187–95. Another fine explanation can be found in Smolin, *Life of the Cosmos,* pp. 51–54.

165 **Yang and Mills paper on gauge symmetry:** C. N. Yang and R. L. Mills, *Physical Review* 96, no. 1 (1954): 191–95.

167 **"Are all the particles":** MGM and T. Rosenbaum, "Elementary Particles," *Scientific American,* July 1957, p. 88.

167–8 **"Suppose that the eight baryons are really,"** etc.: *High Energy Nuclear Physics,* Proceedings of the Seventh Rochester Conference, Section IX, p. 11.

168 **Schwinger says he is working on the same idea:** ibid., p. 15.
"Model of the Strong Couplings" was published in *Physical Review* 106, no. 6 (1957): 1296–300.
"grave doubts," etc.: ibid., p. 1296.
"If a child grows up to be a scientist": letter dated 1/27/58 to Marianne Besser of Cincinnati, Ohio.
"metallic colors," etc.: letter dated 7/29/58 from Alvarez to MGM.
MGM's personality change: This assessment is based on interviews with colleagues who knew him at Yale, MIT, the Institute for Advanced Study, and the University of Chicago. The change was noticed in particular by Goldberger (4/11/96 interview), who probably knew him best.

169 **Norton's impressions of MGM:** Norton interview, 4/2/97.
The martini incident: interviews with Norton (4/2/97) and MGM (4/17/97).
Feynman teasing MGM at theoretical seminar: Norton interview, 4/2/97.
Feynman knew little about anything but physics: This commonly held opinion was expressed in particularly strong terms in an interview with Yuval Ne'eman (4/1/97): "Feynman was an ignoramus whereas Murray knew everything."
MGM's mixed feelings about Feynman: expressed in various interviews with MGM (1/22/97, 2/19/97, 4/17/97, 8/23/97, and 12/3/97) and in his article "Dick Feynman—The Guy in the Office Down the Hall," *Physics Today,* February 1989, pp. 50–54.

170 The uxyl scheme is described by MGM in "Particle Theory from S-Matrix to Quarks," p. 484. See chapter 5 notes for complete citation.
MGM's account of red and blue neutrinos and the Stanford meeting: MGM, "Particle Theory from S-Matrix to Quarks," p. 484; MGM interview (2/19/97). According to the *American Physical Society Bulletin* (Series II, vol. 2, no. 8, 19 December 1957), Lederman was at the meeting. He said in an interview (1/19/98) that he did not recall MGM's talk or anything about red and blue neutrinos.
two-neutrino discovery: The experiment by Lederman, Schwartz, and Steinberger is described in Crease and Mann, *Second Creation,* pp. 286–90.

171 **schizons:** MGM, "Particle Theory from S-Matrix to Quarks," pp. 484–85.
Schwinger's electroweak theory: "A Theory of the Fundamental Interactions," *Annals of Physics* 2, no. 5 (1957). The theory is described in Crease and Mann, *Second Creation,* pp. 215–18.
MGM's paper on weak magnetism: "Test of the Nature of the Vector Interaction in Beta Decay," *Physical Review* 111, no. 1 (1958): 363–65. Correction published in *Physical Review* 112, no. 6 (1958): 2139.

172 **Landau's doubts about field theory:** Crease and Mann, *Second Creation,* p. 238.
Goldberger and MGM talks at Rochester 6 (1956): See notes for chapter 5.

MGM describes his argument with Landau in "Particle Theory from S-Matrix to Quarks," p. 479.

"We all have our doubts about field theory," etc.: Rochester 7 Proceedings, Section IX, p. 7; also in Polkinghorne, *Rochester Roundabout*, p. 68.

173 **"trying to show that the methods of dispersion theory," etc:** "Statement of Professional History and Goals," attached to an application to the National Science Foundation for a senior postdoctoral fellowship, dated 9/27/58.

Details of 1959 Kiev conference: Polkinghorne, *Rochester Roundabout*, pp. 78–85; Crease and Mann, *Second Creation*, p. 238; Hildebrand interview (11/4/96).

"Aren't you ashamed of yourself": The incident is described in Crease and Mann, *Second Creation*, p. 238.

MGM's impressions of Landau: MGM, "Particle Theory from S-Matrix to Quarks," pp. 479–81; MGM interview (1/22/97).

Details of the Heineman Prize were reported 16 April 1959 in the *New York Times*. MGM's account is from an interview (4/17/97).

"Perhaps you are still paralyzed with joy": letter to MGM from Karl K. Darrow of the American Physical Society (3/3/59) following his initial announcement letter (2/21/59).

174 **"The photo is certainly a splendid one":** letter from MGM to American Institute of Physics (5/18/59).

"Well done, old boy!": letter from Samuel P. Huntington (4/23/59).

Yale honorary degree anecdote from interviews with Morowitz (4/24/95) and MGM (4/17/97).

CHAPTER 9 *The Magic Eightball*

175 The incident in Michel's office is described in Bernstein, *Life It Brings*, pp. 161–62.

Bernstein's impressions of MGM at Columbia Grammar: ibid, pp. 19–20, 141–43; interview with Bernstein (2/21/94).

MGM lectures at Harvard: Bernstein, *Life It Brings*, pp. 142–43.

176 **the paper that impressed Bernstein:** MGM, "Test of the Nature of the Vector Interaction in Beta Decay" (cited in the notes for chapter 8).

"That's interesting," etc.: Bernstein, *Life It Brings*, pp. 141–44.

MGM's plan for Paris is described in his National Science Foundation application (cited in the notes for chapter 8).

"cut the Gordian knot": ibid.

176–7 **experiences living in Paris:** MGM interview, 4/17/97.

177 **Gell-Mann loses sleep over a grammatical error:** He told the story in the documentary *Strangeness Minus Three* (cited in the notes to chapter 7).

177–8 **Feynman first agreed to put his name on the paper with MGM and Lévy:** letter from Feynman to MGM (1/8/60). **"I am in one of those activity lows," etc.:** letter from Feynman to MGM (2/21/60). **"Listen, I may sound like a fickle woman":** letter from Feynman to MGM (3/2/60). **"But I find gossip like that passes quickly":** letter from Feynman to MGM (4/18/60). All are in MGM's personal files. An early typescript of the paper, "The Axial Vector Current in Beta Decay," with Feynman listed as a coauthor, has found its way onto the shelves of the Oppenheimer Library at Los Alamos, New Mexico.

178 **The papers MGM wrote in Paris:** "On the Renormalization of the Axial Vector Coupling Constant in β-Decay" (with J. Bernstein and L. Michel) *Nuovo Cimento* 16, no. 3 (1960): 560–68; "The Axial Vector Current in Beta Decay" (with M. Lévy) *Nuovo Cimento* 16, no. 4 (1960): 705–25.

how MGM did physics: Bernstein, *Life It Brings,* pp. 163–64; Bernstein interview (2/21/94).

"The currents are commuting like angular momenta," etc.: Bernstein, *Life It Brings,* pp. 163–64.

179 **"Where have you been?":** ibid., p. 164.

Glashow's visit to Paris was described to me in interviews with Glashow (3/8/94) and MGM (4/17/97). It is also described in MGM's "Particle Physics from S-Matrix to Quarks" (cited in the notes for chapter 5), p. 488, and Glashow's (maddeningly unindexed) *Interactions,* pp. 143–44.

180 Glashow's family background is described in Glashow, *Interactions,* pp. 11–31, and in Crease and Mann, *Second Creation,* pp. 218–20.

"What you're doing is good": Crease and Mann, *Second Creation,* p. 225.

description of MGM's work in Paris on the strong force: MGM, "Particle Physics from S-Matrix to Quarks," pp. 488–89; MGM interview (4/17/97); Crease and Mann *Second Creation,* pp. 265–66.

181 **"The hell with it":** MGM interview, 4/17/97.

MGM described his encounter with the French relativist in "Nature Conformable to Herself," *Complexity* 1, no. 4 (1995/96): 9.

182 **"If we go to Africa, can I wear a safari jacket?":** MGM interview, 4/17/97.

MGM asks oil company executive to help him rent a Land Rover: letter dated 6/20/60 to G. G. Biggar, vice president for public relations at Shell Oil Company, the American counterpart of Royal Dutch/Shell.

182–3 **description of Africa trip:** MGM interview (4/17/97); Berland, *Scientific Life,* pp. 68-69; *Los Angeles Times* interview with Margaret Gell-Mann, 9 November 1969.

183 **MGM's bird-watching list:** Berland, *Scientific Life,* p. 51.

MGM's fascination with diversity: ibid., p. 69.

negative impressions of Africa: ibid., pp. 50, 68–69; MGM interview (4/17/97).

MGM's Altadena house: MGM interview (4/17/97); Berland, *Scientific Life,* p. 57.

184 **A journalist visiting the Gell-Manns around that time:** This was Theodore Berland, author of *The Scientific Life.*

"Isn't the food here vile": ibid., p. 58.

MGM described the visit to Feynman's house in "Dick Feynman—The Guy in the Office Down the Hall" (cited in the notes to chapter 7), p. 50.

184–5 **scene with Zachariasen:** Berland, *Scientific Life,* p. 58.

185 **"Q-square omega Q-square," etc.:** ibid., p. 58.

MGM's parents move to Pasadena: MGM interviews, 4/17/97 and 3/20/98.

186 **Historical background on Jason:** Goldberger interview (3/31/97); Brueckner interview (4/10/97); unpublished American Institute of Physics oral history interviews with Edwin Salpeter (3/30/78, pp. 61–68) and Gordon MacDonald (4/16/86, pp. 10–35); Kevles, *The Physicists,* pp. 402–3; Deborah Shapley, "Jason Division: Defense Consultants Who Are Also Professors Attacked," *Science* 179, no. 4072: 459–62, 505.

"Usually some of us": letter from Goldberger to MGM dated 7/23/57. Jason was started in 1959.

The early Jason membership list, dated 12/6/60, is from MGM's files.

MGM's interests outside physics: Berland, *Scientific Life,* p. 64.

186–7 **Gell-Mann's research leading to the Eightfold Way:** MGM interview (4/17/97); MGM, "Particle Theory from S-Matrix to Quarks," p. 490; Glashow interview (3/8/94); Glashow, *Interactions,* pp. 150–59; Crease and Mann, *Second Creation,* p. 266.

187 **Feynman was in on the early discussions:** Zachariasen interview, 4/2/97. In Lee Edson, "Two Men in Search of the Quark," *New York Times Magazine,* 8 October 1967, p. 68, MGM describes doing early work on SU(3) with Feynman. Years later he said he couldn't recall any collaboration (4/17/97 interview).

"Yang-Mills trick": MGM, "Particle Theory from S-Matrix to Quarks," p. 490.

MGM expressed his opinions of Glashow's strengths and weaknesses as a physicist in recommendation letters to the Sloan Foundation (4/10/61) and the University of California at Berkeley (3/21/62).

MGM's arguments with Glashow and Feynman over the Eightfold Way: Zachariasen interview, 4/2/97.

hanging out at the Beachcomber: Glashow interview, 3/8/94.

trips to Baja: interviews with Glashow (3/8/94) and Zachariasen (4/1/94).

187ff. **MGM's encounter with Richard Block and his epiphany about group theory:** MGM, "Particle Theory from S-Matrix to Quarks," p. 490; MGM interview (4/17/97); Crease and Mann, *Second Creation,* pp. 266–69.

190–2 **Final unfolding of Eightfold Way:** MGM, "Particle Theory from S-Matrix to Quarks," pp. 490–92; MGM interview (4/17/97). Good popular accounts of the work can be found in Riordan, *Hunting of the Quark,* pp. 87–93, and Crease and Mann, *Second Creation,* pp. 266–69. Bernstein provides a nice description of the organizing scheme in *The Tenth Dimension,* pp. 69–73. In his original formulation of the scheme, MGM didn't actually use charge and strangeness on the axes but rather isospin and a closely related quantity called hypercharge.

192 **"Now this, O monks, is noble truth":** in G. F. Chew, M. Gell-Mann, and A. Rosenfeld, "Strongly Interacting Particles," *Scientific American,* February 1964, p. 89.

MGM's SU(3) preprint: MGM, "The Eightfold Way: A Theory of Strong Interaction Symmetry," Caltech Synchrotron Laboratory Report 20 (1961), unpublished. Later reprinted in Gell-Mann and Ne'eman, *The Eightfold Way,* pp. 11–57.

192–3 **Glashow thought his name should also be on the paper:** Zachariasen interview (4/2/97); Glashow, personal communication (4/2/97).

193 **Gell-Mann and Glashow's attempt at unifying strong and electroweak forces:** MGM, "Particle Theory from S-Matrix to Quarks," p. 490. The work was published as "Gauge Theories of Vector Particles," *Annals of Physics* 15 (1961): 437–60.

"The model we have discussed:" ibid., p. 459.

The Berkeley coffee cup incident: Crease and Mann, *Second Creation,* p. 459, footnote 35; MGM interview (8/23/97).

"Theory of Strong Interaction Symmetry": A. Herschman, the acting editor of the *Physical Review* used this title in a letter to MGM dated 9/27/63. MGM described the contents in "Particle Theory from S-Matrix to Quarks," p. 491.

193–4 **encounter with Fowler:** MGM interview (8/23/97); unpublished MGM interview by Anderson and Else for *New Scientist,* summer 1994.

194 Fowler's accomplishments with strange particles are described in articles by William Chinowsky and William Fowler, in Brown et al., eds., *Pions to Quarks,* pp. 331–42, 342–47.

"You're not smart enough": Zweig interview, 8/20/97.

MGM's waffling over the Eightfold Way: MGM, "Particle Theory from S-Matrix to Quarks," p. 491; MGM interviews (4/17/97 and 8/23/97); unpublished interview with *New Scientist* (cited above).

MGM described his hopes for the India visit in a letter dated 1/10/61 to M. G. K. Menon at the Tata Institute in Bombay, organizer of the summer school. He also talked about the trip in an interview (4/17/97).

194–5 **"My dearest love," etc.:** letter dated 6/19/61 from Margaret Gell-Mann.

195 The La Jolla "Conference on Strong and Weak Interactions" in June 1961 is described in MGM, "Particle Theory from S-Matrix to Quarks," pp. 481–82, and MGM interview (1/22/97). I also interviewed participants Geoffrey Chew (4/1/97) and Keith Brueckner (4/10/97). Though there were no proceedings, an unpublished copy of MGM's presentation, "Symmetries, Currents, and Resonances in the Theory of Baryons and Mesons," is in his files.

"Sir, all our wines are at least five years old": MGM, "Particle Theory from S-Matrix to Quarks," p. 481.

195–6 **Gell-Mann's impressions of Chew's talk and of the bootstrap:** MGM interview (1/22/97); MGM, "Particle Theory from S-Matrix to Quarks," p. 481.

196 **"hadronic egalitarianism":** MGM interview, 12/3/97.

MGM spoke favorably of the bootstrap: MGM, "Symmetries, Currents, and Resonances in the Theory of Baryons and Mesons," pp. 3, 13.

MGM nominated Chew for the Heineman Prize: letter dated 8/17/62 to John Wheeler.

"Most of the predictions": letter to S. A. Goudsmit dated 9/11/61.

197 After all the agonizing, the Eightfold paper finally appeared as "Symmetries of Baryons and Mesons," *Physical Review* 125, no. 3 (1962): 1067–84.

"that the apparatus of field theory may be a misleading encumbrance": ibid., p. 1068.

MGM cited Chew's talk at La Jolla: ibid.

CHAPTER 10 *Holy Trinity*

198 Biographical material on Ne'eman and the story of his discovery of SU(3) can be found in his historical reminiscence, "Hadron Symmetry, Classification and Compositeness," in *Symmetries in Physics (1600–1980)* (cited in the notes to chapter 5), pp. 501–40. I also relied on Crease and Mann, *Second Creation,* pp. 269–72, and an interview with Ne'eman (4/1/97), followed by E-mail correspondence with him.

199 **"You are embarking":** Ne'eman, "Hadron Symmetry," p. 513.

200 **Ne'eman was disappointed to hear about the Sakata group:** letter from Ne'eman to Gell-Mann dated 3/7/61.

"Derivation of Strong Interactions from a Gauge Invariance" was published in *Nuclear Physics* 26 (1961): 222–29.

"Look at that!" etc.: Ne'eman, "Hadron Symmetry," p. 515. Salam also described his reaction to the Eightfold Way in a letter to MGM dated 8/31/61.

Ne'eman immediately wrote to Gell-Mann: letter from Ne'eman to Gell-Mann dated 3/7/61.

Ne'eman and Gell-Mann at Imperial College: Ne'eman interview (4/1/97); MGM interview (8/23/97).

201 Wigner's essay, "The Unreasonable Effectiveness of Mathematics in the Natural Sciences," was published in *Communications in Pure and Applied Mathematics,* 13, no. 1 (1960).

MGM's early ideas on mathematical elegance are described in Berland, *Scientific Life,* p. 67.

202 The Sakata model is described in Crease and Mann, *Second Creation,* pp. 261–62, and Riordan, *Hunting of the Quark,* pp. 81–82.

MGM influenced by Sakata-like triplets: MGM, "Particle Theory from S-Matrix to Quarks" (cited in the notes for chapter 5), p. 491.

MGM speaks about Sakata triplets at La Jolla: MGM, "Symmetries, Currents, and Resonances in the Theory of Baryons and Mesons" (cited in notes for chapter 9), p. 22.

203 **"fictitious" leptons, etc.:** "The Eightfold Way," pp. 17–18 (as reprinted in Gell-Mann and Ne'eman, *The Eightfold Way*). Of course, one now talks about quarks as being distinct from leptons, particles like the electron and neutrino that do not feel the strong force.

"Could a baryon be made of three leptons?": Ne'eman, "Hadron Symmetry," p. 518.

Ne'eman's early ideas on triplets: ibid., pp. 517–20; Ne'eman interview (4/1/97).

The Ne'eman and Goldberg paper was published in January 1963 as "Baryon Charge and R-Inversion in the Octet Model," *Nuovo Cimento* 27, no. 1 (1963). Reprinted in *Symmetries in Physics*, pp. 531–32.

204 **Pauline's death:** interview with MGM, 3/20/98. Arthur's reaction is described in his letter to the Gell-Manns in Geneva, dated 6/7/62.

"The shock of her death": ibid.

The summer in Geneva: MGM interview, 3/20/98.

205 **"except, of course, for a few orders of magnitude":** letter from Gell-Mann to Morowitz dated 11/7/61; he also compared the sixties to the twenties in a letter to David Pines, dated 11/7/62.

MGM's early work with Zachariasen on Regge poles was written up as S. C. Frautschi, F. Zachariasen, and M. Gell-Mann, "Experimental Consequences of the Hypothesis of Regge Poles," *Physical Review* 126, no. 6 (1962): 2204–18.

MGM talk on Regge trajectories at Geneva conference: *High Energy Physics*, Proceedings of the 11th Rochester Conference, pp. 533–42. The conference is also described in Polkinghorne, *Rochester Roundabout*, pp. 94–106, and Crease and Mann, *Second Creation*, pp. 273–74.

206–7 Ne'eman describes his version of the omega-minus discovery in "Hadron Symmetry," pp. 520–22. MGM gives his in "Particle Theory from S-Matrix to Quarks," p. 492. Gerson Goldhaber described his and Sula Goldhaber's meeting with Ne'eman in "The Encounter on the Bus," in E. Gotsman and G. Tauber, eds., *From SU(3) to Gravity* (Cambridge, England: Cambridge University Press, 1985), pp. 107–12. I also relied on the BBC documentary "Strangeness Minus Three," cited in notes for chapter 7, which includes interviews with MGM, Ne'eman, and Samios, and on my own interviews with Ne'eman (4/1/97) and MGM (8/23/97).

207 **MGM mentioned the omega-minus in 1955:** "The Interpretation of the New Particles as Displaced Charge Multiplets," *Nuovo Cimento Supplement* 4 (1956): appendix, p. 865.

208 **"Did you tell him about your prediction?":** Ne'eman interview, 4/1/97.

MGM's statement at Geneva: Rochester 11 Proceedings, p. 805. His slip-up is not reproduced there.

208 Samios described his thoughts about the omega-minus and the lunch encounter with MGM in "Strangeness Minus Three."

Dyson's reaction to Eightfold Way: Dyson interview, 1/15/98.

209 **details on work with Regge poles:** interviews with Goldberger (8/26/97), Zachariasen (4/2/97), and MGM (8/23/97). General descriptions of Reggeism can be found in G. Chew, M. Gell-Mann, A. Rosenfeld, "Strongly Interacting Particles," *Scientific American*, February 1964, pp.74–93; Riordan, *Hunting of the Quark*, pp. 85–86; and Pickering, *Constructing Quarks*, pp.

75–78. Four articles eventually came out of the collaboration: "Elementary Particles of Conventional Field Theory as Regge Poles" (with M. Goldberger), *Physical Review Letters* 9 (1962): 275–77; "Elementary Particles of Conventional Field Theory as Regge Poles II" (with M. Goldberger, F. Low, and F. Zachariasen), *Physics Letters* 4 (1963): 265–67; "Elementary Particles of Conventional Field Theory as Regge Poles III" (with M. Goldberger, F. Low, E. Marx, and F. Zachariasen), *Physical Review B* 133 (1964): 145–60; "Elementary Particles of Conventional Field Theory as Regge Poles IV" (with M. Goldberger, F. Low, V. Singh, and F. Zachariasen), *Physical Review B* 133 (1964): 161–74.

210 The gourmet dinner at Schwinger's house was described in interviews with Goldberger (8/26/97) and Low (8/6/96).

MGM is promised a corner office with a view: letter from Charles Zemach dated 9/30/63.

"coup to end all coups," etc.: letter dated 5/6/63 from Chew to MGM, Goldberger, and Low.

211 **"the Billy Graham of physics":** letter from Goldberger to MGM dated 1/27/62.

The story of the encounter with Serber is told in MGM, "Particle Theory from S-Matrix to Quarks," p. 493, and in his talk "Quarks, Color, and QCD," in Hoddeson et al., *Rise of the Standard Model*, p. 625. Serber gave his account in his memoirs, *Peace and War*. I am grateful to his editor, Robert Crease, for making the relevant portion of the manuscript available. In the account (pp. 119–201), Serber says that MGM had not thought of the triplet scheme ("It was news to Murray") until Serber confronted him with the idea at Columbia, and that some time later MGM "told Murph Goldberger that he had never thought of it." Goldberger did not recall such a conversation (8/26/97 interview), and Gell-Mann has insisted in all his accounts and interviews (see earlier references in this chapter) that he was already playing with the idea at MIT and that Serber merely jogged him into taking it more seriously. Considering that Gell-Mann used fictitious triplets in his original Eightfold Way preprint and that it was mathematically evident that the fundamental representation of SU(3) was a triplet, Serber may have overestimated his influence.

T. D. Lee's reaction to triplets: Riordan, *Hunting of the Quark*, p. 101. He cites an interview with Lee.

212 **origin of "kworks":** Serber remembered the name coming up at the coffee after the colloquium and wrote that Dr. Bacqui Beg of Rockefeller University, who was present, recalled MGM saying "the existence of such a particle would be a strange quirk of nature, and quirk was jokingly transformed into quark" (Serber memoir, p. 200). MGM doesn't remember when he first uttered the word (8/23/97 interview). In Edson's 1967 *New York Times Magazine* article (cited in the notes to chapter 9), p. 54, MGM says the name first came up in a bout of wordplay between him and Feynman.

Details of Jason's fall meeting at Woods Hole are described in numerous memos from the Institute for Defense Analysis filed with MGM's 1963 papers.

213 **Ne'eman at Caltech with MGM and Feynman:** Ne'eman interview, 4/1/97.

Ne'eman discussed his own triplet scheme with MGM: personal communication from Ne'eman, 8/28/97.

MGM's attitude toward Jews and Israel: Several of his friends and colleagues mentioned this, but not for attribution: "One word that he would avoid like the black plague was the word 'Jew,' " one told me.

his friend simply refused to be categorized: Glashow interview, 3/8/94.

213–14 **"Jews and science?" etc.:** The story is from an interview with Ne'eman (4/1/97).

214 **"Three quarks for Muster Mark":** The line is on page 383 of the 1939 Viking edition.

MGM's version of the quark tale is told in MGM, "Particle Theory from S-Matrix to Quarks," pp. 493–97, and MGM, *Quark and Jaguar,* pp. 180–81.

MGM on the pronunciation of "quark": ibid.

"Please, Murray," etc.: MGM, "Particle Theory from S-Matrix to Quarks," p. 495.

the quark paper: MGM, "A Schematic Model of Baryons and Mesons," *Physics Letters* 8, no. 3 (1964): 214–15.

Mandelstam's reaction to quarks: MGM, "Particle Theory from S-Matrix to Quarks," p. 496.

215 **"It is fun to speculate," etc.:** MGM, "A Schematic Model of Baryons and Mesons," p. 215.

216 **The editor of *Physics Letters* was pressing MGM to submit a paper:** Crease and Mann, *Second Creation,* p. 284.

Ne'eman's Coral Gables story is told by him in "Strangeness Minus Three," cited in notes for chapter 7.

217 The discovery of the omega-minus is recounted by Samios in "Early Baryon and Meson Spectrosopy Culminating in the Discovery of the Omega-Minus and Charmed Baryons," in Hoddeson et al., *Rise of the Standard Model,* pp. 525–41, in "Strangeness Minus Three," and in Crease, *Making Physics,* pp. 304–14. Fine popular accounts are given in Riordan, *Hunting of the Quark,* pp. 95–98; Crease and Mann, *Second Creation,* pp. 274–79; and Close et al., *The Particle Explosion,* pp. 141–42.

218 **MGM hears the good news about omega-minus:** Crease and Mann, *Second Creation,* p. 279; MGM interview (8/23/97).

"Please excuse the oversight": quoted in Ne'eman, "Hadron Symmetry," p. 522.

Yang didn't believe finding the omega-minus was decisive: R. J. Oakes and C. N. Yang, "Meson-Baryon Resonances and the Mass Formula," *Physical Review Letters* 12 (1964): 174–78.

Feynman's wisecrack about the omega-minus: Ne'eman, "Hadron Symmetry," p. 522.

Serber kicked himself for not publishing the quark idea: Serber, *Peace and War,* p. 201.

CHAPTER 11   *Aces and Quarks*

219 **Gell-Mann's comments on precociousness in particle physics:** Berland, *Scientific Life,* p. 52.

220 Zweig told his own story in "Origins of the Quark Model," later published in *Baryon '80,* Proceedings of the Fourth International Conference on Baryon Resonances, July 14–16, 1980 (Toronto: University of Toronto, 1981). Page numbers here refer to the original typescript. I also relied on an interview with Zweig (8/20/97), and on Riordan, *Hunting of the Quark,* pp. 103–8.

220–1 **Zweig's early impressions of MGM:** Zweig, "Origins of the Quark Model," p. 17.

221 **how graduate students thought of him:** Robert Weinman, personal communication, 8/18/98.

the Mayan incident: ibid.

being interrupted by MGM: Lederman interview, 1/19/98.

Feynman could be nasty too: interviews with former colleagues, including Zachariasen (4/2/97) and Zweig (8/20/97).

Weinberg gets the Caltech treatment: Weinberg interview (1/15/98); Zachariasen interview (4/2/97).

221–2 **Weinberg and MGM at Gatlinburg:** Weinberg interview (1/15/98); MGM interview (3/20/98).

222 **"Does the vanishing of the denominator":** Weinberg's question and Gell-Mann's terse response are recorded in *High Energy Physics,* Proceedings of the 11th Rochester Conference, p. 542.

MGM gave his opinion of Weinberg's physics in a recommendation letter, dated 11/6/61, to the University of California at Berkeley.

223 Yang and Lee's breakup is documented in Regis, *Who Got Einstein's Office,* pp. 144–47.

225–6 **Zweig's aces papers:** "An SU(3) Model for Strong Interaction Symmetry and Its Breaking," CERN preprint 8182/TH.401, 17 January 1964, and "An SU(3) Model for Strong Interaction Symmetry and Its Breaking II," CERN preprint 8419/TH.412, 21 February 1964.

226 **Zweig felt sick when he read MGM's quark paper:** Zweig, personal communication, 10/12/97.

"We will work with these aces as fundamental units": first Zweig preprint, p. 2.

"Are aces particles?" etc.: second Zweig preprint, pp. 40–42.

227 **These were the words of an iconoclast, a romantic:** Zweig, "Origins of the Quark Model," p. 35.

228 **Zweig's talk at Erice:** A. Zichichi, ed., *Symmetries in Elementary Particle Physics,* Proceedings of the 1964 summer school at the Ettore Majorana Centre in Erice, Sicily (New York: Academic Press, 1965).

The description of Erice is from Glashow, *Interactions,* p. 209, and various travel guides.

MGM's diverse interests beyond physics: He discussed this as early as 1967 in a television interview, "Particles Are a Family Affair" (cited in the notes to chapter 2).

MGM's mid-sixties speculations on psychology: MGM, "Personal Statement" (cited in the notes to chapter 1), pp. 8–9.

228–9 **"Dear Murray":** letter from Goldberger dated 1/27/62.

229 **MGM's effort to expand Caltech's interests:** MGM interview (12/3/97); Berland, *Scientific Life,* p. 64; MGM, *Quark and Jaguar,* p. 116; Goldberger, "Remarks on the Occasion of Murray Gell-Mann's More or Less 60th Birthday," in J. Schwarz, ed., *Elementary Particles and the Universe,* p. 212. MGM made the "barber college" remark to several colleagues, including Glashow (interview 3/8/94).

"Murray, you don't have to go": quoted in Goldberger, "Remarks on the Occasion of Murray Gell-Mann's More or Less 60th Birthday," p. 212.

230 **"I became frightened":** letter from MGM to Weisskopf.

MGM's efforts to get a joint MIT-Harvard appointment are documented in letters from MGM to Weisskopf (7/15/63) and Weisskopf to MGM (5/8/63 and 5/12/63).

colleagues joked about MGM's demands on Harvard: Edson, "Two Men in Search of the Quark," *New York Times Magazine,* 8 October 1967, p. 56.

"I like diversity,": ibid., p. 74.

The history of the Pugwash conferences is described throughout Rotblat, *Science and World Affairs.* Details of the meeting in Udaipur are in Rotblat, *Scientists in the Quest for Peace,* pp. 58–60.

230–1 MGM's impressions of the Udaipur meeting are from MGM, "Personal

Statement," pp. 10–12, and interviews with him (8/23/97) and Ruina (11/11/98). I also relied on a letter to me from Ruina (6/21/95).

231 **MGM and Ruina's paper:** "Ballistic Missile Defence and the Arms Race," Proceedings of the Twelfth Pugwash Conference on Science and World Affairs: Current Problems of Disarmament and World Security, Udaipur, India, January 27 to February 1, 1964, pp. 232–5.

The meeting where Millionshchikov sought out MGM was the Dubna conference, described later in the chapter. The anecdote is from MGM, "Personal Statement," p. 11.

MGM's visit to Kyoto: "Particle Theory from S-Matrix to Quarks" (cited in the notes to chapter 5), p. 494; MGM interview (8/23/97).

233 The Dubna meeting is described in Polkinghorne, *Rochester Roundabout*, p. 107; Glashow, *Interactions*, pp. 191–92; MGM interview (8/23/97); and *High Energy Physics*, Proceedings of the 12th Rochester Conference.

234 **quarks went over "like a lead balloon":** Crease and Mann, *Second Creation*, p. 283; they cite their interview with Gell-Mann on 3/3/83.

Salam's talk: "Symmetry of Strong Interactions," Rochester 12 Proceedings, pp. 834–47.

"more impressive than one has any right to expect," etc.: Salam, pp. 838–39.

Schwinger's paper: "Field Theory of Matter," *Physical Review* 135, no. 3B (1964): 816–30.

"He introduces particles of fractional charge": ibid., p. 817, footnote 5.

T. D. Lee's triplet scheme is described on p. 839 of the Rochester 12 Proceedings.

Glashow and Bjorken's fourth quark: Bjorken and Glashow, *Physics Letters* 11, no. 3 (1964): 255–57. The story of the theory can be found in Crease and Mann, *Second Creation*, pp. 291–92, and Glashow, *Interactions*, pp. 189–91.

"Some people do not know when to stop": Rochester 12 Proceedings, p. 841.

235 **Salam's cartoon:** ibid., pp. 834–47.

MGM's talk at Dubna: MGM, "Possible Triplets in the Eightfold Way," Rochester 12 Proceedings, pp. 809–10.

"Field theories involving triplets may be very useful": ibid., p. 809 (correcting typos).

"which may or may not have anything to do with reality": MGM, "The symmetry Group of Vector and Axial Currents," p. 198.

the veal-and-pheasant paper: MGM, "The Symmetry Group of Vector and Axial Currents," *Physics* 1 (1964) preprint: 63–75; reprinted in MGM and Ne'eman, *The Eightfold Way*, pp. 172–206. MGM attributes the recipe to Telegdi in "Progress in Elementary Particle Theory," Brown et al., *Pions to Quarks*, p. 709. Though there are all kinds of elaborate concoctions described in the 1941 American edition of *The Escoffier Cookbook and Guide to the Fine Art of Cookery*, including pheasant stuffed with duck foie gras and truffles, and veal roast stuffed with larks, quails, or thrushes, I could find no mention of pheasant cooked between slices of veal. Escoffier recommends roasting the birds by wrapping them with ordinary bacon, though he allows that sometimes veal or beef fat may be used (p. 117, section 252). Maybe Telegdi misread the passage. He denies (personal communication, 2/6/98) that he was pulling MGM's leg.

236 **Gell-Mann called quarks "fictitious":** For example, in the 1966 Rochester conference at Berkeley, he referred to "three hypothetical and probably fictitious spin ½ quarks" (*High Energy Physics*, Proceedings of the 13th Rochester Conference, p. 5). As late as 1972, he was still using the term "fictitious" inter-

changeably with "mathematical" when describing quarks: MGM, "Quarks," *Acta Physica Austriaca,* suppl. 9 (1972): 738, 746.

"If quarks are not found, remember I never said they would be": Polkinghorne, *Rochester Roundabout,* p. 110.

The trip to Delhi and Kathmandu: Zachariaesen, personal communication (1/9/98); MGM interview (12/3/97).

238 **MGM and Ne'eman anticipate the statistics problem:** MGM, "Quarks, Color, and QCD," p. 624. Ne'eman corroborated this in an e-mail dated 9/28/97. The article they consulted was by S. Kamefuchi.

Pickering elaborates on the history of parastatistics and color in *Constructing Quarks,* pp. 218–19.

Greenberg's principal parastatistics paper was published in *Physical Review Letters* 13, no. 20 (1964): 598–602.

Han and Nambu's paper: "Three-Triplet Model with Double SU(3) Symmetry," *Physical Review* 139, no. 4B (1965): 1006–10.

239 **how a theorist should be judged:** MGM, "Particle Theory from S-Matrix to Quarks," p. 476.

Zweig called a charlatan: Zweig, "Origins of the Quark Model," p. 36. He hasn't revealed the identity of his accuser, whom he describes here as "one of the most respected spokesmen for all of theoretical physics."

Gell-Mann thought Zweig was outstanding: He expressed this opinion in recommendations to the National Academy of Science (1/7/63) and the Sloan Foundation (11/18/65).

"concrete block model": Riordan, *Hunting of the Quark,* p. 108; Zweig interview.

"quarkonics industry": MGM, "Particle Theory from S-Matrix to Quarks," p. 495.

"a chief criterion," etc.: MGM, "Particles and Principles," reprinted in *Physics Today,* November 1964, pp. 22–28.

CHAPTER 12   *The Swedish Prize*

241 The account of Feynman's Nobel Prize is from Gleick, *Genius,* pp. 377–83.

"sweeping them under the rug": ibid, p. 378.

"a dippy process," "hocus-pocus": Feynman, *QED,* p. 128.

242 **MGM's reaction to Feynman's prize:** Zweig interview, 8/20/97.

The fruitless search for quarks is described in Riordan, *Hunting of the Quark,* pp. 112–16, and Polkinghorne, *Rochester Roundabout,* p. 110.

243 **looking for quarks in oysters:** MGM, "The Elementary Particles of Matter," *Engineering and Science,* January 1967, p. 25.

Australian researcher is "99 percent sure": reported in a UPI dispatch in the *New York Times,* 4 September 1969, and in *Time,* 12 September 1969.

244 **"the best feature about quarks," etc.:** The exchange is recorded in A. Zichichi, ed., *Strong and Weak Interactions: Present Problems,* Proceedings of the 1966 summer school at the Ettore Majorana Centre in Erice, Sicily (New York: Academic Press, 1966), pp. 786–87.

Berkeley conference organizers decided to let MGM give whole presentation: Goldberger told this story in *Science* 166, no. 3906 (1969): 720–22. The conference is described in Polkinghorne, *Rochester Roundabout,* pp. 116–21, and in *High Energy Physics,* Proceedings of the 13th Rochester Conference.

"In some respects" etc.: Rochester 13 Proceedings, p. 3.

MGM thought field theory and the S-matrix were just different perspectives: MGM, "Present Status of the Fundamental Interactions," in A. Zichichi, ed., *Hadrons and Their Interactions,* Proceedings of the 1967 summer school at the

Ettore Majorana Centre in Erice, Sicily (New York: Academic Press, 1968), p. 695.
245 **"chimeron":** Rochester 13 Proceedings, p. 9.
"I think that it's not discouraging": ibid.
"probably fictitious," etc.: ibid., p. 5.
MGM's early notion that quarks are permanently trapped: ibid. Here is how he put it: "One may think of mathematical quarks as the limit of real light quarks confined by a barrier, as the barrier goes to an infinitely high one."
MGM tells the Dirac story in "Present Status of the Fundamental Interactions," in A. Zichichi, *Hadrons and Their Interactions,* p. 693.
246 A film of MGM's Caltech anniversary lecture, "About the Elementary Particles of Physics," October 1966, has been transferred to videotape and is in the school's archives. A slightly edited version, "The Elementary Particles of Matter," was later published in the Caltech magazine, *Engineering and Science,* January 1967, pp. 20–25. MGM gave essentially the same lecture at the British Royal Institution: *Proceedings of the Royal Institution* 41, no. 189 (1966): 150–61.
"Cosmic Gall" appeared in *The New Yorker* in 1960. From *Collected Poems 1953–1993* by John Updike. Copyright © 1993 by John Updike. Reprinted by permission of Alfred A. Knopf Inc.
246–7 "Perils of Modern Living," copyright 1956, originally published in *The New Yorker.*
248 **"purely illusory," etc.:** MGM, "Present Status of the Fundamental Interactions," p. 694.
"The hottest properties," etc.: Edson, "Two Men in Search of the Quark," *New York Times Magazine,* 8 October 1967, p. 74.
249 **history of SLAC and the deep inelastic scattering experiments:** Riordan, *Hunting of the Quark,* pp. 136–68; Crease and Mann, *Second Creation,* pp. 299–307; Gleick, *Genius,* pp. 391–96; Pickering, *Constructing Quarks,* pp. 125–43. The work is described in detail in essays by two experimenters published in Hoddeson et al., *Rise of the Standard Model:* Jerome Friedman, "Deep-Inelastic Scattering and the Discovery of Quarks," pp. 566–88, and James Bjorken, "Deep-Inelastic Scattering: From Current Algebra to Partons," pp. 589–99.
"The peach didn't seem to have a pit": Crease and Mann, *Second Creation,* p. 303.
250 **"I would also like to disassociate myself":** *Third International Symposium on Electron and Photon Interactions at High Energy,* Stanford, Calif., September 5–9, 1967, p. 126.
bootstrappers' explanation of SLAC data: Riordan, *Hunting of the Quark,* p. 147.
250–1 **Feynman descends on SLAC:** described in an untranscribed tape recording of an American Institute of Physics oral history interview with Feynman conducted 2/4/73 by Charles Weiner; in Gleick, *Genius,* pp. 391–96; Crease and Mann, *Second Creation,* pp. 304–7; Riordan, *Hunting of the Quark,* pp. 148–54; Gribben and Gribben, *Richard Feynman,* pp. 196–99; and Bjorken, "Feynman and Partons," in Brown et al., eds., *Most of the Good Stuff,* pp. 89–96.
251 **Feynman was embarrassed by his ignorance of the strong force:** AIP oral history interview with Feynman by Charles Weiner.
"I've always taken the attitude": ibid.
"One of the reasons why I haven't done anything," etc.: from a transcript of Feynman's talk in the Caltech archives: "Feynman Papers," Box 2, folder 19.
Feynman tries to turn down a Marshak invitation: documented in correspondence between Feynman and Marshak dated 10/21/66, 12/14/66, 1/16/67, and 1/26/67; "Feynman Papers," Box 2, folder 19.

Feynman and Zweig at lunch: Zweig interview, 8/20/97.

253 **"Of course you must know this":** Bjorken, "Feyman and Partons," in *Most of the Good Stuff,* p. 91.

The Vienna conference is described in Riordan, *Hunting of the Quark,* pp. 154–55; Polkinghorne, *Rochester Roundabout,* pp. 122–29; and *High Energy Physics,* Proceedings of the 14th Rochester Conference.

MGM didn't follow SLAC developments closely: MGM interview, 12/3/97.

The Gell-Manns' life in Princeton: ibid.

253–4 The *Los Angeles Times* story about Pasadena appeared 27 April 1969.

254 **Feynman at the topless bar:** Mehra, *Beat of a Different Drum,* pp. 583–84; Gleick, *Genius,* p. 412.

"That's Dick's view of civil rights": Telegdi interview, 3/31/94.

MGM's reaction to partons: MGM interview, 12/3/97. Also see the accounts in Gleick, *Genius;* Crease and Mann, *Second Creation;* and Riordan, *Hunting of the Quark.*

MGM was put off by the lack of a cooperative spirit at the Institute: Regis, *Who Got Einstein's Office?,* p. 144.

the Institute as a place of Machiavellian intrigue: ibid., pp. 202–7; MGM interview (3/20/98). The controversy over Carl Kaysen's appointment was widely covered at the time in the national press.

255 **MGM's decision to turn down full-time Institute job:** MGM interviews (12/3/97 and 3/20/98); Regis, *Who Got Einstein's Office?,* p. 283.

MGM rejects overtures for Lucasian chair: documented in various letters from early 1968 in MGM's personal files.

The paper on the weak interaction: MGM, M. Goldberger, N. M. Kroll, F. E. Low, "Amelioration of Divergence Difficulties in the Theory of Weak Interactions," *Physical Review* 179 (1969): 1518–27.

"an absolutely disgusting model": David Politzer, quoted in Crease and Mann, *Second Creation,* p. 322.

"a piece of trash": MGM interview, 3/20/98.

trip back to Pasadena, reaction to RFK's death: MGM interview, 12/3/97.

256 **Jason and Vietnam:** interviews with MGM (3/20/98), Dyson (1/15/98), and Weinberg (1/15/98). Also see general Jason references in the notes to chapter 9.

257 **MGM and the Nixon administration:** MGM interview, 12/3/97. His meetings with Rockefeller and Kissinger and his work on Nixon's science advisory committee are documented in various letters in his personal files.

training the dog to grimace at the name "Nixon": Telegdi interview (3/31/94); MGM interview (3/20/98).

deciding to sell the Altadena house: MGM interview, 3/20/98.

258 **description of the Armada house:** *Pasadena Star-News* interview with Margaret Gell-Mann, 4 November 1969; MGM interview (3/20/98); personal visit to neighborhood, spring 1994.

life in the new neighborhood: interviews with MGM (12/3/97), Zachariasen (4/1/94), Paul Macready (5/19/95), Judy Macready (11/11/98), and Nick Gell-Mann (5/26/98); school reports from Pasadena Polytechnic in MGM's personal files.

259 **the husky story:** Berreby interview with MGM for *New York Times Magazine* profile (cited in the notes to chapter 1).

Gell-Mann's problem moving from Pasadena: MGM interview, 12/3/97.

260 **wanting to build a log cabin:** Profiles of Margaret Gell-Mann in *Los Angeles Times,* 7 November 1969, and *Pasadena Star-News,* 4 November 1969.

reasons for buying the Armada house: MGM interview, 12/3/97.

possible Nobel hint from T. D. Lee: ibid.

260–1 **"Standard physics cocktail party conversation":** Goldberger, *Science* 166 (1969): 720.

261 **Delbrück and the film crew:** MGM interview, 12/3/97.

MGM's expectations concerning the prize: *Newsweek,* 10 November 1969, p. 70. Details of his first day as a laureate are also from MGM interview, 12/3/97; *Time,* 7 November 1969; p. 52; *Los Angeles Times,* 31 October 1969; *California Tech,* special edition, 31 October 1969; and *New York Times,* 31 October 1969.

"Dr. Gell-Mann has contributed probably more than anyone": quoted in "Gell-Mann: Order Out of Chaos," *Engineering and Science,* November 1969, p. 14.

"This event marks the public recognition," etc.: ibid.

262 **MGM brings in champagne for the theory group:** personal communication from Zachariasen, 1/9/98.

celebrating at the Beachcomber: MGM interview, 12/3/97.

"The quark is just a notion," etc.: quoted in the *Los Angeles Times,* 31 October 1969.

MGM's brother Ben sent to Stockholm: MGM interview, 12/3/97.

The description of the Nobel events is based on articles in the Swedish press (*Svenska Dagbladet,* 12 December 1969) and material from the files of the Nobel Foundation, including the program for the awards ceremony, the menu from the banquet, and MGM's schedule for the week. He described his own memories of the events in interviews (12/3/97 and 3/20/98). The presentation speech by Ivar Waller is reproduced in *Nobel Lectures: Physics 1963–1970* (Amsterdam: Elsevier, 1972), pp. 295–98. Further details of the award ceremony are in Frank N. Magill, ed., *The Nobel Prize Winners: Physics,* vol. 3, 1968–1988 (Pasadena: Salem Press, 1989), pp. 946–51. Swedish Science Radio provided a tape recording of the Nobel banquet, including MGM's acceptance speech, which was reprinted, slightly edited, in *Adventures in Experimental Physics,* vol. gamma, Princeton, N.J.: World Science Communications, 1972), p. iv. I also found a typescript of the speech in MGM's files, and relied on interviews with him (12/3/97 and 3/20/98) and a personal visit to the 1997 Nobel festivities. (Those who have attended Nobel ceremonies after 1973 may be surprised to read that the banquet was held upstairs in the Golden Hall rather than downstairs in the more spacious Blue Hall.) Marie Granmar gave me her impressions of Gell-Mann's Swedish.

263 **"It has not yet been possible to find individual quarks," etc.:** Waller speech, *Nobel Lectures,* pp. 295–98.

264 **MGM's thoughts on economics prize:** MGM interview, 12/3/97.

265 **"Your Majesty, Your Royal Highnesses":** lecture as recorded by Swedish Science Radio.

CHAPTER 13 *Quantum Chromodynamics*

267 Gell-Mann spoke of the warning from Hoyle in a 1967 NET documentary called "Particles Are a Family Affair"; a copy is in the Caltech archives.

"Please feel free": undated letter to K. T. Mahanthappa, circa 1970, in MGM's files.

"Unfortunately when I make promises": letter dated 8/11/70 to E. L. Feinberg and V. L. Ginzburg, Lebedev Physical Institute, Moscow.

267–8 **"It would be a great pleasure":** letter dated 1/10/69 to Marcello Conversi.

268 **"Manuscript contains many serious distortions":** letter dated 2/29/68 to J. Geheniau, Université Libre de Bruxelles.

MGM's recycled Nobel lecture: interview with MGM, 7/30/98. The versions delivered at Caltech and in London are cited in the notes to chapter 12. Though he never submitted a lecture for publication by the Nobel committee, a Swedish translation of the talk published in the proceedings of the British Royal Institution appeared a year later: *Kosmos* 47 (1970): 11–17.

268–9 **"TERRIBLY SORRY":** telegram dated 1/19/70, to Nobel Foundation; **"EDITOR TROUBLED":** 3/3/70 from Nobel Foundation; **"VERY SORRY TO BE EVEN MORE DELINQUENT":** 3/9/70 to Nobel Foundation; **"FURTHER ABJECT APOLOGIES":** 5/4/70 to Nobel Foundation; **"ASSUME YOUR UNPRECEDENTED DELAY":** 6/2/70 from Nobel Foundation; **"VERY MUCH REGRET FAILURE":** 6/5/70 to Nobel Foundation.

270 For a description of Aspen's social millieu, see Ted Conover's *Whiteout: Lost in Aspen.*

270–1 **Lisa in Aspen:** interviews with Margaret Low Smith (3/2/95), MGM (6/30/98), and Nick Gell-Mann (5/96/98).

271 The 1969 Aspen symposium, "Progress in a Living Environment," was covered by the *San Francisco Chronicle* (22 and 23 September 1969), the *New York Times* (22 September 1969), the *Christian Science Monitor* (27 September 1969), *Newsday* (22 September 1969), the *Guardian* (23 September 1969), and *New Scientist* (9 October 1969).

"The less of us there are": quoted in Victory McElheny, "Aspen Technology Conference Ends in Chaos," *Science* 169, no. 3951, p. 1187.

"the world is indeed beset": ibid.

272 The various 1970 conferences are documented in letters in MGM's files.

MGM's thoughts on Santa Barbara and bird-watching: "Talk Given by Murray Gell-Mann at Dedication of New Physics Building at University of California at Santa Barbara, Monday, October 26, 1970," typescript in MGM's files. A slightly edited version of the talk was later published as "How Scientists Can Really Help," *Physics Today,* May 1971, pp. 23–25.

"For me the two things are inseparable," etc.: ibid.

273 **"How many people did Jason kill last year?":** The incident was described in the *Denver Post,* 30 July 1969.

274 Gell-Mann's work with Fritzsch is described in MGM, "Quarks, Colors, and QCD," in Brown et al., *Rise of the Standard Model,* pp. 629–33; Fritzsch, "The Development of Quantum Chromodynamics," in Doncel et al., *Symmetries in Physics (1600–1980)* (cited in the notes to chapter 5), pp. 595–608; Pickering, *Constructing Quarks,* pp. 222–23; Crease and Mann, *Second Creation,* pp. 327–29; and Riordan, *Hunting of the Quark,* pp. 228–54. I also relied on interviews with Gell-Mann (12/3/97) and Fritzsch (4/3/98), followed by several E-mail exchanges.

Full citations to the SLAC experiments indicating that partons had the characteristics of quarks can be found in the histories cited above. The work was later described in a lecture by Feynman, published as "Structure of the Proton," *Science* 183, no. 4125 (1974): 601–10.

"Our eyes alone": Riordan, *Hunting of the Quark,* p. 188.

275 **"Subatomic Tests Suggest a New Layer of Matter":** The *Times* article appeared 25 April 1971, followed by the editorial on 1 May 1971.

"Write about Zweig": The comments, in MGM's handwriting, are on a copy of the editorial in his personal files.

"Hadrons act as if they are made": from "Quarks," a lecture given by MGM at a conference in Schladming, Austria, held 21 February to 4 March 1972; published in *Acta Physica Austriaca,* Suppl. 9 (1972): 733–61; the quote is on p. 746.

276 Gell-Mann's opinion of Fritzsch can be found in his recommendation letters dated 2/11/71 to the Max Planck Institute for Physics and Astrophysics, Munich, and 12/26/73 to the University of Hamburg.

MGM's first two papers with Fritzsch: "Scale Invariance and the Light Cone," *Proceedings of the 1971 Coral Gables Conference on Fundamental Interactions at High Energy* (New York: Gordon & Breach, 1971), pp. 1–53; "Light Cone Current Algebra," in E. Gotsman, ed., *Proceedings of the International Conference on Duality and Symmetry in Hadron Physics* (Weizmann Science Press of Israel, 1971), pp. 317–74.

276–7 **"to get the benefits without the embarrassments":** Polkinghorne, *Rochester Roundabout,* p. 139.

277 **"Nature reads in the book of free field theory":** Fritzsch, "The Development of Quantum Chromodynamics," p. 600.

The story of electroweak unification, greatly compressed here, is described in more detail in Crease and Mann, *Second Creation,* pp. 231–49; Riordan, *Hunting of the Quark,* pp. 198–204; and Sidney Coleman, "The 1979 Nobel Prize in Physics," *Science* 206, no. 4424 (1979): 1290–92.

Salam's analogy between symmetry-breaking and choosing a salad plate is in Riordan, *Hunting of the Quark,* p. 204.

279 **"such an extraordinarily ad hoc and ugly theory":** Thomas Kibble, quoted in Crease and Mann, *Second Creation,* p. 248.

"an absolutely disgusting model": The paper and the comment are cited in the notes to chapter 12.

279–80 The SLAC experiments giving hints of gluons were also described in Feynman, "Structure of the Proton." Full citations to the experiments can be found in the histories cited earlier in this chapter.

280 **Origin of "gluon":** MGM used it unattributed on page 1073 of "Symmetries of Baryons and Mesons" (cited in the notes to chapter 9). It was in an interview (12/3/97), that he speculated that Teller coined the term for use in an entirely different context.

The Kuti and Weisskopf paper: "Inelastic Lepton-Proton Scattering and Lepton Pair Production in the Relativistic Quark-Parton Model," *Physical Review D* 4 (1971): 3418–39.

280–1 **"unnecessary scaffolding," etc.:** Quoted in Bjorken, "Feynman and Partons," *Most of the Good Stuff,* p. 94. He cites Feynman, *Photon-Hadron Interactions* (Reading, Mass: W. A. Benjamin, 1972).

281 **Feynman felt like a fly:** untranscribed AIP interview (2/4/73) with Feynman by Charles Weiner (cited in the notes to Chapter 12).

MGM described his Aspen real estate dealings in an interview (12/3/97).

MGM arranged financing for Geneva: documented in letters in his files.

281–2 Gell-Mann's return to the statistics problem and the road to color was described in his "Quarks, Colors, and QCD," and in Fritzsch, "The Development of Quantum Chromodynamics." The story of color is also told in Crease and Mann, *Second Creation,* pp. 297–99; Pickering, *Constructing Quarks,* pp. 218–19; and Riordan, *Hunting of the Quark,* pp. 228–30.

282 **"the quark hypothesis is strongly excluded":** Adler is quoted in Crease and Mann, *Second Creation,* p. 328.

MGM and Fritszch dealt with pion decay in their paper, "Light Cone Algebra," where they used Greenberg's term "parastatistics of rank three" on p. 361.

The Indiana physicist who used "color" was D. B. Lichtenberg, in *Unitary Symmetry and Elementary Particles,* (New York: Academic Press, 1970).

MGM's paper with Fritzsch and Bardeen: "Light Cone Algebra, π^0 decay and e$^+$e$^-$ Annihilation," in R. Gatto, ed., *Scale and Conformal Symmetry in Hadron Physics* (New York: Wiley, 1973), pp. 139–51.

283 Nambu's paper with Han is cited in the notes to chapter 11. He elaborated on his theory in an E-mail to me dated 4/15/98.

284 MGM gave his opinion of Nambu in letters of recommendation to the University of Pennsylvania (2/20/61) and the University of Colorado (4/21/61). He told the story of missing Nambu's second paper in "Quarks, Color, and QCD," pp. 628–29, and in an interview (8/23/97).

The Nambu paper Murray didn't read: "A Systematics of Hadrons in Subnuclear Physics," in A. De-Shalit, H. Feshbach, and L. Van Hove, eds., *Preludes in Theoretical Physics: In Honor of V. F. Weisskopf* (Amsterdam: North Holland: 1966), pp. 133–43.

"a very drastic and far-reaching": Dalitz, "Symmetries and the Strong Interactions," *High Energy Physics,* Proceedings of the 13th Rochester Conference, p. 232.

285 **"If the quarks are only mathematical entities,"** etc.: MGM's exchanges with Wigner and Nambu are from a typewritten transcript of the session, sent to MGM by Jay Melosh (7/29/72).

Veneziano's work is described in Kaku and Trainer, *Beyond Einstein,* pp. 89–90. Gell-Mann's early thoughts about duality and string theory are in his paper "Quarks, Color, and QCD," p. 631, and his 1972 Fermilab talk with Fritzsch: "Current Algebra: Quarks and What Else?" in *High Energy Physics,* Proceedings of the 16th Rochester Conference, Section II, p. 137. He elaborated on the subject in an interview (7/30/98).

286 The Paris demonstrations against MGM were described in a short news item, "Gell-Mann Protested in Paris," *Science* 177, no. 4047 (1972): 414, and in an interview (3/20/98).

287 **"For more than a decade, we particle theorists":** MGM and H. Fritsch, "Current Algebra," in Rochester 16 Proceedings, Section II, p. 136.

MGM's recollections of the Fermilab talk are in his "Quarks, Color, and QCD," pp. 631–32. David Gross gave a somewhat different account of the talk in "Asymptotic Freedom and the Emergence of QCD," in Hoddeson et al., *Rise of the Standard Model,* p. 222.

287–8 **MGM's Tower of Babel remarks:** "General Status: Summary and Outlook," Rochester 16 Proceedings, Section IV, pp. 337–38.

288 The story of asymptotic freedom is described in a number of accounts, including Crease and Mann, *Second Creation,* pp. 329–35, and Pickering, *Constructing Quarks,* pp. 207–14. Wilczek gave a personal account in his book with Betsy Devine, *Longing for the Harmonies,* pp. 207–17. Also see Gross, "Asymptotic Freedom and the Emergence of QCD," pp. 216–23.

MGM and Low's paper on the renormalization group is cited in the notes to chapter 6.

289 **"It was like you're sure there's no God":** in Crease and Mann, *Second Creation,* p. 333.

290 **Pagels's paper first using the term "quantum chromodynamics":** W. Marciano and H. Pagels, "Quantum Chromodynamics," *Physics Reports* C36 (1978): 137–276.

Feynman a dedicated "quarkerian": Weiner interviews.

Feynman nominated MGM and Zweig for a Nobel: Gleick, *Genius,* p. 396.

"Protons are not fundamental particles": Feynman, "Structure of the Proton," p. 601.

CHAPTER 14   *Superphysics*

291 The history of the Standard Model is detailed in the many historical reminiscences in Hoddeson et al., *Rise of the Standard Model,* and in Part V of Crease

and Mann, *Second Creation,* which also includes citations for the important papers.

"quantum flavor dynamics": Gell-Mann used the term in, for example, his 1980 Wolfson lecture, "Questions for the Future," in J. H. Mulvey, ed., *The Nature of Matter* (Oxford, England: Clarendon Press, 1981), p. 176.

292 The dramatic discovery of the weak bosons is recounted in Taubes, *Nobel Dreams.*

Gell-Mann lobbied for Glashow to share the Nobel: MGM interview, 12/3/97.

293 **"worse than the disease":** Bernstein, *Tenth Dimension,* p. 90.

"It was as if anthropologists": Close et al., *The Particle Explosion,* p. 178.

296 I dig more deeply into the construction of the Standard Model in chapter 2 of my book *Fire in the Mind* (New York: Alfred A. Knopf, 1995), pp. 29–59.

297 **high and low quarks:** Experiments in 1989 at SLAC and CERN are now taken as establishing that there are no more than three families of particles.

297–8 The implications of a universe with more than four dimensions are explored in Kaku, *Hyperspace.*

298 Popular descriptions of string theory can be found in Greene, *The Elegant Universe;* MGM, *Quark and Jaguar,* pp. 127–29, 199–213; Weinberg, *Dreams of a Final Theory,* pp. 211–19; Morris, *Edges of Science,* pp. 153–66; Kaku, *Hyperspace,* pp. 151–177; Kaku and Trainer, *Beyond Einstein;* and Crease and Mann, *Second Creation,* pp. 414–16.

299 **MGM on the origins of "slanguage":** "From Renormalizability to Calculability?" in Jackiw et al., eds., *Shelter Island II,* p. 13, cited in notes to chapter 3. He also spoke about this in an interview (7/30/98).

299–300 **MGM's flatlander example:** in Telegdi Festschrift paper, pp. 18–19.

MGM's work with Ramond and Slansky: "Color Embedding, Charge Assignments, and Proton Stability in Unified Gauge Theories," *Review of Modern Physics,* 50 (1978): 721–44.

300 **"nature reserve for an endangered species":** ibid., p. 19.

Some of the inventors of supergravity were Peter van Nieuwenhuizen, Dan Freedman, Bruno Zumino, and Julius Weiss. See Kaku and Trainer, *Beyond Einstein,* pp. 128–30.

300–1 Descriptions of what it was like to be a postdoc working with MGM in the seventies were given in interviews with Brink (7/27/98), Ramond (7/29/98), and Glennys Farrar (8/11/98). Barton Zwiebach related almost identical experiences about working with him in the early 1980s (interview 11/4/98).

The Shelter Island II talk and a later paper in a Festschrift for Telegdi go into detail about MGM's views on the development of superstrings: "Is the Whole World Composed of Superstrings?" in K. Winter, ed., *Festi-Val: Festschrift for Val Telegdi* (Amsterdam: Elsevier, 1988), pp. 119–40. In my notes, I use the page numbers from the original typescript.

301 **Papers on supergravity MGM wrote with Brink, Ramond, and Schwarz:** "Extended Supergravity as Geometry of Superspace," *Physics Letters* 76B (1978): 417–22; "Prepotentials in a Superspace Formulation of Supergravity," *Nuclear Physics B* 145 (1978): 93–109. He also wrote many reviews and popular talks on the subject in addition to those cited earlier.

302–3 **Hawking declared the end of physics was in sight:** mentioned, for example, in Kaku, *Hyperspace,* pp. 147–48.

303 **"Eventually, there would be a limit to human patience":** MGM, "Questions for the Future," p. 186.

"The ground of physics is littered": Dyson, *Disturbing the Universe,* p. 62.

MGM's talk at the Einstein symposium was published in Ne'eman, ed.,

To Fulfill a Vision: Jerusalem Einstein Centennial Symposium on Gauge Theories and Unification of Physical Forces (Reading, Mass.: Addison-Wesley, 1981), pp. 257–64.

MGM on Einstein's unification equations: ibid., p.257.

"While cultivating successful ideas," etc.: ibid., p. 264.

306 **"Let the ruling class tremble":** quoted in Scott McLemee, "Nothing to Be Done," *In These Times*, 21 March 1994. Other information about the MLP came from Harvey Klehr, *Far Left of Center: The American Radical Left Today* (New Brunswick, N.J.: Trans-Action Books, 1988), pp. 125–26, and from the MLP's own documents, available from Marxist-Leninist Books in Chicago. I also benefited from discussions with Scott McLemee and Chip Berlet. Lisa Gell-Mann declined, through her brother Nick, to be interviewed.

"Though encircled by NATO and Warsaw," etc.: Quotes from the various songs are from *Down with Ronald Reagan, Chieftain of Capitalist Reaction and Other Songs of Revolutionary Struggle and Socialism* (Chicago: Marxist-Leninist Publications, 1982).

307 The course of Margaret Gell-Mann's illness was described to me in interviews with MGM (7/30/98), Nick Gell-Mann (5/26/98), Sydney Meshkov (6/8/98), Lydia Matthews (7/21/98), Fred Zachariasen (4/1/94), Helen Tuck (3/31/94), David Pines (11/13/96), Marvin Goldberger (4/11/96 and 3/31/97), and Robert Walker (11/4/98).

308 **Pais offers MGM a place to stay:** Pais interview, 2/28/94.

Margaret told Lisa how she wanted to spend her life: interview with Nick Gell-Mann.

309 Working with MGM after Margaret's death: interview with Lars Brink, 7/27/98.

CHAPTER 15   *From the Simple to the Complex*

310 **Gell-Mann felt like a farmer or a troubador:** MGM, "Particle Theory from S-Matrix to Quarks," in M. G. Doncel, A. Hermann, L. Michel, and A. Pais, eds., *Symmetries in Physics (1600–1980),* Proceedings of the First International Meeting on the History of Scientific Ideas (Barcelona: Belaterra, 1987), p. 475; MGM, "Progress in Elementary Particle Theory, 1950–1964," in Brown et al., *Pions to Quarks,* p. 694.

"orgy of reminiscence": MGM, "Progress in Elementary Particle Theory," p. 694.

MGM's Paris talk: "Strangeness," in *Proceedings of an International Conference on the History of Particle Physics,* Paris, July 1982, pp. 395–408; cited in notes to chapter 6.

311 **"As you ramble":** MGM, "Particle Theory from S-Matrix to Quarks," p. 475.

"I tend to keep my eye on the hole": ibid.

312 **MGM seemed to be saying he had discovered everything:** This opinion was expressed in previously cited interviews with colleagues, including Zachariasen, Zweig, and Telegdi.

MGM's blowup over Zweig's paper: Zweig interview, 8/20/97. The published result: Zweig, "Origins of the Quark Model," in *Baryon '80* (cited in the notes to chapter 11).

313 **Serber never stopped believing:** Serber, *Peace and War,* pp. 199–201.

MGM's behavior: interviews with numerous colleagues.

MGM almost evicted from planes: Lydia Matthews interview, 7/21/98.

MGM's inability to collaborate with Feynman: "Dick Feynman—The Guy in the Office Down the Hall" (cited in the notes to chapter 9), pp. 50–1.

Feynman at Esalen: Gleick, *Genius,* p. 407; Feynman, *"Surely You're Joking,"* pp. 338–39.

MGM thought Feynman preoccupied with generating anecdotes: "Dick Feynman—The Guy in the Office Down the Hall," p. 50.

314 **Feynman's tea story:** Feynman, *"Surely You're Joking,"* pp. 59–61.

"I was very excited": ibid., p. 250.

Murray threatens to sue Feynman: Gleick, *Genius,* p. 411.

"Dick's joke book": Crease and Mann, interview transcript for *Second Creation.*

Feynman and the *Challenger:* Gleick, *Genius,* pp. 415–24; Mehra, *Beat of a Different Drum,* pp. 594–99.

315 **MGM emerging from depression:** MGM interview, 7/30/98.

The loneliness of the years after Margaret: interviews with Lydia Matthews (7/21/98), Marcia Southwick (9/28/98), Nick Gell-Mann (5/26/98), Rick Lipkin (9/24/98), Helen Tuck (3/31/94), and MGM (7/30/98).

MGM's speculations on the connection between simple particles and complex creatures: ibid., pp. 8–9.

315–16 **the condors:** MGM, *Quark and Jaguar,* pp. 7–8.

317 The history of superstring theory in the 1980s is described in the books by Kaku and others cited in the notes to chapter 14.

MGM's papers with Zwiebach: "Curling up Two Spatial Dimensions with SU(1.1)/U(1)," *Physics Letters* B 147, 111–14 (1984); "Space-Time Compactification Due to Scalars," *Physics Letters* B 141, 333–36 (1984); "Dimensional Reduction of Space-Time Induced by Non-linear Scalar Dynamics and Noncompact Extra Dimensions," *Nuclear Physics* B 260, 569–92 (1985).

318 **"There are certainly some indications":** "Is the Whole World Composed of Superstrings?" (cited in notes to chapter 14), p. 22.

some believed it would take extraterrestrials to explain superstrings: Kaku, *Hyperspace,* p. 189.

MGM's early beliefs about complexity were laid out in several lectures, including "The Concept of the Institute" (talk given at the School of American Research, Santa Fe, N.M., November 1984, published in Pines, ed., *Emerging Syntheses in Science,* pp. 1–15) and "Patterns of Convergence in Contemporary Science" (keynote address at the annual meeting of the American Association for the Advancement of Science, Los Angeles, Calif., 26 May 1985). Later talks include "The Santa Fe Institute" (talk given in Santa Fe, N.M., 9 January 1990); "Visions of a Sustainable World" (introductory talk to a meeting on the subject in Santa Fe, N.M., May 1990); "Remarks on Complex Adaptive Systems" (talk given at the Santa Fe Institute Workshop on Integrative Themes, 8 July 1992, reprinted in Cowan et al., eds., *Complexity: Metaphors, Models, and Reality*), and "What Is Complexity?" in *Complexity* 1, no. 1 (1995): 16–19. Most of the ideas in these talks are developed throughout MGM, *Quark and Jaguar.*

319 **"a crude look at the whole":** MGM, *Quark and Jaguar,* p. xiv.

The story of the beginning of the Santa Fe Institute is told in Waldrop, *Complexity,* pp. 52–98. Gell-Mann gave me his version in an interview (8/31/98).

321 **"The ability to reduce everything":** P. W. Anderson, "More Is Different," *Science,* 177, no. 4047, p. 393.

"emerging syntheses": MGM, "The Concept of the Institute."

322 **MGM's difficulties as an administrator:** Waldrop, *Complexity,* pp. 89–90.

The Santa Fe conferences mentioned here are documented in Anderson et al., eds., *The Economy as an Evolving Complex System;* Langton, ed., *Artificial Life;* and Zurek, ed., *Complexity, Entropy, and the Physics of Information.*

323 **"squalid-state physics":** Gell-Mann has used this term so frequently that it has become part of the folklore of solid-state physics.

Cowan threatens to quit: Waldrop, *Complexity,* pp. 348–49.

323–4 **MGM and Kauffman, etc.:** My information on Gell-Mann's recent relationship with the Institute comes from conversations since 1992 with many of his Santa Fe colleagues, including Bruce Abell, Philip Anderson, Brian Arthur, John Casti, Jim Crutchfield, George Cowan, J. Doyne Farmer, James Hartle, Stuart Kauffman, Christopher Langton, Seth Lloyd, John Miller, Melanie Mitchell, Mike Simmons, Joseph Traub, and Wojciech Zurek, and with members of the staff. I am also drawing on my own observations from regularly attending Institute lectures and events.

324 **plectics:** "Let's Call It Plectics," *Complexity* 1, no. 5 (1995–96): 3.

complex adaptive systems: See earlier references to MGM's papers on complexity; the ideas are also presented in Part 1 of MGM, *Quark and Jaguar.*

325 **Seth Lloyd's encounter with MGM:** Lloyd, personal communication, 1/14/99.

326 **MGM's work with Lloyd:** "Information Measures, Effective Complexity, and Total Information," *Complexity* 2, no. 1 (1996): 44–52.

"Since quantum mechanics is correct": Fermi, quoted in MGM, *Quark and Jaguar,* p. 149.

"Quantum flapdoodle": ibid., pp. 167–75.

326–7 **MGM's and Hartle's work on decoherence:** "Quantum Mechanics in Light of Quantum Cosmology," in W. Zurek, ed., *Complexity, Entropy and the Physics of Information,* pp. 425–58.

327 For more on decoherence, see Lindley, *Where Does the Weirdness Go?,* pp. 193–226, and Johnson, *Fire in the Mind,* pp. 156–77.

328 **"Goblin worlds":** MGM, *Quark and Jaguar,* p. 165.

MGM and the Tesuques: This story came from a conversation I had with Alfonso Ortiz in 1993, when I was writing *Fire in the Mind.*

CHAPTER 16 *The Quark and the Jaguar*

329 Feynman's death is described in Gleick, *Genius,* pp. 437–38, and Mehra, *Beat of a Different Drum,* pp. 606–8.

Feynman memorial service: The program and videotapes of both sessions are in the Caltech archives.

330 The Peruvian artifacts story is documented in Kirkpatrick, *Lords of Sipan;* Carl Nagin, "The Peruvian Gold Rush," *Art & Antiquities,* May 1990, pp. 98–145; and *United States v. David Swetnam* (CV 88-00914) and *Peru v. Johnson* (CV 88-6990). MGM gave his account in an interview (7/30/98).

"Smugglers' Cove": Nagin, "Peruvian Gold Rush," p. 142.

331 MGM's Festschrift is documented in Schwarz, ed., *Elementary Particles and the Universe: Essays in Honor of Gell-Mann,* and "Where Are Our Efforts Leading?" *Engineering and Science,* Spring 1989, pp. 11–21. I also relied on previously cited interviews with Telegdi and Low.

332–3 **Goldberger on MGM's Chinese:** "Where Are Our Efforts Leading?," p. 21. MGM's linguistic abilities: MGM interview (11/2/98); Berreby transcripts (cited in the notes to chapter 1).

333 **MGM and music:** Telegdi interview, 3/31/94.

334 **Sze gives Murray book of poems:** Sze interview, 9/17/98.

Sze's poem: Sze, *River River,* p. 14. Copyright 1998 by Arthur Sze.

335 **almost seeing a jaguar:** MGM, *Quark and Jaguar,* pp. 3–5.

MGM asks Bernstein about book-writing: Zachariasen interview, 4/1/94.

Background on Brockman can be found in Eric Konigsberg, "Science Made

Easy," *New Republic*, 13 March 1996, pp. 17–19; James Gorman, "Nimble Deal-Maker for Stars of Science," *New York Times*, 14 October 1997, p. F-1; and Thomas Weyr, "John Brockman and the Science of Deal-Making," *Publishers Weekly*, 7 February 1994, pp. 33–35. Brockman declined to be interviewed.
Brockman's representation of Capra: This was after Capra wrote *The Tao of Physics*.

336–7 The genesis of the *Quark and Jaguar* deal was described to me in interviews with Leslie Meredith (9/22/98), Roger Lewin (9/16/98), and MGM (11/2/98). I also relied on the following published accounts: Herbert R. Lottman, "Reunification Frankfurt," *Publishers Weekly*, 2 November 1990; Paul Nathan, "Rights," *Publishers Weekly*, 23 November 1990; and Alun Anderson and Tim Lincoln, "Million-Dollar Quark," *Nature*, 348, no. 6297, p. 102.

337 **"I am embarked upon a personal voyage," etc.:** MGM book proposal, obtained by author.

338 **Gell-Mann's agony over writing the book:** MGM, *Quark and Jaguar*, pp. xv–xviii; MGM interview (11/2/98).

339 **"We found him on the aircraft":** Crease and Mann, *Second Creation*, p. 263.
MGM berates Robert Crease: Crease and Mann, interview transcripts for *Second Creation*.
Norwegian lighthouse keeper: MGM, *Quark and Jaguar*, pp. xvi–xviii.
Murray meets Marcia Southwick: interviews with MGM (7/30/98 and 11/3/98) and Marcia Southwick (9/28/98). They also described their meeting in Marsha McEuen, "Santa Fe's Brain Trust," *Santa Fean*, April 1998.

341ff. **The saga of writing *Quark and Jaguar*:** interviews with MGM (11/2/98), Leslie Meredith, Marcia Southwick, Rick Lipkin (9/24/98), and Jeremiah Lyons (9/17/98). Ed Dobb declined to be interviewed but provided the following statement (9/23/98): "[Gell-Mann] was very generous in the deal upfront, and for reasons I don't care to go into we didn't work very well as a partnership, and so he needed to find someone else to complete the project."

343 **the wedding:** interviews with MGM (7/30/98) and Marcia Southwick.

345 **MGM on page one:** William J. Broad, "Top Quark, Last Piece in Puzzle of Matter Appears to Be in Place," *New York Times*, 26 April 1994.
the reading on the Upper West Side: I was in the audience.
some of the *Quark and Jaguar* reviews: *Publishers Weekly*, 21 February 1994; *Kirkus Reviews*, 1 March 1994; *Washington Post*, 12 June 1994; *Science* 264, no. 5164, pp. 1480–81; *New York Times Book Review*, 1 May 1994.

346 **MGM had never worked so hard on anything:** MGM, *Quark and Jaguar*, p. xvii.

EPILOGUE *Valentine's Day, 1997*

349 **Glashow compares string theory to theology:** in, for example, Sheldon Glashow and Paul Ginsparg, "Desperately Seeking Superstrings?" *Physics Today* (May 1986), pp. 7–9.

350 **Supercollider site considered for a mushroom farm:** Associated Press, "Well-Bred Mushrooms from $2 Billion Beds?" in *New York Times*, 2 March 1994.

350 For more on recent twists in superstring theory, see George Johnson, "Almost in Awe, Physicists Ponder 'Ultimate' Theory," *New York Times*, 22 September 1998.
"We no longer are what we once were," etc.: *Workers' Advocate*, 28 November 1993, p. 1.
MGM and his son growing closer: interview with Nick Gell-Mann, 5/26/98.

350–1 **the Gell-Manns' Manhattan apartment:** MGM interview, 11/2/98.

351 The symposium Murray and Marcia collaborated on was held at SITE Santa
Fe, 4–5 December 1998.

The roboticist and video artist were Alan Rath and Pipilotti Rist.

MGM's goals as a bird-watcher: ibid.

351–2 **Zweig's feelings about MGM:** Zweig interview, 8/20/97.

352 **life annotated by Gell-Mann:** interview with Jeremiah Lyons, 9/17/98.

353 **"In our work we are always":** cited at the very beginning of these notes.

Anderson, Philip W., Kenneth J. Arrow, David Pines, eds. *The Economy as an Evolving Complex System*. Redwood City, Calif.: Addison-Wesley, 1988.

Barrow, John D. *Theories of Everything: The Quest for Ultimate Explanation*. New York: Oxford University Press, 1991.

———. *The World Within the World*. New York: Oxford University Press, 1991.

Bell, J. S. *Speakable and Unspeakable in Quantum Mechanics: Collected Papers on Quantum Philosophy*. Cambridge, Mass.: Cambridge University Press, 1987.

Beller, Steven. *Vienna and the Jews, 1867–1938: A Cultural History*. Cambridge, England: Cambridge University Press, 1989.

Berland, Theodore. *The Scientific Life*. New York: Coward-McCann, 1962.

Bernstein, Jeremy. *A Comprehensible World: On Modern Science and Its Origins*. New York: Random House, 1967.

———. *The Life It Brings: One Physicist's Beginnings*. New York: Ticknor & Fields, 1987.

———. *The Tenth Dimension: An Informal History of High Energy Physics*. New York: McGraw-Hill, 1989.

Brockman, John. *Einstein, Gertrude Stein, Wittgenstein & Frankenstein: Re-Inventing the Universe*. New York: Viking, 1986.

———. *The Third Culture: Beyond the Scientific Revolution*. New York: Simon & Schuster, 1995.

Brower, Kenneth. *The Starship and the Canoe*. New York: Harper & Row, 1983.

Brown, Laurie M., and Lillian Hoddeson, eds. *The Birth of Particle Physics*. Cambridge, England: Cambridge University Press, 1983.

———, Max Dresden, and Lillian Hoddeson, eds. *Pions to Quarks: Particle Physics in the 1950s*. Cambridge, England: Cambridge University Press, 1989.

———, and John S. Rigden, eds. *Most of the Good Stuff: Memories of Richard Feynman*. New York: American Institute of Physics, 1993.

Buckley, William F. *God and Man at Yale: The Superstitions of Academic Freedom*. Chicago: Regnery, 1951.

Casti, John L. *Complexification: Explaining a Paradoxical World Through the Science of Surprise*. New York: HarperCollins, 1994.

Chalfen, Israel. *Paul Celan: A Biography of His Youth*. Trans. Maximilian Bleyleben. New York: Persea Books, 1991.

Christianson, Gale E. *Edwin Hubble: Mariner of the Nebulae*. Chicago: University of Chicago Press, 1995.

Cline, Barbara Lovett. *Men Who Made a New Physics: Physicists and the Quantum Theory*. New York: Thomas Y. Crowell, 1965.

Close, Frank, Michael Marten, and Christine Sutton. *The Particle Explosion.* New York: Oxford University Press, 1987.

Conover, Ted. *Whiteout: Lost in Aspen.* New York: Random House, 1991.

Cooper, Leon N. *An Introduction to the Meaning and Structure of Physics.* Short ed. New York: Harper & Row, 1970.

Coughlan, G. D., and J. E. Dodd. *The Ideas of Particle Physics: An Introduction for Scientists.* Cambridge, England: Cambridge University Press, 1991.

Cowan, George, David Pines, and David Meltzer, eds. *Complexity: Metaphors, Models, and Reality.* Santa Fe Institute Studies in the Science of Complexity, vol. 19. Reading, Mass.: Addison-Wesley, 1994.

Crease, Robert P. *Making Physics: A Biography of Brookhaven National Laboratory, 1946–1972.* Chicago: University of Chicago Press, 1999.

————, and Charles C. Mann. *The Second Creation: Makers of the Revolution in 20th-Century Physics.* Rev. ed. Rutgers: Rutgers University Press, 1996.

Davies, P. C. W., and Julian Brown. *Superstrings: A Theory of Everything?* Cambridge, England: Cambridge University Press, 1992 (first published 1988).

DeWitt, Bryce S., and Neill Graham, eds. *The Many-Worlds Interpretation of Quantum Mechanics.* Princeton, N.J.: Princeton University Press, 1973.

Dyson, Freeman. *Disturbing the Universe.* New York: Harper & Row, 1979.

Escoffier, Auguste. *The Escoffier Cookbook and Guide to the Fine Art of Cookery.* New York: Crown, 1941.

Feinberg, Gerald, ed. *T. D. Lee, Selected Papers, vol. 3: Random Lattices to Gravity.* Boston: Birkhauser, 1986.

Fermi, Laura. *Atoms in the Family.* Chicago: University of Chicago Press, 1954.

Ferris, Timothy. *The Whole Shebang: A State-of-the-Universe(s) Report.* New York: Simon & Schuster, 1997.

Feynman, Richard P. *The Character of Physical Law.* Cambridge, Mass.: MIT Press, 1967.

————. *QED: The Strange Theory of Light and Matter.* Princeton, N.J.: Princeton University Press, 1985.

————. *"Surely You're Joking, Mr. Feynman!": Adventures of a Curious Character.* New York: W. W. Norton, 1985.

————. *What Do You Care What Other People Think?" Further Adventures of a Curious Character.* New York: W. W. Norton, 1988.

————, Robert B. Leighton, and Matthew Sands. *The Feynman Lectures on Physics.* 3 vols. Reading, Mass.: Addison-Wesley, 1963–65.

Fritzsch, Harald. *Quarks: The Stuff of Matter.* New York: Basic Books, 1983.

Gamow, George. *Thirty Years That Shook Physics: The Story of Quantum Theory.* Garden City, N.Y.: Doubleday, 1966.

Gardner, Martin. *The New Ambidextrous Universe: Symmetry and Asymmetry from Mirror Reflections to Superstrings.* Rev. ed. New York: W. H. Freeman, 1990.

Gell-Mann, Murray. Afterword to *Last of the Curfews,* by Fred Bodsworth. Washington, D.C.: Counterpoint, 1995.

————. *The Quark and the Jaguar: Adventures in the Simple and the Complex.* New York: W. H. Freeman, 1994.

————, and Yuval Ne'eman. *The Eightfold Way: A Review, with a Collection of Reprints.* New York: W. A. Benjamin, 1964.

Glashow, Sheldon. *The Charm of Physics.* New York. Simon & Schuster, 1991.

————, with Ben Bova. *Interactions: A Journey Through the Mind of a Particle Physicist and the Matter of This World.* New York: Warner Books, 1988.

Gleick, James. *Chaos: Making a New Science.* New York: Viking, 1987.

————. *Genius: The Life and Science of Richard Feynman.* New York: Pantheon, 1992.

Goldsmith, Emanuel S. *Architects of Yiddishism at the Beginning of the Twentieth Cen-*

tury: A Study in Jewish Cultural History. Rutherford/Madison/Teaneck, N.J.: Fairleigh Dickinson University Press, 1976.

Greene, Brian. *The Elegant Universe: Superstrings, Hidden Dimensions, and the Quest for the Ultimate Theory.* New York: Norton, 1999.

Gribben, John, and Mary Gribben. *Richard Feynman: A Life in Science.* New York: Dutton, 1997.

Griffiths, David. *Introduction to Elementary Particles.* New York: Wiley, 1987.

Halliday, David, and Robert Resnick. *Physics: Parts I and II.* New York: Wiley, 1966.

Hapgood, Fred. *Up the Infinite Corridor: MIT and the Technical Imagination.* New York: Addison-Wesley, 1992.

Hawking, Stephen W. *A Brief History of Time: From the Big Bang to Black Holes.* New York: Bantam Books, 1988.

Heisenberg, Werner. *Physics and Beyond: Encounters and Conversations.* New York: Harper & Row, 1971.

———. *Physics and Philosophy: The Revolution in Modern Science.* New York: Harper & Row, 1958.

Herbert, Nick. *Quantum Reality: Beyond the New Physics.* New York: Anchor/Doubleday, 1985.

Hoddeson, Lillian, Laurie Brown, Michael Riordan, and Max Dresden. *The Rise of the Standard Model: Particle Physics in the 1960s and 1970s.* Cambridge, England: Cambridge University Press, 1997.

Horgan, John. *The End of Science: Facing the Limits of Knowledge in the Twilight of the Scientific Age.* Reading, Mass. Addison-Wesley, 1996.

Howe, Irving. *World of Our Fathers: The Journey of the East European Jews to America and the Life They Found and Made.* New York: Harcourt Brace Jovanovich, 1976.

Janik, Allan, and Stephen Toulmin. *Wittgenstein's Vienna.* New York: Simon & Schuster, 1973.

Joyce, James. *Finnegans Wake.* New York: Viking, 1939.

Kaku, Michio. *Hyperspace: A Scientific Odyssey Through Parallel Universes, Time Warps, and the 10th Dimension.* New York: Oxford University Press, 1994.

———, and Jennifer Trainer. *Beyond Einstein: The Cosmic Quest for the Theory of the Universe.* New York: Bantam Books, 1987.

Kauffman, Stuart A. *The Origins of Order: Self-Organization and Selection in Evolution.* New York: Oxford University Press, 1963.

———. *At Home in the Universe: The Search for Laws of Self-Organization and Complexity.* New York: Oxford University Press, 1995.

Kevles, Daniel J. *The Physicists: The History of a Scientific Community in Modern America.* New York: Alfred A. Knopf, 1977.

Kirkpatrick, Sidney D. *Lords of Sipan: A Tale of Pre-Inca Tombs, Archeology, and Crime.* New York: William Morrow, 1992.

Langton, Christopher G., ed. *Artificial Life.* Santa Fe Institute Studies in the Science of Complexity, vol. 6. Redwood City, Calif.: Addison-Wesley, 1989.

Lederman, Leon, with Dick Teresi. *The God Particle: If the Universe Is the Answer, What Is the Question?* Boston: Houghton Mifflin, 1993.

Lewin, Roger. *Complexity: Life at the Edge of Chaos.* New York: Macmillan, 1992.

Lindley, David. *The End of Physics: The Myth of a Unified Theory.* New York: Basic Books, 1993.

———. *Where Does the Weirdness Go?: Why Quantum Mechanics Is Strange, but Not as Strange as You Think.* New York: Basic Books, 1996.

Mason, Stephen F. *A History of the Sciences.* New York: Collier, 1962.

Mehra, Jagdish. *The Beat of a Different Drum: The Life and Science of Richard Feynman.* Oxford: Oxford University Press, 1994.

Oren, Dan A. *Joining the Club: A History of Jews and Yale.* New Haven: Yale University Press, 1985.

Pagels, Heinz R. *Perfect Symmetry: The Search for the Beginning of Time.* New York: Simon & Schuster, 1985.

Pais, Abraham. *Inward Bound: Of Matter and Forces in the Physical World.* New York: Oxford University Press, 1986.

————. *A Tale of Two Continents: A Physicist's Life in a Turbulent World.* Princeton, N.J.: Princeton University Press, 1997.

————. *"Subtle Is the Lord . . .:" The Science and the Life of Albert Einstein.* New York: Oxford University Press, 1982.

Pickering, Andrew. *Constructing Quarks: A Sociological History of Particle Physics.* Edinburgh: Edinburgh University Press, 1984.

Pines, David, ed. *Emerging Syntheses in Science.* Proceedings of the Founding Workshops of the Santa Fe Institute, vol. 1. Redwood City, Calif.: Addison-Wesley, 1988.

Polkinghorne, John. *Rochester Roundabout: The Story of High Energy Physics.* Longman, 1989.

Regis, Ed. *Who Got Einstein's Office? Eccentricity and Genius at the Institute for Advanced Study.* New York: Addison-Wesley, 1987.

Rhodes, Richard. *The Making of the Atomic Bomb.* New York: Simon & Schuster, 1987.

Riordan, Michael. *The Hunting of the Quark: A True Story of Modern Physics.* New York: Simon & Schuster, 1987.

Rotblat, Joseph. *Science and World Affairs: History of the Pugwash Conferences.* London: Dawsons of Pall Mall, 1962.

————. *Scientists in the Quest for Peace.* Cambridge, Mass.: MIT Press, 1972.

Schorske, Carl E. *Fin-de-Siècle Vienna: Politics and Culture.* New York: Alfred A. Knopf, 1980.

Schwarz, John H., ed. *Elementary Particles and the Universe: Essays in Honor of Gell-Mann.* Cambridge: Cambridge University Press, 1991.

Schweber, Silvan S. *QED and the Men Who Made It: Dyson, Feynman, Schwinger, and Tomonaga.* Princeton, N.J.: Princeton University Press, 1994.

Segrè, Emilio. *Enrico Fermi, Physicist.* Chicago: University of Chicago Press, 1970.

Serber, Robert. *Peace and War: Reminiscences of a Life on the Frontiers of Science.* Ed. Robert P. Crease. New York: Columbia University Press, 1998.

Smolin, Lee. *The Life of the Cosmos.* New York: Oxford University Press, 1997.

Southwick, Marcia. *The Night Won't Save Anyone: Poems.* Athens, Georgia: University of Georgia Press, 1980.

————. *A Saturday Night at the Flying Dog & Other Poems.* (Field Poetry Series). Oberlin, Ohio: Oberlin College Press, 1999.

————. *Why the River Disappears.* Ithaca, N.Y.: Cornell University Press, 1990.

Starr, Kevin. *Inventing the Dream: California Through the Progressive Era.* New York: Oxford University Press, 1985.

Stenberg, Peter. *Journey to Oblivion: The End of the East European Yiddish and German Worlds in the Mirror of Literature.* Toronto: University of Toronto Press, 1991.

Sze, Arthur. *River River.* Providence, R.I.: Lost Roads, 1987.

————. *The Redshifting Web: Poems 1970–1998.* Port Townsend, Washington: Copper Canyon, 1998.

Taubes, Gary. *Nobel Dreams: Power, Deceit and the Ultimate Experiment.* New York: Random House, 1986.

Traweek, Sharon. *Beamtimes and Lifetimes: The World of High Energy Physicists.* Cambridge, Mass.: Harvard University Press, 1988.

Waldrop, M. Mitchell. *Complexity: The Emerging Science at the Edge of Order and Chaos.* New York: Simon & Schuster, 1992.

Weinberg, Steven. *The Discovery of Subatomic Particles.* New York: W. H. Freeman, 1983.

————. *The First Three Minutes: A Modern View of the Origin of the Universe.* Rev. ed. New York: Basic Books, 1988.

————. *Dreams of a Final Theory.* New York: Pantheon, 1992.

Weisskopf, Victor. *The Joy of Insight: Passions of a Physicist.* New York: Basic Books, 1991.

Weyl, Hermann. *Symmetry.* Princeton, N.J.: Princeton University Press, 1969.

Wheeler, John Archibald. *Geons, Black Holes & Quantum Foam: A Life in Physics.* New York: Norton, 1998.

————, and Wojciech Hubert Zurek, eds. *Quantum Theory and Measurement.* Princeton, N.J.: Princeton University Press, 1983.

Wilczek, Frank, and Betsy Devine. *Longing for the Harmonies: Themes and Variations from Modern Physics.* New York: W. W. Norton, 1987.

Yang, Chen Ning. *Selected Papers, 1945–1980, With Commentary.* San Francisco: W. H. Freeman, 1983.

Zee, A. *Fearful Symmetry: The Search for Beauty in Modern Physics.* New York: Macmillan, 1986.

Zurek, Wojciech H., ed. *Complexity, Entropy, and the Physics of Information.* Santa Fe Institute Studies in the Science of Complexity, vol. 8. Redwood City, Calif.: Addison-Wesley, 1990.

A NOTE ABOUT THE AUTHOR

George Johnson is a writer for the *New York Times* and the author of *Fire in the Mind: Science, Faith, and the Search for Order.* His other books include *In the Palaces of Memory: How We Build the Worlds Inside Our Heads, Machinery of the Mind: Inside the New Science of Artificial Intelligence,* and *Architects of Fear: Conspiracy Theories and Paranoia in American Politics.* He lives in Santa Fe, New Mexico.

A NOTE ON THE TYPE

This book was set in Baskerville, a facsimile of the type cast from matrices designed by John Baskerville (1706–1775) of Birmingham, England. Baskerville's original face was one of the forerunners of the type style known to printers as "modern face"—a "modern" of the period A.D. 1800.

Composed by Creative Graphics,
Allentown, Pennsylvania
Printed and bound by Quebecor,
Martinsburg, West Virginia
Diagrams by Arlene Lee
Designed by Anthea Lingeman